Characterization, Epidemiology, and Management

Phytoplasma Diseases in Asian Countries
Characterization, Epidemiology, and Management
Volume 3

Edited by

A.K. Tiwari

Kenro Oshima

Amit Yadav

Seyyed Alireza Esmaeilzadeh-Hosseini

Yupa Hanboonsong

Suman Lakhanpaul

Academic Press is an imprint of Elsevier
125 London Wall, London EC2Y 5AS, United Kingdom
525 B Street, Suite 1650, San Diego, CA 92101, United States
50 Hampshire Street, 5th Floor, Cambridge, MA 02139, United States
The Boulevard, Langford Lane, Kidlington, Oxford OX5 1GB, United Kingdom

Copyright © 2023 Elsevier Inc. All rights reserved.

No part of this publication may be reproduced or transmitted in any form or by any means, electronic or mechanical, including photocopying, recording, or any information storage and retrieval system, without permission in writing from the publisher. Details on how to seek permission, further information about the Publisher's permissions policies and our arrangements with organizations such as the Copyright Clearance Center and the Copyright Licensing Agency, can be found at our website: www.elsevier.com/permissions.

This book and the individual contributions contained in it are protected under copyright by the Publisher (other than as may be noted herein).

Notices

Knowledge and best practice in this field are constantly changing. As new research and experience broaden our understanding, changes in research methods, professional practices, or medical treatment may become necessary.

Practitioners and researchers must always rely on their own experience and knowledge in evaluating and using any information, methods, compounds, or experiments described herein. In using such information or methods they should be mindful of their own safety and the safety of others, including parties for whom they have a professional responsibility.

To the fullest extent of the law, neither the Publisher nor the authors, contributors, or editors, assume any liability for any injury and/or damage to persons or property as a matter of products liability, negligence or otherwise, or from any use or operation of any methods, products, instructions, or ideas contained in the material herein.

ISBN: 978-0-323-91671-4

For information on all Academic Press publications visit our website at https://www.elsevier.com/books-and-journals

Publisher: Nikki P. Levy
Acquisitions Editor: Nancy Maragioglio
Editorial Project Manager: Lena Sparks
Production Project Manager: Swapna Srinivasan
Cover Designer: Christian J. Bilbow

Typeset by TNQ Technologies

Dedicated to Dr. Govind Pratap Rao

Dr. Govind Pratap Rao is working as a Director, Institute of Agriculture & Natural Sciences, DDU Gorakhpur University, Gorakhpur, Uttar Pradesh, India. Dr. Rao was born on January 15, 1960, to the parents of an educated middle-class peasantry in Kushinagar District of Uttar Pradesh, India. He graduated in 1979 and postgraduated in MSc (Botany) from the University of Gorakhpur, India, in 1981. He obtained PhD degree in Botany (Plant Virology) from Gorakhpur University in 1986. As a student, he had been very attentive, indulgent, and single minded. He did Postdoc at the University of Urbana-Champaign, Illinois, USA, with Prof. R. E. Ford on Sugarcane and Maize Viruses in 1994 and with Dr. Phillippe Rott at CIRAD, Montpellier, France in 1998. Dr. Rao has joined UP Council of Sugarcane Research as Scientific Officer (Plant Pathology) in 1988. He has been actively involved and engaged in research on different aspects of characterization, epidemiology, and management of sugarcane diseases. He has been served in different capacities as Scientific Officer, Sr. Scientific Officer, Head of Plant Pathology Division, and Station In-Charge at different centers of UP Council of Sugarcane Research, Shahjahanpur till 2010. In 2010, Dr. Rao joined the Division of Plant Pathology, Indian Agricultural Research Institute, New Delhi, India, as a Principal Scientist. With only his support and initiatives, now 20 different centers/institutions have been involved in phytoplasma research in India. Dr. Rao has completed nearly 10 research projects on different aspects of characterization, epidemiology, and management of phytoplasmas on different agricultural crops in India. His work on phytoplasma diseases associated with sugarcane, brinjal, sesame, fruits, vegetables, medicinal crops, and floriculture crops are well recognized. Because of his sincere efforts,

phytoplasma research has become integral part of plant pathology research all over India and an awareness has been created among farmers, students, and scientists all over India. He has been invited in national/international conferences to deliver invited, plenary, and keynote lectures. He has also chaired technical sessions in several international conferences in India and abroad.

Dr. Rao has 33 years of research experience on plant pathology especially on plant virology and phytoplasmas. He did significant contributions in characterization of plant viruses and phytoplasmas of various crop species and published over 200 research publications, nearly 10 textbooks, and 16 edited books to his credit. Dr. Rao has recently published a series of edited books with international eminent scientists on updated research on phytoplasmas as Phytoplasma: Plant Pathogenic Bacteria *Vol. I, II, and III with Springer. Dr. Rao is the Editor-in-Chief of* Phytopathogenic Mollicutes, *an international journal of phloem-limited microorganisms.*

Over 20 students have been awarded PhD degree under Dr. Rao's supervision. Apart from that, Dr. Rao has helped and supervised many students of various universities and institutions of India by various ways, especially economically weaker section students and never taken credits of the same. Dr. Rao has given his support to several academicians for writing articles, preparing project proposals, and literature supports in the early 20th century when internet access was limited and dependency was on hard copies. He has been awarded several prestigious awards to his credit. The most important ones are as follows: National Biotechnology Associateship Award (1991−1992), DBT, Govt. of India; Young Scientist Award (1994−1995) from DST, Govt. of India; Overseas BOYSCAST Award (1996) from DST, Govt. of India; President Award, Society for General Microbiology, UK, 1998; Best U. P. Agriculture Scientist Award (UPCAR), Govt. of Uttar Pradesh in 2002; Vigyan Ratna Award by CST, Govt. of UP for the year 2003−2004; Jin Xiu Qiu Award in 2006 by People's Govt. of Guangxi Province, Nanning, China; Global Award of Excellence, IS 2008, Al-Ahrish, Egypt; Dr. Ram Badan Singh Vishisht Krishi Vaigyanik Puraskar-2014 by UPCAR, Lucknow, India; Leadership Excellence Award in Sugarcane Crop Protection by Thailand Society of Sugar Cane Technologists, Bangkok, in 2016; S. N. Dasgupta Memorial Award by Indian Phytopathology Society 2019.

Dr. Rao is a member of several prestigious scientific societies and organizations related to plant pathology and microbiology. Besides, Dr. Rao has visited over 30 countries as visiting scientists, for invited talk, postdoc fellow, research training, panel discussion, and for attending workshop and conferences. Dr. Rao has been President, Secretary General, and Secretary of various renowned academic societies: Technology Society for Basic and Applied Sciences, Gorakhpur; Society for Conservation and Resource Development of Medicinal Plants, New Delhi; Society for

Sugar Research and Promotion, New Delhi; Indian Virological Society, New Delhi; Integrated Association of Professional in Sugarcane Technologies, Nanning, China, etc. He has also been elected as fellow of several Academic Societies in India and abroad. He is also Editor-in-Chief of international scientific journals published from Springer-Nature and Indian Journals.

Dr. Rao has made his presence marked as a dedicated, devoted and disciplined teacher, researcher, and guide. His simplicity and sagacity always endeared him to his students and colleagues. He has been a man of few words, frank, straight forward, enlightened, forward looking, mentally alert, and agile, always full of new creative ideas in science. His planning and execution of research projects have been exemplary. Coming from a modest educated family, he scaled glorious academic heights by shear hard work, resilience, and catholic devotion. He is a world known scientist working on phytoplasma's research and popularly known as **"Phytoplasma Man of India."**

At present, Dr. Rao is working on characterization, epidemiology, and management of virus-infected cereal crops, millets, and maize and phytoplasmas infecting important agriculture and horticultural crops in India. He has described over 50 new phytoplasma records and also reported several potential weeds and vectors as natural reservoirs for plant viruses and phytoplasmas. His work in the establishment of phytoplasma in India and abroad is well recognized. Dr. Rao retired in January 2022 and he is still very active in phytoplasma research and has collaborations with many academic institutions. It's really hard to compile and express his work and dedication toward research in Uttar Pradesh in words here.

Thank you for everything you have done for the Phytoplasma Research in India.

Contents

List of contributors xv
Authors biography xix
Foreword xxiii
Characterization, Epidemiology, and Management xxv

CHAPTER 1 Novel methods of phytoplasma detection of phytoplasma in Asian countries 1
S.M.K. Widana Gamage, Nguyen Ngoc Bao Chau, Nguyen Bao Quoc, Saman Abeysinghe and Ajay Kumar Tiwari
  1. Introduction 1
  2. Historical background 2
  3. Symptomatic diagnosis 3
  4. Microscopic detection 4
  5. Light microscopy 4
  6. Fluorescence microscopy 5
    6.1 Epifluorescence 5
    6.2 Immunofluorescence 5
  7. Transmission electron microscopy 6
    7.1 Thin section procedure 6
    7.2 Immunosorbent and immunoelectron microscopy 7
    7.3 Scanning electron microscopy 7
    7.4 Critical-point drying technique 7
    7.5 Cryofixation 7
    7.6 Cryoscanning microscopy 8
  8. Molecular detection 8
    8.1 DNA extraction 8
    8.2 Primer selection 9
    8.3 PCR techniques 10
    8.4 Nested PCR with universal or specific primers 10
    8.5 Quantitative PCR for routine detection and quantification 11
    8.6 Reverse transcription-PCR with crude sap extracts 11
    8.7 Loop-mediated isothermal amplification for field pathogen detection 11
    8.8 Microarray 12
    8.9 DNA barcoding of phytoplasmas with 16S rRNA and *tuf* gene sequences 13
  References 13

CHAPTER 2 Graft and vegetative transmission of phytoplasma-associated diseases in Asia and their management 21
Kadriye Caglayan, Elia Choueiri and Govind Pratap Rao
1. Introduction 21
2. Graft transmission 22
   2.1 Graft transmission in woody plants 22
   2.2 Graft transmission in herbaceous plants 25
   2.3 Transmission by micropropagation 29
   2.4 Dodder transmission 29
3. Management 30
4. Conclusions and perspectives 32
References 32

CHAPTER 3 Transmission of lime witches' broom 37
Mohammad Mehdi Faghihi, Abdolnabi Bagheri, Mohammad Salehi and Majid Siampour
1. Introduction 37
2. Vector transmission 37
   2.1 *Hishimonus phycitis* 37
   2.2 *Diaphorina citri* 39
3. Seed transmission 40
4. Graft transmission and experimental hosts 40
5. Conclusion and perspectives 41
References 41

CHAPTER 4 Major insect vectors of phytoplasma diseases in Asia 45
Chamran Hemmati, Mehrnoosh Nikooei and Abdullah Mohammed Al-Sadi
1. Introduction 45
2. Insect acquisition and transmission of phytoplasmas 46
3. How the phytoplasma insect vectors can be identified? 46
4. Major insect vectors of phytoplasmas 49
   4.1 *Hishimonus* spp. 49
   4.2 *Orosius albicinctus* Distant, 1918 54
   4.3 *Neoaliturus* spp. 56
5. Some minor phytoplasma insect vectors in Asia 58
6. General management 59
7. Conclusion 60
References 60
Further reading 66

CHAPTER 5 Genomic studies on Asian phytoplasmas 67
Ching-Ting Huang, Shen-Chian Pei and Chih-Horng Kuo
  1. Background 67
  2. Historical perspective 69
  3. Biological insights 73
     3.1 Gene content and metabolism 73
     3.2 Effectors and putative secreted proteins 74
     3.3 Mobile genetic elements 75
  4. Linking genomics to taxonomy 76
  5. Conclusions and perspectives 77
  References 77

CHAPTER 6 Cross-boundary movement of phytoplasmas in Asia and status of plant quarantine 85
V. Celia Chalam, Pooja Kumari, D.D. Deepika, Priya Yadav, K. Kalaiponmani and A.K. Maurya
  1. Introduction 85
  2. Exclusion of phytoplasma through quarantine 86
     2.1 International scenario: imports and exports 86
     2.2 Status of phytoplasmas as regulated pest in Asia 89
     2.3 National scenario: imports 89
     2.4 National scenario: exports 93
     2.5 National domestic quarantine 93
  3. Perspectives 94
  References 95

CHAPTER 7 Updates on phytoplasma diseases management 97
Nursen Ustun, Maryam Ghayeb Zamharir and Abdullah Mohammed Al-Sadi
  1. Introduction 97
  2. Quarantine 98
  3. Healthy planting material 100
  4. Antibiotics 101
  5. Insect vector control strategies 102
  6. Other management practices 103
  7. Host resistance and/or tolerance? 104
  8. Recovery 108
  9. Induced resistance 108
  10. Case study 1: witches' broom diseases of lime 109
  11. Case study 2: grapevine yellows 110

- 12. Biotic resistance inducers 111
- 13. Biocontrol 112
- 14. Conclusions 112
- References 113
- Further reading 123

CHAPTER 8 Management of insect vectors associated with phytoplasma diseases 125
Chamran Hemmati, Mehrnoosh Nikooei, Ajay Kumar Tiwari and Abdullah Mohammed Al-Sadi

- 1. Introduction 125
- 2. Vector taxonomy and ecology 125
- 3. Insect vector managements 127
  - 3.1 Insecticides 127
  - 3.2 Clean propagation material 128
  - 3.3 Resistant plants 128
  - 3.4 Alternative host's and weeds control 129
  - 3.5 Habitat management 130
  - 3.6 Covering plants 130
  - 3.7 Biological controls 131
  - 3.8 Symbiotic control 132
- 4. Conclusion 133
- References 133
- Further reading 136

CHAPTER 9 Elimination of phytoplasmas: an effective control perspective 137
Chamran Hemmati, Mehrnoosh Nikooei and Ajay Kumar Tiwari

- 1. Introduction 137
- 2. In vivo methods 138
  - 2.1 Chemotherapy 138
  - 2.2 Thermotherapy by hot water or hot air 139
- 3. In vitro culture 140
  - 3.1 In vitro meristem preparation in combination with in vitro thermotherapy 141
- 4. In vitro chemotherapy 141
- 5. Conclusion 142
- References 143
- Further reading 146

CHAPTER 10 Phytoplasma resistance 147
Isil Tulum and Kadriye Caglayan

- 1. Introduction 147

- 2. Natural (genetic) resistance 148
  - 2.1 Natural resistance in temperate fruit trees 149
  - 2.2 Natural resistance in economically important crops 152
- 3. Induced resistance 153
  - 3.1 Resistance inducers 153
  - 3.2 The "recovery" phenomenon 154
  - 3.3 Effectors and plant immunity 155
- 4. Transgenic resistance 155
- 5. Current status of phytoplasma resistance in India, Iran, Japan, and Turkey 156
  - 5.1 India 156
  - 5.2 Iran 156
  - 5.3 Japan 157
  - 5.4 Turkey 158
- 6. Conclusion and perspectives 159
- References 159

CHAPTER 11 microRNAs role in phytoplasma-associated developmental alterations 167

Sapna Kumari, Amrita Singh and Suman Lakhanpaul

- 1. Introduction 167
  - 1.1 Role of noncoding RNAs 167
  - 1.2 Discovery of microRNAs 167
  - 1.3 Biogenesis of microRNAs 168
- 2. Role of microRNAs in vegetative and reproductive development in plants 168
  - 2.1 Leaf development 169
  - 2.2 Root and shoot development in plants 169
  - 2.3 Vascular development 169
  - 2.4 Flower and fruit development 169
  - 2.5 Abiotic and biotic stress 170
- 3. microRNAs associated with phytoplasma-induced diseases 170
  - 3.1 Role of primary microRNAs involved in phytoplasma infection 170
- 4. Putative mechanism(s) of interaction between miRNAs and their target genes in response to phytoplasma infection 176
  - 4.1 Defective development of anthers 176
  - 4.2 Anthocyanin accumulation in leaves (leaf margins and petioles) ... 176
  - 4.3 Witches' broom and phyllody 178
  - 4.4 Dwarfism or stunting 179
- 5. Conclusion 181
- 6. Future directions 181
- Acknowledgments 181
- References 182

CHAPTER 12 Characteristic features of genome and pathogenic factors of phytoplasmas 187
Ai Endo and Kenro Oshima
   1. Characteristic features of phytoplasma genomes 187
   2. Characteristic features of pathogenic factors 188
   References 190

Index 195

List of contributors

Saman Abeysinghe
Department of Botany, Faculty of Science, University of Ruhuna, Matara, Sri Lanka

Abdullah Mohammed Al-Sadi
Department of Plant Sciences, College of Agriculture and Marine Sciences, Sultan Qaboos University, Muscat, Oman; Department of Crop Sciences, College of Agricultural and Marine Sciences, Sultan Qaboos University, Muscat, Oman

Abdolnabi Bagheri
Plant Protection Research Department, Hormozgan Agricultural and Natural Resources Research and Education Centre, AREEO, Bandar Abbas, Iran

Nguyen Ngoc Bao Chau
Faculty of Biotechnology, Ho Chi Minh City Open University, Ho Chi Minh City, Vietnam

Kadriye Caglayan
Plant Protection Department, Faculty of Agriculture, Hatay Mustafa Kemal University, Antakya, Turkey

V. Celia Chalam
Division of Plant Quarantine, ICAR-National Bureau of Plant Genetic Resources, New Delhi, Delhi, India

Elia Choueiri
Department of Plant Protection, Lebanese Agricultural Research Institute, Tal Amara, Zahlé, Lebanon

D.D. Deepika
Division of Plant Quarantine, ICAR-National Bureau of Plant Genetic Resources, New Delhi, Delhi, India

Ai Endo
Department of Clinical Plant Science, Faculty of Bioscience and Applied Chemistry, Hosei University, Koganei, Tokyo, Japan

Mohammad Mehdi Faghihi
Plant Protection Research Department, Fars Agricultural and Natural Resources Research and Education Centre, AREEO, Zarghan, Iran

Chamran Hemmati
Department of Agriculture, Minab Higher Education Center, University of Hormozgan, Bandar Abbas, Iran; Plant Protection Research Group, University of Hormozgan, Bandar Abbas, Iran

Ching-Ting Huang
Institute of Plant and Microbial Biology, Academia Sinica, Taipei, Taiwan

K. Kalaiponmani
Division of Plant Quarantine, ICAR-National Bureau of Plant Genetic Resources, New Delhi, Delhi, India

Sapna Kumari
Department of Botany, University of Delhi, New Delhi, Delhi, India

Pooja Kumari
Division of Plant Quarantine, ICAR-National Bureau of Plant Genetic Resources, New Delhi, Delhi, India

Chih-Horng Kuo
Institute of Plant and Microbial Biology, Academia Sinica, Taipei, Taiwan

Suman Lakhanpaul
Department of Botany, University of Delhi, New Delhi, Delhi, India

A.K. Maurya
Division of Plant Quarantine, ICAR-National Bureau of Plant Genetic Resources, New Delhi, Delhi, India

Mehrnoosh Nikooei
Department of Agriculture, Minab Higher Education Center, University of Hormozgan, Bandar Abbas, Iran; Plant Protection Research Group, University of Hormozgan, Bandar Abbas, Iran

Kenro Oshima
Department of Clinical Plant Science, Faculty of Bioscience and Applied Chemistry, Hosei University, Koganei, Tokyo, Japan

Shen-Chian Pei
Institute of Plant and Microbial Biology, Academia Sinica, Taipei, Taiwan; Department of Plant Pathology and Microbiology, National Taiwan University, Taipei, Taiwan

Nguyen Bao Quoc
Research Institute for Biotechnology and Environment, Nong Lam University, Ho Chi Minh City, Vietnam; Faculty of Biological Sciences, Nong Lam University, Ho Chi Minh City, Vietnam

Govind Pratap Rao
Division of Plant Pathology, Indian Agricultural Research Institute, New Delhi, Delhi, India

Mohammad Salehi
Plant Protection Research Department, Fars Agricultural and Natural Resources Research and Education Centre, AREEO, Zarghan, Iran

Majid Siampour
Department of Plant Protection, College of Agriculture, Shahrekord University, Shahrekord, Iran

Amrita Singh
Department of Botany, Gargi College, University of Delhi, New Delhi, Delhi, India

Ajay Kumar Tiwari
UPCSR-Sugarcane Research and Seed Multiplication Center, Gola, Uttar Pradesh, India

Isil Tulum
Istanbul University, Faculty of Science, Department of Botany, Istanbul, Turkey; Istanbul University, Centre for Plant and Herbal Products Research-Development, Istanbul, Turkey

Nursen Ustun
Laboratory of Bacteriology, Plant Diseases Deparment, Plant Protection Research Institute, Bornova, Izmir, Turkey

S.M.K. Widana Gamage
Department of Botany, Faculty of Science, University of Ruhuna, Matara, Sri Lanka

Priya Yadav
Division of Plant Quarantine, ICAR-National Bureau of Plant Genetic Resources, New Delhi, Delhi, India

Maryam Ghayeb Zamharir
Plant Diseases Department, Iranian Research Institute of Plant Protection, Agricultural Research, Education and Extension Organization (AREEO), Tehran, Iran

Authors biography

Dr. A. K. Tiwari

Email: ajju1985@gmail.com

Ajay Kumar Tiwari, PhD, is serving as Officer-In-Charge at UPCSR-Sugarcane Research and Seed Multiplication Center, Gola, Khiri, UP, India. He completed his PhD on Cucurbit Viruses at the CCS University, Meerut, UP, India, in 2011. He has published 100 research articles in reputed national and international journals. He has also authored eight books published by Springer, Taylor & Francis, and Nova. He has submitted more than 300 plant pathogen nucleotide sequences to GenBank. Dr. Tiwari is a regular reviewer and member of the editorial boards of several international journals and is Managing Editor of Sugar Tech, Editor of *Annals of Applied Biology*, and Chief Editor of the *Agrica* journal. He received the CIPAM Young Researcher Award in 2011 and the DST-SERB Young Scientist Award. He received the Young Scientist Award by the Chief Minister of the State Government of UP for his outstanding contributions in the area of plant pathology. He has attended several conferences, workshops, and delivered invited talk in India and abroad. He is currently involved in research on the characterization and management of agricultural plant pathogens especially phytoplasma and bacterial diseases of sugarcane and the production of healthy sugarcane seed materials and their distributions in UP through cane societies.

Professor Kenro Oshima

Email: kenro@hosei.ac.jp

Dr. Kenro Oshima is a Plant Pathology Professor at the Hosei University, Japan. He has 20 years of research experience on plant pathology especially on the genomes and pathogenicity of phytoplasmas. He has more than 60 peer-reviewed international publications and has delivered numerous oral and posters presentations in national and international meetings. He and his group determined the first complete genome sequence of phytoplasma and found that the phytoplasma has lost many genes of metabolic pathway in the process of evolution. They also discovered the virulence factors of phytoplasmas such as TENGU and PHYL1 and clarified its mechanism of action. He is a member of the International Phytoplasma Working Group (IPWG) and has participated in the revision of the '*Candidatus* Phytoplasma' species description guidelines in 2022. He is an Associate Editor of *Frontiers in Microbiology* for the section of Evolutionary and Genomic Microbiology. He received the Kitamoto Award from the Japanese Society of Mycoplasmology in 2018 and the Society Award from the Phytopathological Society of Japan in 2021.

Dr. Amit Yadav

Email: amityadav@nccs.res.in

Amit Yadav, PhD, is a Scientist at National Centre for Cell Science at Pune, India. He carried out his doctoral studies on 'Sugarcane Grassy Shoot' (SCGS) phytoplasma from the University of Pune, Pune. Currently, his research group is involved in perusing research on '*Candidatus* Phytoplasma' involving the functional genomics, epidemiology, detection and diagnosis, plant—insect—phytoplasma interactions, and taxonomy. '*Candidatus* Phytoplasma' is a 'not-yet-cultured' plant pathogenic bacterium involved in devastating diseases in economically important plants across the globe. Dr. Yadav is currently working on epidemiology of Sandalwood Spike Disease related to '*Ca*. P. asteris' and genome characterization of peanut witches' broom (PWB) phytoplasma associated with *Glycine max* and *Parthenium hysteroporus*. He has published 58 peer-reviewed research articles, which include the reports on phytoplasma diseases on different plant species, phytoplasma genome sequencing, and genome characterizations. He is review editor of specialty section in *Frontiers in Microbiology*: Microbe and Virus Interactions with Plants and Evolutionary and Genomic Microbiology (specialty section of '*Frontiers in Microbiology*,' '*Frontiers in Genetics*,' and '*Frontiers in Ecology and Evolution*'). Also, he is an editorial member of '*Phytopathogenic Mollicutes*,' a journal dedicated to phytoplasma studies. Dr. Yadav is a member of Bergey's International Society for Microbial Systematics (BISMiS), Association of Microbiologists of India (AMI), Indian Association of Mycoplasmologists (IAM), International Organization of Mycoplasmology (IOM), and a member of board of studies in Botany at Sir Parashurambhau College, Pune.

Dr. Seyyed Alireza Esmaeilzadeh-Hosseini

E-mail: phytoplasma.iran@gmail.com

Seyyed Alireza Esmaeilzadeh-Hosseini, PhD, started working as a researcher in the Plant Protection Research Department, Yazd Agricultural and Natural Resources Research and Education Center, A-REEO, Yazd, Iran, since 2000. Currently, he is a member of the academic staff and the head of the center. He has also taught plant virology and phytopathogenic prokaryotes courses in the universities of Yazd province. Dr. Esmaeilzadeh-Hosseini is reviewer of several international and national journals. He has participated in more than 52 research projects in the field of plant pathology, especially phytoplasma diseases. He has participated in the publication of more than 61 research articles in reputed national and international journals and also contributed to 76 paper presentations in national and international scientific congresses. Dr. Esmaeilzadeh-Hosseini has won the top position among the researchers of the universities of Yazd province and the researchers of the Iranian Research Institute of Plant Protection (IRIPP) several times. He is a member of the International Phytoplasma Working Group (IPWG) and has participated and presented papers in the second to fourth IPWG Congress. He has introduced some new subgroups of phytoplasmas (16SrXXIX-B, 16SrIX-I, and 16SrII-Z) and new hosts of these diseases in the world. He is currently involved in research on the biological and molecular characteristics of phytoplasmas and the management of these diseases in Iran.

Professor Dr. Yupa Hanboonsong

Email: yupa_han@kku.ac.th, yupaento@gmail.com

Yupa Hanboonsong is a Professor of Entomology in the Department of Entomology and Plant Pathology, KhonKaen University, Thailand. She obtained a PhD in Entomology from Lincoln University, New Zealand. Professor Hanboonsong and her research group is a pioneer on sugarcane white leaf disease caused by phytoplasma. Her research group contributed essential new knowledge and a better understanding about this sugarcane phytoplasma disease and its epidemiology including identification of new disease vectors, mode of transmission of the pathogen, the description of probing activities pattern of the vector, and working out various measures for integrated management of the disease. Her research findings on sugarcane white leaf management expanded to other Asian countries for insect vectors and sugarcane white leaf disease control. Last year (2021) she has published a manual on new norm of seed cane propagation for sugarcane white leaf management and her new method of seed cane production for reducing sugarcane white disease is well known and practiced by sugarcane farmers in Thailand. Dr. Hanboonsong is also a member of the scientific board committee on sugarcane white disease management in Thailand under the office of cane and sugar board of the Ministry of Industry, Thailand. She is also member of the editorial board for the International Journal *Sugar Tech*.

Professor Suman Lakhanpaul

Email: sumanlp2001@yahoo.com

Suman Lakhanpaul is a Senior Professor and presently Head, Department of Botany, University of Delhi, Delhi, India. She did her doctoral work from Delhi University in 1989 on Genome organization and Evolution in Asiatic *Vigna* and postdoctorate at Cornell University. She served ICAR-National Bureau of Plant Genetic Resources, New Delhi, India, during 1987−2002 prior to joining Delhi University. She has worked on molecular basis of genetic diversity using modern approaches with a special focus on crops and their allied taxa. In addition, she is passionate about studying mechanisms underlying multipartner transkingdom interactions in nature such as host plant−lac insect−microbe and phytoplasma−host taxa−insect vector interactions. She has over 100 publications in journals of reputed national and international journals and books. She is a regular reviewer of several scientific journals. She has guided 13 PhD and 9 MPhil students and has been actively teaching Masters students for 3 decades. She has handled several research projects from prestigious funding agencies including the World Bank. Professor Lakhanpaul has widely travelled and has been a visiting scientist to USDA research institutes and delivered lectures in international conferences in Australia, Austria, the United States of America, etc. Currently, her research interest is phytoplasma biology that included identifying new hosts of phytoplasma and unraveling the mechanism underlying phytoplasma-associated developmental alterations in host taxa using genomic approaches.

Foreword

Characterization, Epidemiology, and Management of Phytoplasma Diseases in Asia is the last one of three book series covering the aspects related to phytoplasma-associated diseases in Asian countries. These diseases are limiting the quality and productivity of economically important agriculture crops worldwide and are especially relevant in Asian countries. Their management options are focused on minimizing their spread by insect vectors and propagation materials and on the development of host plant resistance. Annual losses due to phytoplasma diseases may vary, but under the pathogen-favorable condition, phytoplasma disease may lead to disastrous consequences for farming and industry community. The phytoplasma-associated plant diseases were first detected in China during the Song dynasty (960−1227 CE) on green peonies, but important scientific progress about the identification of phytoplasmas only began after 1980s after the discovery carried out in Japan in 1967. Significant advancement in the last decades on biological and molecular properties, epidemiology, host−pathogen−insect interactions as well as the management of phytoplasma-associated diseases has been made. The book is enclosing updated information on phytoplasmas and phytoplasma-associated diseases focused on the Asian countries and provides comprehensive information on recent approaches for diagnostics, transmission, and management aspects. This volume contains 12 chapters contributed by experienced scientist working in Asian countries on different aspects of phytoplasma-associated diseases. These chapters cover novel methods for phytoplasma detection and available information about graft transmission and insect vector transmission with an emphasis on citrus crops. The addition of chapters dedicated to the management of these diseases by controlling their insect transmission and eliminating phytoplasmas from infected plants providing relevant information for the agricultural operators and farmers. The role of quarantine in phytoplasma diseases containment is also presented and focused on the Asian countries' situation. Genomic studies that started in Japan with the first sequence of a phytoplasma published in 2004 are summarized and updated with the Asian-based experience on pathogenicity factors and possible mechanisms of interaction with plants enclosing plant resistance to phytoplasma-associated diseases. The information on various topics is advanced and comprehensive, and provides appropriate information to everyone interested in phytoplasmas and their associated disease management. The book serves as an exhaustive and up-to-date compendium on various aspects of phytoplasmas affecting important crops in Asian countries.

Although phytoplasmas remain the most poorly characterized plant pathogens, the presentation of the knowledge acquired in area of the world will allow reducing the incidence and the losses due to these pathogens in the Asian agriculture and also reducing the impact and worldwide spreading of phytoplasma-associated diseases.

Assunta Bertaccini

Characterization, Epidemiology, and Management

Characterization, Epidemiology, and Management is the last of three volumes in the *Phytoplasma Diseases in Asian Countries* series, covering new research and providing an updated status of developments in characterization, detection, diagnosis, host metabolic interaction, and modes of transmission of phytoplasmas. In the last decades, significant advancement has been made on diagnostic, biological, and molecular properties, epidemiology, host—pathogen—insect interactions, and management of the phytoplasmas. Nowadays, it is necessary to collect an authentic compilation to know the progress of phytoplasmas' characterization. This book also provides an update on genomics, effectors, and pathogenicity factors toward a better understanding of phytoplasma—host metabolic interactions. It offers a comprehensive overview on biological, serological, and molecular characterization of the phytoplasmas, including recently developed approaches in diagnostics such as transcriptomics studies. Also included is information available on full genome sequencing of important phytoplasma strains from the Asian continent.

As there is no cure for these diseases, all effective and promising management strategies have been included in this volume, with a particular focus on exclusion, minimizing their spread by insect vectors and propagation materials, and developing host plant resistance. This book discusses the latest information on the epidemiology and management of phytoplasma-associated diseases, providing a comprehensive, up-to-date overview of distribution, occurrence, and identification of the phytoplasmas, recent diagnostics approaches, transmission, losses, and geographical distribution, as well as management aspects. This book is a valuable resource for plant pathologists, researchers in agriculture, and PhD students.

Editors

A.K. Tiwari

Kenro Oshima

Amit Yadav

S.A. Esmaeilzadeh-Hosseini

Yupa Hanboonsong

Suman Lakhanpaul

CHAPTER 1

Novel methods of phytoplasma detection of phytoplasma in Asian countries

S.M.K. Widana Gamage[1], Nguyen Ngoc Bao Chau[2], Nguyen Bao Quoc[3,4], Saman Abeysinghe[1] and Ajay Kumar Tiwari[5]

[1]Department of Botany, Faculty of Science, University of Ruhuna, Matara, Sri Lanka; [2]Faculty of Biotechnology, Ho Chi Minh City Open University, Ho Chi Minh City, Vietnam; [3]Research Institute for Biotechnology and Environment, Nong Lam University, Ho Chi Minh City, Vietnam; [4]Faculty of Biological Sciences, Nong Lam University, Ho Chi Minh City, Vietnam; [5]UPCSR-Sugarcane Research and Seed Multiplication Center, Gola, Uttar Pradesh, India

1. Introduction

Prevalence of diseases associated with phytoplasmas has been reported in Asia since two centuries. Before the discovery of phytoplasmas in 1967, the agents of several diseases remain unknown. Phytoplasmas were initially named mycoplasma-like organisms (MLOs). They are prokaryotes that lack cell wall, and are localized in sieve elements of the phloem tissue. Phytoplasmas are pleomorphic and have been observed in many shapes and sizes range from 0.1 to 2 μm (Doi et al., 1967; Lebsky and Poghosyan, 2014). Because of this unique morphology phytoplasmas can be easily identified from other intracellular pathogens. As a result, mere presence of phytoplasmas in infected tissues can be used as a preliminary tool for their detection. Phytoplasmas accumulate in plant parts where functional phloem tissues are present such as immature shoots, mid-rib and petiole of young leaves, and are ideal plant parts for sampling. Microscopic techniques especially transmission electron microscopy (TEM) and scanning electron microscopy (SEM) have been employed to observe ultrastructural details of phytoplasmas. In addition, other microscopic techniques could also be used to detect the presence of phytoplasmas in symptomatic plant tissues. However, the presence of low titer and inconsistent distribution of phytoplasmas in infected tissues (Siddique et al., 1998) often hamper their accurate detection.

Presence of phytoplasmas induces a wide range of disease symptoms in plants such as little leaf, witches'broom, phyllody, dwarfing, stunting, virescence. Therefore, disease symptoms are used as a preliminary phytoplasma detection tool. However, often phytoplasma symptom identification is combined with another diagnostic tool for more reliable disease diagnosis because some of the phytoplasma disease symptoms are also associated with the presence of other pathogens such as viruses.

Presently, molecular-based disease detection is the most widely used tool worldwide, including Asia. Polymerase chain reaction (PCR) offers platform for almost all molecular-based detection tools in practice today. PCR offers opportunities for detection, quantification, and classification of phytoplasmas. There are different PCR techniques each with its own pros and cons. The nested PCR is the most widely used technique and may provide information for phytoplasma differentiation into ribosomal groups and subgroups. Much of the research efforts in the recent past were focused on searching specific target genes or sequences in the phytoplasma genome in order to obtain finer resolution in their differentiation. With the discovery of loop-mediated isothermal amplification (LAMP) (Notomi et al., 2000), PCR-based detection of phytoplasmas conducted so far in the laboratory was brought to the field. This novel technological advancement made it feasible to pathogen detection in a sample in less than an hour.

In this chapter, historical background of phytoplasma disease detection in Asia and techniques that are currently used in the detection and quantification of phytoplasmas are described. The underlying technology behind each technique and their advantages and limitations are also briefly discussed.

2. Historical background

Diseases associated with phytoplasmas are reported date back to 17th century in Asia. However, until the discovery of small pleomorphic particles that resemble mycoplasmas in infected phloem tissues, those diseases were thought to be caused by a known source or virus. For instance, diseases that existed in Japan since two centuries back such as mulberry dwarf, rice yellow dwarf, and paulownia witches'broom were thought be caused by a known source. Later, when the plant viruses were identified, those diseases were attributed to virus because of virus-like symptoms and their insect transmissibility (Maejima et al., 2014). In 1967, for the first time, mycoplasma-like pleomorphic particles were discovered from ultrathin sections of phloem tissues infected with diseases such as mulberry dwarf disease and rice yellow dwarf (Doi et al., 1967). Those particles were named as MLOs due to their morphological similarity to animal infecting mycoplasmas and sensitivity to tetracycline.

In India, the history of phytoplasma diseases dates back to more than 100 years. Similar to the situation in other countries, some early diseases thought to be caused by viruses were later identified as phytoplasma diseases after confirming the presence of phytoplasmas in infected tissues. For example, brinjal little leaf disease that prevailed in India earlier thought to be a virus disease later confirmed a phytoplasma disease after observing phytoplasma particles in tissues (Varma et al., 1969; Shantha and Lakshmanan, 1984).

When the descriptions on particle morphology of phytoplasmas became available, diseases associated with phytoplasmas were mostly detected in early days by using TEM due to their extremely small dimension. Although TEM served as a promising diagnostic tool, it had limited applicability because of high initial investment for the instrument and technical expertise required for sample preparation and instrument operation. Hence many laboratories tend to use light microscopy with appropriate staining methods when TEM facility was not available (Purohit et al., 1978). Therefore, in 1980s, simple diagnostic techniques such as direct fluorescence detection (Namba et al., 1981) of phloem cells and staining of phytoplasmas with fluorescent dyes such as 4′,6′-diamidino-2-phenylindole (DAPI) (Hiruki and da Rocha, 1986) were developed. Later in the same decade, enzyme-linked immunosorbent assay (ELISA) which at that time widely used as a plant virus diagnostic tools was also tested for the detection of phytoplasmas. However, since then it has been rarely used in phytoplasma disease detection due to the absence of high-quality antibodies against phytoplasmas because of the difficulty in purifying phytoplasmas for antibody preparation. But there are several successful cases (Lin and Chen, 1985; Viswanathan, 1997).

Later in 1990s, with the advancement of molecular biological techniques, PCR became a widely used diagnostic tool for phytoplasmas (Lee et al., 1993) around the world including Asia. PCR-amplified 16S rRNA gene sequence became the gold standard in the detection and classification of phytoplasmas. Recently, LAMP has been extensively used in Asia as a novel diagnostic tool for the detection of phytoplasmas belonging to different ribosomal groups and subgroups (Nair et al., 2016; Sugawara et al., 2012; Maejima et al., 2014; Quoc et al., 2021; Yu et al., 2020). LAMP became an extremely useful technique because it is a rapid in-field diagnostic tool for which expensive equipment for phytoplasma DNA extraction and PCR are not required. The first LAMP-based universal phytoplasma detection kit has been commercially available in Japan since 2011.

3. Symptomatic diagnosis

Phytoplasma infection in plants associates with a range of diverse symptoms including chlorosis, leaf yellowing, stunting, witches'broom, little leaf, phyllody, big bud, shortened internodes, giant calyx, virescence, vascular discoloration, vein clearing, leaf curl, and distortion (Fig. 1.1). Interestingly, most

FIGURE 1.1

Symptoms associated with phytoplasmas in horticultural crops and sugarcane. (A) Witches'broom disease in brinjal, (B) little leaf disease in brinjal, (C) close up image of little leaf in brinjal, (D) white leaf disease in sugarcane, (E) witches'broom in chili, and (F) close up image of witches'broom in chili. All images were from the fields of Sri Lanka.

of the phytoplasmas-induced symptoms are similar to those caused by plant virus and viroid infections such as chlorosis, leaf yellowing, stunting, little leaf, vascular discoloration, vein clearing, leaf curl, and distortion. As a result, conventional symptomatic diagnosis of phytoplasmas is often confusing and unreliable. Hence, additional support such microscopic evidence for the presence of phytoplasma particles in the intracellular environment may be required. Further, symptomatic diagnosis of phytoplasmas is becoming complicated with the discovery of new phytoplasmas and diseases associated with them. Phytoplasmas may induce different symptoms in different host plant species, and/or distinct phytoplasmas may induce similar symptoms in different host plant species. For instance, witches' broom in brinjal is associated with phytoplasmas belonging to different subgroups, where it is 16SrII-D in Oman and China (Al-Subhi et al., 2011; Li et al., 2019) and 16SrV in India (Snehi et al., 2021).

4. Microscopic detection

Microscopic techniques are used in the visualization of particle morphology, intracellular localization, and studying histological changes induced by the infection in host plants. High-resolution microscopy has shown that phytoplasmas are pleomorphic single cells, lacking a cell wall. As a result, phytoplasmas appear in many shapes and sizes and hence enables more accurate diagnosis among the other intracellular microorganisms in the host tissues. In general, phytoplasmas are localized in the sieve elements of phloem and occasionally in parenchyma cells in the plant. They induce histological changes in plant tissues where they are localized and associated cells and tissues such as deposition of callose, suberin, lignin, and polyphenolic compounds that can be observed with the aid of a microscope. These unique features of phytoplasmas and associated histological alterations make microscopy a powerful preliminary tool in the detection of phytoplasmas even today amid the other modern diagnostic tools.

5. Light microscopy

Light microscopy has been widely used in Asia in the past and even present day phytoplasma disease detection as indirect evidence. Electron microscopic studies have revealed that the size of phytoplasmas (0.1–2 µm in diameter) falls within the resolving power of the light microscope (Horne, 1970; Waters and Hunt, 1980; Lebsky and Poghosyan, 2014). Hence, light microscopic techniques are continuing to be used in the localization and visualization of phytoplasmas in infected tissues. Moreover, light microscopic techniques are relatively fast and less expensive in contrast to other microscopic techniques and therefore have a diagnostic value as a preliminary method to detect phytoplasmas. More often, semi-thin sections and staining give sufficient information on the detection and localization of phytoplasmas in cells containing high concentrations of these bacteria (Musetti and Favali, 2004).

Studies have shown successful visualization of phytoplasmas in plant tissues stained with several basic dyes such as methylene green, Feulgen, Dienes' stain, toluidine blue, thionin, and acridine orange. The phloem tissues of phytoplasma-infected stems stain dark blue with Dienes' stain (Deeley et al., 1979). It stains xylem in blue-green and cortex in light blue.

Toluidine blue–stained phytoplasma-infected tissues can be seen as blue or violet accumulations of phytoplasmas inside sieve elements whereas cell walls are stained in blue (Wang and Valkonen,

2008). Thionin/acridine orange, another dye gives clear contrast between phytoplasma and non-phytoplasma areas in which phytoplasmas are stained as purple spots scattered along the walls of sieve elements whereas cellulose cell walls are stained in yellow (Cousin et al., 1986).

Although light microscopy has long been used as a diagnostic tool in phytoplasmas studies, its success depends on several factors such as the accumulation of high concentration of phytoplasmas in the tissue, appropriate staining procedure, and quality of the cross section.

6. Fluorescence microscopy

Fluorescence microscopy has become an essential tool in many fields in biology including plant pathology due to the attributes that are not available in traditional light microscopy. The application of a wide range of fluorochromes enables to identify cells, cellular components, and molecules with a high degree of specificity. It can detect the presence of even a single molecule. Hence, fluorescence microscopy could be used in the detection of intracellular pathogens including phytoplasmas with very low titer in host cells. Therefore, fluorescence microscopy has been used as a diagnostic tool in the detection of phytoplasmas (Ghosh et al., 1999; Gopinath et al., 2016; Ong et al., 2021).

Due to the autofluorescence of cell walls, chlorophyll, and some other cellular components, autofluorescence could not be used to detect phytoplasmas in plant tissues. Therefore, fluorescence for phytoplasmas has to be obtained using specific fluorophores. Fluorophores are stains that specifically attaches to specific targets and are excited by specific wavelength of irradiating light and emit light of defined intensity. Fluorescence microscope separates emitted fluorescence from the excitation light resulting superimposed high contrast images of target structures or molecules.

6.1 Epifluorescence

Epifluorescence with fluorescent stains such as DAPI and SYTO13 is a simple application of fluorescence microscopy to detect phytoplasmas in fresh plant tissues (Lee et al., 2012; Buxa et al., 2013). DAPI and SYTO13 specifically attaches to DNA. The DNA of phytoplasmas stained with dyes can be seen as fluorescent spots localized in sieve elements which are absent in healthy tissues (Andrade and Arismendi, 2013; Buxa et al., 2013). Therefore, DAPI staining gives precise localization of phytoplasmas. DAPI staining techniques have been used as a phytoplasma diagnostic tool in both herbaceous and woody host plant (Musetti et al., 2000; Russell et al., 1975; Gatineau et al., 2002; Schaper and Seemüller, 1982). Staining with fluorescent dyes is simple which only requires hand-cut or microtome sections of infected plant tissues and incubation time for the staining (Pasternak et al., 2015; Buxa et al., 2013). Therefore, it appears a quick, easy, and inexpensive method that can be used in the detection phytoplasmas in infected plant tissues.

6.2 Immunofluorescence

Although the simplicity and robustness of DAPI staining, it could not provide specific identification of different phytoplasmas. As a result, one phytoplasma could not be distinguished from the other. Therefore, application of DAPI has been limited to confirm the presence or absence of phytoplasmas in infected plant tissues. With the development of poly- and monoclonal antibodies against phytoplasmas, they could be more reliably and accurately identified. As a result, DAPI has been replaced by immunofluorescent probes that were specifically developed for the identification of a specific phytoplasma or a group in recent phytoplasma diagnostics (Ong et al., 2021; Gopinath et al., 2016).

Immunofluorescence for the detection of phytoplasmas can be used in either direct or indirect methods like in other serology-based techniques, ELISA, for example. In the direct method, the phytoplasma-infected plant tissue is stained with a solution of antibody produced against the phytoplasma antigen. The antibody is conjugated with a fluorophore, such a fluorescein isothiocyanate. In the indirect method, firstly phytoplasma-infected plant tissue is stained by incubation in a solution of unlabeled antibody (primary antibody) produced against phytoplasma antigen. In the second step, plant tissue is incubated with a secondary antibody conjugated with a fluorophore produced against the primary antibody. Therefore, fluorescence signal provides evidence for the presence and precise localization of phytoplasma in the plant tissue. Immunologically detected plant tissues are then visualized using a fluorescent microscope.

Since the reaction of antigen and antibody is specific and the procedure is technically simple and little time-consuming, immunofluorescence appears an ideal diagnostic tool for phytoplasma diseases. However, production of specific antisera viz., monoclonal antibodies against phytoplasmas is not an easy task since phytoplasma are difficult to be cultured *in vitro* for the purpose of purification in artificial media (Brzin et al., 2003; Loi et al., 2002; Hodgetts et al., 2014). Therefore, more often only partially purified antisera which also contain antibodies produced against the host viz., polyclonal antibodies could be obtained from infected host plants. As a result, immunofluorescence has limited application as a diagnostic tool in phytoplasma-associated plant diseases.

7. Transmission electron microscopy

There is no argument that high-resolution microscopes with high magnification are required for better visualization of phytoplasmas not only because they are small but also because of their pleomorphic nature. Electron microscope (EM) offers visualization of ultrastructural details of phytoplasmas. This section describes applications of TEM for the detection of phytoplasmas. There are several preparation techniques that have been used in TEM.

7.1 Thin section procedure

For the first time in 1967, the ultrastructure of phytoplasmas with TEM were described in mulberry, potato tissue, and aster sections taken from plants showing symptoms of dwarf, witches'broom, and yellowing, respectively (Doi et al., 1967). They described the presence of specific pleomorphic particles with spherical or irregularly ellipsoidal in shape in the size range of 80–800 µm and found consistently present in sieve elements and sometimes in phloem parenchyma cells. Since then, many other studies have observed and described ultrastructure of different phytoplasmas (Li et al., 2013; Rao, 2018). The procedure involves several steps for the preparation of plant material for preparing thin sections. Those steps are chemical fixation, dehydration, and resin embedding. Thin sections can be obtained as microtome sections of resin-embedded materials. Cryosections prepared by rapid freezing of thin sections of phytoplasma-infected materials give better preserved ultrastructural details of phytoplasma than the resin-embedded plant materials (Musetti and Favali, 2004).

In addition to the phytoplasma morphological descriptions, many studies have also reported changes in host cellular structure such as thickened and distorted phloem cell walls and callose deposits in the sieve tube as a consequence of the phytoplasmas on the plant tissue (Bertaccini and Marani, 1980, 1982; Musetti et al., 2000, 2004).

7.2 Immunosorbent and immunoelectron microscopy

Immunology coupled with electron microscopy could be performed with either crude sap preparations (immunosorbent electron microscopy, ISEM) or thin sections prepared (immunoelectron microscopy, IEM) from phytoplasma infected plant tissues. ISEM gives specific identification of phytoplasmas in crude sap preparations of infected plant materials in the background of host-derived tissue debris which always hamper better observation of phytoplasmas in EM (Vera and Milne, 1994). ISEM also allows observation of immunologically unrelated phytoplasmas in the sample (Musetti et al., 2002; Gurr and Gurr, 2007). Vera and Milne (1994) for the first time developed a rapid and specific protocol for immunogold labeling of phytoplasmas. They trapped phytoplasmas in crude sap on grids coated with antibodies prepared against phytoplasmas. Phytoplasmas were detected using gold-labeled antibodies. This method allowed them to distinguish morphologically similar but serologically unrelated phytoplasmas from each other.

7.3 Scanning electron microscopy

The understanding on phytoplasma particle morphology and their intracellular localization in host plant and insect tissues has been greatly improved though the applications of electron microscopy, both TEM and SEM. The presence of phytoplasmas in infected tissues of areca palm (Kanatiwela-de Silva et al., 2015) and coconut (Kanatiwela-de Silva et al., 2019) has been reported. There are several techniques in sample preparation for SEM analyses. Nowadays, techniques such as critical-point drying (CPD), cryofixation, and cryoscanning are becoming popular.

7.4 Critical-point drying technique

Presently in many laboratories, phytoplasma tissue samples are prepared for SEM using the CPD technique (Lebsky and Poghosyan, 2014; Anderson, 1951). This method clears sieve elements by removing cell contents while attaching phytoplasma particles firmly to the cell wall and pores of the sieve plate (Haggis and Sinha, 1978; Poghosyan et al., 2004). This method was able to overcome the previously experienced difficulties in visualizing phytoplasmas using SEM due to the interference driven by phloem contents.

7.5 Cryofixation

In cryofixation, the specimen is frozen quickly without an intermediary ice crystal phase (vitrification). There are several methods to achieve vitrification such as plunge freezing, slam freezing, or high-pressure freezing. In plunge freezing, specimen is plunged into liquid nitrogen. Since the process is performed quickly (within milliseconds), specimen is preserved and successful imaging can be obtained with tissue sections obtained by using a cryoultramicrotome cooled with liquid nitrogen. In high-pressure freezing, cryoprotectants such as yeast paste or 20% bovine serum albumin are used to lower the freezing point of intracellular water. Freezing specimen is brought into room temperature over a period of three to four days while replacing frozen water in the tissue with acetone. At room temperature, specimen is dehydrated and resin-embedded for sectioning.

In cryofixation, specimen is perfectly preserved in its native state while preventing changes to the cell membranes and shrinkage. However, only a small volume of the specimen can be successfully vitrified, generally less than 100 µm from the edge of the sample. This mean cryofixation gives perfect preservation of specimens only for ultrathin sections. Cryofixed samples can either be visualized using a cryo-EM. In the absence of cryo-EM facility, specimens can be visualized after freeze substitution and chemical fixation. Cryofixation has been used in the phytoplasma diagnostics (Devonshire, 2013).

7.6 Cryoscanning microscopy

In this technique, the specimen is frozen and hydrated. There is no need for stains or fixatives. Therefore, integrity and external features of samples are preserved and internal contents can be observed intact. During imaging, microscopic stage is cooled to the temperatures, below which ice crystals are formed, usually with liquid nitrogen. Freeze fracturing or cryoultramicrotome sectioning enables visualization of cells and ultrastructure beneath the surface. Cryoscanning microscopy has intrinsic limitation such as time-consuming and requirement of special equipment and conditions for specimen preparation, storage, holding, and imaging.

8. Molecular detection

Although their presence and particle morphology can be identified by using microscopic techniques, phytoplasma identification relies heavily on PCR-based molecular biological techniques. Success in PCR-based diagnosis depends on the quality and quantity of template DNA, selection of appropriate primers, and PCR technique applied.

8.1 DNA extraction

Obtaining high quality and sufficient quantity of template DNA is essential for any molecular biological technique. There is no universal sampling protocol available for DNA extraction from phytoplasma-infected plant tissues. Generally, phytoplasmas are localized in tissues with functional phloem like actively growing plant parts such as young shoots and roots, buds, immature leaf petioles, midrib, sapwood scrapings of woody perennials (Oropeza et al., 2011; Cordova et al., 2000; Vázquez-Euán et al., 2011). However, unlike in nonwoody annual plants, detection of phytoplasmas in woody perennial plants is often difficult and inconsistent due to their irregular distribution and low titer in tissues (Oropeza et al., 2011; Boudon-Padieu et al., 2003). The decision taken on the extraction protocol, tissue selection, and timing of sample collection is critical for accurate pathogen detection. Therefore, pooling multiple samples from different tissues and collected at different time points is often recommended (Boudon-Padieu et al., 2003; Palmano, 2001).

Often, DNA extraction from phytoplasma-infected herbaceous plant tissues using established DNA extraction protocols for plant genomic DNA such as cetyltrimetyl ammonium bromide (CTAB) method has successfully been used in DNA extraction from infected plant tissues (Sunpapao, 2014; Ong et al., 2021).

However, DNA extraction from woody trunks such as coconut and palms, trunk boring or drilling is often used to reach phloem tissue (Nipah et al., 2007; Harrison et al., 2013). DNA is then extracted by grinding in liquid nitrogen or directly with the CTAB method. Sometimes, DNA is extracted from sawdust of the woody stem through prolonged incubation in the CTAB extraction buffer when the facility of trunk boring is unavailable (Nipah et al., 2007).

Since the CTAB method is laborious and time-consuming, it is not an ideal method for extracting DNA from bulk samples and extracting DNA in the field. Using an automated DNA extraction protocol, DNA from a large number of grapevine and fruit tree phytoplasmas in a single extraction has been achieved with quality and quantity comparable with CTAB method (Mehle et al., 2013; Marcone et al., 2022). The tissue lysate is extracted using a tissue homogenizer in which collisions of a matrix and the plant material provided with lysis buffer release DNA in less than 40 seconds. Lysate is then centrifuged to remove tissue debris. DNA is purified from the cleared lysate mixed with magnetic beads in a fully automated DNA extraction machine. This is a rapid method of extracting DNA from a large number of samples with minimum possibilities for cross contamination, therefore, ideal for the laboratories where large number of samples are tested at one time. Purified DNA is suitable for downstream applications such as PCR, reverse transcription-PCR (RT-PCR), and LAMP. However, there is going to be high initial cost for automated homogenizer and DNA purification system.

For instances, when plant materials have to be tested for the presence of phytoplasmas in the field, Danks and Boonham (2007) developed a novel method in which DNA is extracted into the nitrocellulose membrane of a lateral flow device. This method was later modified for the *in situ* extraction of DNA from phytoplasma-infected materials from the field as a rapid method of DNA extraction and coupled with LAMP, discussed later in this chapter) procedure for the phytoplasma detection (Tomlinson et al., 2010).

8.2 Primer selection

Phytoplasmas are currently classified within the provisional genus '*Candidatus* Phytoplasma'. Members are placed on the genus based on their 16S rRNA gene sequence identity. A threshold of 98.65% identity with any previously described species is used to propose new taxons. Moreover due to the large number of 'Ca. Phytoplasma' species (49), a supporting approach has been developed based on the use of a computerized system for the simulation of restriction fragment length polymorphism (RFLP) analyses of 16S rRNA gene sequence (Lee et al., 1998a, 1998b; Wei et al., 2007, 2008).

A range of PCR primers is used in present phytoplasma detection. Several universal or group specific primers have been developed based on the 16S rRNA gene sequence (Duduk et al., 2013; Gundersen and Lee, 1996). Those primers can be used in identification of phytoplasmas. Among the primers, P1/P7 and R16F2/R16R2 primers are most widely used in phytoplasma diagnostics since RLFP patterns for both amplicons are conserved. Therefore, those two primer pairs have been extensively used in the identification of phytoplasmas (Rao, 2018; Raj et al., 2008; Biswas et al., 2014; Al-Saady et al., 2006; Bertaccini et al., 2019).

However, due to the highly conserved nature of the 16S rRNA gene sequence, recently less conserved and more variable single copy genes have been developed as genetic markers for more finer strain differentiation including the genes of ribosomal proteins (*rpl22* and *rpS3*), *secY*, *secA*, *tuf*, and *groEL* (Lee et al., 2006; Marcone et al., 2000; Mitrović et al., 2011; Martini et al., 2007).

Ribosomal protein (*rp*) genes are moderately conserved genes. They are able to group phytoplasmas further into subgroups which could not be resolved by using 16S rRNA gene sequences (Lee et al., 2004a, 2004b; Martini et al., 2002; Langer and Maixner, 2004). The presence of distinct biological and ecological properties in those newly formed subgroups provide evidence for the validity for subgrouping (Martini and Lee, 2013; Martini et al., 2007; Ashwathappa et al., 2021).

Similarly, *tuf* gene which code a translation elongation factor, E-Tu, has been used as a genetic marker for differentiation and ecological studies of phytoplasmas (Schneider and Gibb, 1997; Contaldo et al., 2011). Since there is a specific association of *tuf* gene with specific host plants and insect vector, it is also used as a genetic marker in epidemiological studies (Johannesen et al., 2012).

The *secY* gene is used as a genetic marker to differentiate phytoplasma strains in 16SrI, 16SrV, 16SrIX, 16SrX, and 16SrXII-A and other groups (Danet et al., 2011; Arnaud et al., 2007; Clair et al., 2003; Lee et al., 2010, 2012).

Primers designed for the *secA* gene amplify all major 16Sr groups including 16SrII (Dickinson and Hodgetts, 2013). The *secA* gene encodes ATP-dependent energy generator in the bacterial protein translocation cascade system. Primers designed for *secA* gene for nested PCR (Hodgetts et al., 2008) give a specific amplicon for reliable differentiation of phytoplasmas in different groups and subgroups (Mehdi et al., 2012; Bekele et al., 2011a; Jardim et al., 2021; Kumar et al., 2017). Abeysinghe et al. (2016) reported that potential of use of a new set of universal phytoplasma primers that amplify approximately 1 kb of the leucyl transfer RNA synthetase (*leuS*) gene and have been validated on a broad range of phytoplasmas.

Although a wide range of primers that are based on genes such as 16S rRNA, ribosomal protein gene and nonribosomal genes such as *tuf, secY, secA* are available, yet only the 16S rRNA gene could differentiate all phytoplasmas into '*Candidatus*' species (Bertaccini et al., 2022).

8.3 PCR techniques

A range of PCR techniques from PCR to LAMP are being used in the phytoplasma presence detection. The choice of PCR technique mainly depends on the available laboratory facilities; conventional thermal cycler or quantitative thermal cycler and the purpose; detection, identification, quantification.

8.4 Nested PCR with universal or specific primers

Nested PCR is the PCR assay used by many laboratories for the detection and identification of phytoplasmas. Nested PCR with universal primers for 16S rRNA, P1/P7, and R16F2/R16R2 and primers based on more variable single copy genes, *tuf*, *secY*, and *secA* has long been the choice for the detection of phytoplasmas from various host—phytoplasma combinations. In Asia, a large number of phytoplasmas have been described by nested PCR and subsequent sequence analyses (Rao et al., 2017; Maejima et al., 2014).

8.5 Quantitative PCR for routine detection and quantification

Despite being sensitive, nested PCR increases the possibility for cross contamination that could happen during the processing which ultimately resulting in false-positive detection of phytoplasmas (Christensen et al., 2013). On the other hand, phytoplasma titer could not be quantified in nested PCR. Under these circumstances, quantitative PCR has been identified as a promising technology that offers similar sensitivity to nested PCR but with less chances for contamination. In addition, quantitative PCR provides quantification. Phytoplasma titer is known to vary in different plant tissues and with the season (Christensen et al., 2013). Quantification is essential during the screening host plants for resistance against phytoplasmas. There are situations where quantitative PCR has been used, sometimes with modifications to increase sensitivity, specificity, and reduce the time-consuming in processing (Manimekalai et al., 2010; Nair et al., 2014).

8.6 Reverse transcription-PCR with crude sap extracts

PCR-based assays can either use DNA or RNA as template nucleic acid. DNA has been used in most phytoplasma diagnostic assays. However, this requires purified DNA and hence is relatively time-consuming. Alternatively, RT-PCR-based phytoplasma diagnostic assays have been developed as a rapid method with similar sensitivity and specificity to nested PCR (Margaria and Palmano, 2013). Since the assay can be performed with crude sap extracts, the protocol could have applicability in large-scale screening of plant materials for phytoplasma presence and epidemiological studies. Further, the developers of the assay confirm the possibility of using this method in combination with protocols for the simultaneous detection of other pathogens such as RNA viruses using the same crude sap extract. In Asia, research efforts to detect phytoplasmas using a reverse transcription assay are very rare to date.

8.7 Loop-mediated isothermal amplification for field pathogen detection

Recently, LAMP assays have been considered as an alternative technique for in-field detection of phytoplasmas (Dickinson, 2015; Hodgetts, 2018). This approach requires the activity of *Bst* DNA polymerase and a set of four to six special primers designed to recognize six distinct regions on the target gene, and the completion of the LAMP reaction is usually within 30 min at a single temperature between 60 and 65°C (Notomi et al., 2000; Obura et al., 2011). The advantages of LAMP assays over PCR methods are due to cost-effectiveness, analytical specificity, and sensitivity. Therefore, LAMP-based assays have been studied and developed as a rapid, simple, and low-cost diagnostic tool for phytoplasmas within one hour of sampling in the field that is appropriate to poor and developing countries (Hodgetts et al., 2011; Tomlinson et al., 2010). The real-time LAMP system has been successfully demonstrated for rapid detection of the phytoplasma-associated diseases in plants (Tomlinson et al., 2013; Bekele et al., 2011; Kogovsek et al., 2015, 2017; Nair et al., 2016; Vu et al., 2016; Sugawara et al., 2012; Quoc et al., 2020, 2021a, 2021b). Another approach that can be used for in-field detection of phytoplasma-associated diseases is the recombinase polymerase amplification (RPA). With the advantages of this technique as the rapid, sensitive, and cost-effective tool, RPA has been used to detect '*Ca.* P. oryzae' (Wambua et al., 2017) and '*Ca.* P. mali' (Valasevich and Schneider, 2017) until now.

8.8 Microarray

Microarray is a versatile technique for the detection of a wide range of plant pathogens including phytoplasmas. There are microarray formats which analyze single sample to multiple samples up to 96 samples (Nicolaisen et al., 2013). It is a method that uses PCR followed by oligonucleotide probe hybridization. PCR is required to enhance sensitivity of the assay when low titer of pathogen is present as in the case of phytoplasmas. Microarray has been successfully used as a rapid method of detection of phytoplasmas in the groups of 16SrI, 16SrII, 16SrIII, 16SrV, 16SrVI, 16SrVII, 16SrIX, 16SrX, and 16SrXII (Nicolaisen and Bertaccini, 2007). However, microarray has not been a method of choice as diagnostic tool because of unavailability of specific probes and high throughput technology costs of the reader to verify the results.

8.8.1 High-resolution melt analysis and digital PCR

With advances in molecular genetics and technology, new techniques are becoming available that can be used to increase detection sensitivity and reduce the time and resources involved by eliminating many of the post-PCR steps that are involved in RFLP analyses and nested PCR techniques (Harrison et al., 1996). One such technology that has the potential to accomplish a faster and more sensitive diagnostic protocol is qPCR coupled with high-resolution melt analysis (HRMA). This technique allows the confirmation that an amplified product from the qPCR assay is the region of interest based on a signature melting temperature that matches a positive control, either a plant with a known phytoplasma infection or a plasmid with the appropriate insert.

The use of HRMA to detect plant pathogens and differentiate between strains of various pathogens is becoming a common tool in both basic and applied research and could be a valuable tool in rapidly and reliably detecting and identifying phytoplasmas in palm trees (Bahder et al., 2017). The quicker turnaround for results could ultimately reduce the costs for researchers as well as relieve disease pressures due to the faster accumulation of data, which can be essential for effective management strategies to be implemented faster, thus reducing economic loss. Furthermore, the usefulness of digital PCR (dPCR) has been demonstrated in palm phytoplasmas in the Caribbean and significantly increases sensitivity. The newly emerging protocols can not only improve diagnostics of infected palms but also aid in phytoplasma insect vector discovery (Bahder et al., 2019). Chip-based dPCR technology takes a single sample and performs approximately 20,000 PCR replicates of the same sample. Due to the physical nature, this reduces the effect of inhibitors and allows consistent replication of single-copy targets, ultimately making the reactions more sensitive and more accurate for quantification of the target. This technology is suited for both basic and applied projects. Currently, in Florida, nurseries can test many palms at once to confirm they are free of phytoplasma presence. By pooling 20–50 palms in a reaction and still being able to detect a single individual, this technology significantly reduces the cost associated with screening large cohorts of palms. It is also suitable for screening insect vector candidates by testing salivary glands or also screening pooled batches of insect species.

8.9 DNA barcoding of phytoplasmas with 16S rRNA and *tuf* gene sequences

DNA barcoding is relatively an easily adoptable technique for the identification of phytoplasmas. DNA extracted from test samples is amplified with generic primers to obtain short DNA sequences known as DNA barcodes. DNA sequences of amplicons are then compared with sequences in the EPPO Q-bank phytoplasma database (http://www.q-bank.eu/) which contains DNA barcodes to assign '*Ca*. Phytoplasma' species. The DNA barcoding system available for the identification of phytoplasmas is based on 16S rRNA and *tuf* gene sequences (Makarova et al., 2013). Primers for *tuf* gene amplify 420–444 bp barcode, whereas primers for 16S rRNA amplify 600 bp barcode. When sequences of amplicons were searched on the database, both barcodes gave similar results for the identification. This protocol can be used to identify phytoplasmas in many ribosomal groups, 16SrI, 16SrII, 16SrIII, 16SrIV, 16SrV, 16SrVI, 16SrVII, 16SrIX, 16SrX, 16SrXI, 16SrXII, 16SrXV, 16SrXX, and 16SrXXI.

References

Abeysinghe, S., Abeysinghe, P.D., Kanatiwela-de Silva, C., Udagama, P., Warawichanee, K., Aljafar, N., Dickinson, M., 2016. Refinement of the taxonomic structure of 16SrXI and 16SrXIV phytoplasmas of gramineous plants using multilocus sequence typing. Plant Dis. 100 (10), 2001–2010.

Al-Saady, N., AL-Subhi, A., AL-Nabhani, A., Khan, A., 2006. First report of a group 16SrII phytoplasma infecting chickpea in Oman. Plant Dis. 90, 973-973.

Al-Subhi, A., AL-Saady, N., Khan, A., Deadman, M., 2011. First report of a group 16SrII phytoplasma associated with witches'broom of eggplant in Oman. Plant Dis. 95, 360-360.

Anderson, T.F., 1951. Techniques for the preservation of three-dimensional structure in preparing specimens for the electron microscope. Trans. N. Y. Acad. Sci. 13, 130–134.

Andrade, N.M., Arismendi, N.L., 2013. DAPI Staining and Fluorescence Microscopy Techniques for Phytoplasmas. *Phytoplasma*. Springer.

Arnaud, G., Malembic-Maher, S., Salar, P., Bonnet, P., Maixner, M., Marcone, C., Boudon-Padieu, E., Foissac, X., 2007. Multilocus sequence typing confirms the close genetic interrelatedness of three distinct "flavescence dorée" phytoplasma strain clusters and group 16SrV phytoplasmas infecting grapevine and alder in Europe. Appl. Environ. Microbiol. 73, 4001–4010.

Ashwathappa, K.V., Venkataravanappa, V., Cheegatagere, L.R., Manem, K.R., 2021. Multilocus sequence analysis of '*Candidatus* Phytoplasma aurantifolia' associated with phyllody disease of gerbera from India. Acta Virol. 65, 89–96.

Bahder, B.W., Helmick, E.E., Harrison, N.A., 2017. Detecting and differentiating phytoplasmas belonging to subgroups 16SrIV-A and 16SrIV- D associated with lethal declines of palms in Florida using qPCR and high-resolution melt analysis (HRMA). Plant Dis. 101, 1449–1454.

Bahder, B.W., Soto, N., Komondy, L., Mou, D., Humphries, A.R., Helmick, E.E., 2019. Detection and quantification of the 16SrIV-D phytoplasma in leaf tissue of common ornamental palm species in Florida using qPCR and dPCR. Plant Dis. 103, 1918–1922.

Bekele, B., Abeysinghe, S., Hoat, T.X., Hodgetts, J., Dickinson, M., 2011a. Development of specific secA-based diagnostics for the 16SrXI and 16SrXIV phytoplasmas of the Gramineae. Bull. Insectol. 64 (Suppl), S15−S16.

Bekele, B., Hodgetts, J., Tomlinson, J., Boonham, N., Nikolić, P., Swarbrick, P., Dickinson, M., 2011b. Use of a real-time LAMP isothermal assay for detecting 16SrII and XII phytoplasmas in fruit and weeds of the Ethiopian Rift Valley. Plant Pathol. 60, 345−355.

Bertaccini, A., Arocha-Rosete, Y., Contaldo, N., Duduk, B., Fiore, N., Montano, H.G., Kube, M., Kuo, C.-H., Martini, M., Oshima, K., Quaglino, F., Schneider, B., Wei, W., Zamorano, A., 2022. Revision of the 'Candidatus Phytoplasma' species description guidelines. Int. J. Syst. Evol. Microbiol. 72, 005353.

Bertaccini, A., Fiore, N., Zamorano, A., Tiwari, A.K., Rao, G.P., 2019. Molecular and serological approaches in detection of phytoplasmas in plants and insects. Phytoplasmas: Plant Pathogenic Bacteria - III Genomics, Host Pathogen Interactions, and Diagnosis. Springer Nature, Singapore, pp. 105−136.

Bertaccini, A., Marani, F., 1980. Mycoplasma-like organisms in *Gladiolus* sp. with malformed and virescent flowers. Phytopath. Medit. 19, 121−128.

Bertaccini, A., Marani, F., 1982. Electron microscopy of two viruses and mycoplasma-like organisms in lilies with deformed flowers. Phytopath. Medit. 21, 8−14.

Biswas, C., Dey, P., Mandal, K., Mitra, J., Satpathy, S., Karmakar, P., 2014. First report of a 16SrI-B phytoplasma associated with phyllody and stem fasciation of flax (*Linum usitatissimum*) in India. Plant Dis. 98, 1267-1267.

Boudon-Padieu, E., Béjat, A., Clair, D., Larrue, J., Borgo, M., 2003. Grapevine yellows: comparison of different procedures for DNA extraction and amplification with PCR for routine diagnosis of phytoplasmas in grapevine. Vitis 42, 141−149.

Brzin, J., Ermacora, P., Osler, R., Loi, N., Ravnikar, M., 2003. Detection of apple proliferation phytoplasma by ELISA and PCR in growing and dormant apple trees. Z. für Pflanzenkrankh. Pflanzenschutz 110, 476−483.

Buxa, S., Pagliari, L., Musetti, R., 2013. Epifluorescence microscopy imaging of phytoplasmas in embedded leaf tissues using DAPI and SYTO13 fluorochromes. Microscopie 25 (1), 49−56.

Christensen, N.M., Nyskjold, H., Nicolaisen, M., 2013. Real-time PCR for universal phytoplasma detection and quantification. In: Phytoplasma. Springer.

Clair, D., Larrue, J., Aubert, G., Gillet, J., Cloquemin, G., Boudon-Padieu, E., 2003. A multiplex nested-PCR assay for sensitive and simultaneous detection and direct identification of phytoplasma in the elm yellows group and "stolbur" group and its use in survey of grapevine yellows in France. Vitis 42, 151−157.

Contaldo, N., Canel, A., Makarova, O., Paltrinieri, S., Bertaccini, A., Nicolaisen, M., 2011. Use of a fragment of the tuf gene for phytoplasma 16Sr group/subgroup differentiation. Bull. Insectol. 64 (Suppl.), S45−S46.

Cordova, I., Oropeza, C., Almeyda, H., Harrison, N., 2000. First report of a phytoplasma-associated leaf yellowing syndrome of palma jipi plants in southern Mexico. Plant Dis. 84, 807-807.

Cousin, M., Sharma, A., Misra, S., 1986. Correlation between light and electron microscopic observations and identification of mycoplasmalike organisms using consecutive 350 nm thick sections. J. Phytopathol. 115, 368−374.

Danet, J-L., Balakishiyeva, G., Cimerman, A., Sauvion, N., Marie-Jeanne, V., Labonne, G., Laviña, A., Batlle, A., Križanac, I., Škorić, D., 2011. Multilocus sequence analysis reveals the genetic diversity of European fruit tree phytoplasmas and supports the existence of inter-species recombination. Microbiology 157, 438−450.

Danks, C., Boonham, N., 2007. Purification method and kits. Patent WO/2007/104962.

Deeley, J., Stevens, W., Fox, R., 1979. Use of Dienes' stain to detect plant diseases induced by mycoplasmalike organisms. Phytopathology 69, 169−1171.

Devonshire, B.J., 2013. Visualization of Phytoplasmas Using Electron Microscopy. Phytoplasma. Springer.

Dickinson, M., Hodgetts, J., 2013. PCR Analysis of Phytoplasmas Based on the *secA* Gene. Phytoplasma. Springer.

Dickinson, M., 2015. Loop-mediated isothermal amplification (LAMP) for detection of phytoplasmas in the field. Methods Mol. Biol. 1302, 99−111.

References

Doi, Y., Teranaka, M., Yora, K., Asuyama, H., 1967. Mycoplasma-or PLT group-like microorganisms found in the phloem elements of plants infected with mulberry dwarf, potato witches' broom, aster yellows, or paulownia witches' broom. Japanese J. Phytopathol. 33, 259–266.

Duduk, B., Paltrinieri, S., Lee, I-M., Bertaccini, A., 2013. Nested PCR and RFLP analysis based on the 16S rRNA gene. In: Phytoplasma. Springer.

Gatineau, F., Jacob, N., Vautrin, S., Larrue, J., Lherminier, J., Richard-Molard, M., Boudon-Padieu, E., 2002. Association with the Syndrome "basses richesses" of sugar beet of a phytoplasma and a bacterium-like organism transmitted by a *Pentastiridius* sp. Phytopathology 92, 384–392.

Ghosh, D., Das, A., Singh, S., Singh, S., Ahlawat, Y., 1999. Occurrence of witches'broom, a new phytoplasma disease of acid lime (*Citrus aurantifolia*) in India. Plant Dis. 83, 302-302.

Gopinath, R., Deesma, K.P., Soumya, V.P., Mohanan, C., Ramaswamy, M., 2016. Application of light microscopic staining techniques for the detection of phytoplasmas in yellow leaf disease affected arecanut palms in India. Phytopath. Moll. 6, 77–81.

Gundersen, D., Lee, I-M., 1996. Ultrasensitive detection of phytoplasmas by nested-PCR assays using two universal primer pairs. Phytopath. Mediterr. 144–151.

Gurr, G.M., Gurr, G.M., 2007. Australian Lucerne Yellows Disease: Testing and Extension of Disease Management Strategies. Rural Industries Research and Development Corporation.

Haggis, G., Sinha, R., 1978. Scanning electron microscopy of mycoplasmalike organisms after freeze fracture of plant tissues affected with clover phyllody and aster yellows. Phytopathology 68, 677–680.

Harrison, N.A., Davis, R.E., Helmick, E.E., 2013. DNA extraction from arborescent monocots and how to deal with other challenging hosts. In: Phytoplasma. Springer.

Harrison, N.A., Richardson, P.S., Tsai, J.H., 1996. PCR assay for detection of the phytoplasma associated with maize bushy stunt disease. Plant Dis. 80, 263–269.

Hiruki, C., Da Rocha, A., 1986. Histochemical diagnosis of mycoplasma infections in *Catharanthus roseus* by means of a fluorescent DNA-binding agent, 4, 6-diamidino-2-phenylindole-2HCl (DAPI). J. Indian Dent. Assoc. 8, 185–188.

Hodgetts, J., Boonham, N., Mumford, R., Harrison, N., Dickinson, M., 2008. Phytoplasma phylogenetics based on analysis of *secA* and 23S rRNA gene sequences for improved resolution of candidate species of 'Candidatus Phytoplasma'. Int. J. Syst. Evol. Microbiol. 58, 1826–1837.

Hodgetts, J., Tomlinson, J., Boonham, N., Gonzalez-Martin, I., Nikolić, P., Swarbrick, P., Yankey, E.N., Dickinson, M., 2011. Development of rapid in-field loop-mediated isothermal amplification (LAMP) assays for phytoplasmas. Bull. Insectol. 64 (Suppl.), S41–S42.

Hodgetts, J., Johnson, G., Perkins, K., Ostoja-Starzewska, S., Boonham, N., Mumford, R., Dickinson, M., 2014. The development of monoclonal antibodies to the secA protein of cape St. Paul wilt disease phytoplasma and their evaluation as a diagnostic tool. Mol. Biotechnol. 56, 803–813.

Hodgetts, J., 2018. Rapid sample preparation and LAMP for phytoplasma detection. In: Musetti, R., Pagliari, L. (Eds.), Phytoplasmas—Methods and Protocols. Humana Press, pp. 187–201.

Horne, R., 1970. The ultrastructure of mycoplasma and mycoplasma-like organisms. Micron 2, 19–38.

Jardim, B.R., Kinoti, W.M., Tran-Nguyen, L.T., Gambley, C., Rodoni, B., Constable, F.E., 2021. 'Candidatus Phytoplasma stylosanthis', a novel taxon with a diverse host range in Australia, characterised using multilocus sequence analysis of 16S rRNA, *secA*, *tuf*, and *rp* genes. Int. J. Syst. Evol. Microbiol. 71.

Johannesen, J., Foissac, X., Kehrli, P., Maixner, M., 2012. Impact of vector dispersal and host-plant fidelity on the dissemination of an emerging plant pathogen. PLoS One 7, e51809.

Kanatiwela-de Silva, C., Damayanthi, M., de Silva, R., Dickinson, M., de Silva, N., Udagama, P., 2015. Molecular and scanning electron microscopic proof of phytoplasma associated with areca palm yellow leaf disease in Sri Lanka. Plant Dis. 99 (11), 1641.

Kanatiwela-de Silva, C., Damayanthi, M., de Silva, N., Wijesekera, R., Dickinson, M., Weerakoon, D., 2019. Immunological detection of the Weligama coconut leaf wilt disease associated phytoplasma: development and validation of a polyclonal antibody based indirect ELISA. PLoS One 14 (4), e0214983.

Kogovsek, P., Hodgetts, J., Hall, J., Prezelj, N., Nikolic, P., Mehle, N., Lenarcic, R., Rotter, A., Dickinson, M., Boonham, N., Demastia, M., Ravnikar, M., 2015. LAMP assay and rapid sample preparation method for on-site detection of "flavescence dorée" phytoplasma in grapevine. Plant Pathol. 64 (2), 286–296.

Kogovsek, P., Mehle, N., Pugelj, A., Jakomin, T., Schroers, H.J., Ravnikar, M., Dermastia, M., 2017. Rapid loop-mediated isothermal amplification assays for grapevine yellows phytoplasmas on crude leaf-vein homogenate has the same performance as qPCR. Eur. J. Plant Pathol. 148 (1), 75–84.

Kumar, S., Jadon, V.S., Rao, G., 2017. Use of *secA* gene for characterization of phytoplasmas associated with sugarcane grassy shoot disease in India. Sugar Tech 19, 632–637.

Langer, M., Maixner, M., 2004. Molecular characterisation of grapevine yellows associated phytoplasmas of the stolbur-group based on RFLP-analysis of non-ribosomal DNA. Vitis 43, 191–199.

Lebsky, V., Poghosyan, A., 2014. Scanning electron microscopy detection of phytoplasmas and other phloem limiting pathogens associated with emerging diseases of plants. In: Microscopy: Advances in Scientific Research and Education. Formatex Research Center, Barcelona, pp. 78–83.

Lee, I.-M., Bottner-Parker, K., Zhao, Y., Davis, R.E., Harrison, N., 2010. Phylogenetic analysis and delineation of phytoplasmas based on *secY* gene sequences. Int. J. Syst. Evol. Microbiol. 60, 2887–2897.

Lee, I.-M., Bottner-Parker, K.D., Zhao, Y., Bertaccini, A., Davis, R.E., 2012. Differentiation and classification of phytoplasmas in the pigeon pea witches' broom group (16SrIX): an update based on multiple gene sequence analysis. Int. J. Syst. Evol. Microbiol. 62, 2279–2285.

Lee, I.-M., Gundersen-Rindal, D., Davis, R.E., Bottner, K., Marcone, C., Seemüller, E., 2004a. '*Candidatus* Phytoplasma asteris', a novel phytoplasma taxon associated with aster yellows and related diseases. Int. J. Syst. Evol. Microbiol. 54, 1037–1048.

Lee, I.-M., Gundersen-Rindal, D.E., Bertaccini, A., 1998a. Phytoplasma: ecology and genomic diversity. Phytopathology 88, 1359–1366.

Lee, I.-M., Gundersen-Rindal, D.E., Davis, R.E., Bartoszyk, I.M., 1998b. Revised classification scheme of phytoplasmas based on RFLP analyses of 16S rRNA and ribosomal protein gene sequences. Int. J. Syst. Evol. Microbiol. 48, 1153–1169.

Lee, I.-M., Hammond, R., Davis, R.E., Gundersen, D.E., 1993. Universal amplification and analysis of pathogen 16S rDNA for classification and identification of mycoplasmalike organisms. Phytopathology 83, 834–842.

Lee, I.-M., Zhao, Y., Bottner, K., 2006. *SecY* gene sequence analysis for finer differentiation of diversestrains in the aster yellows phytoplasma group. Mol. Cell. Probes 20, 87–91.

Lee, I.-M., Martini, M., Marcone, C., Zhu, S.F., 2004b. Classification of phytoplasma strains in the elm yellows group (16SrV) and proposal of '*Candidatus* Phytoplasma ulmi' for the phytoplasma associated with elm yellows. Int. J. Syst. Evol. Microbiol. 54, 337–347.

Lee, S., Kim, C-E., Cha, B., 2012. Migration and distribution of graft-inoculated jujube witches'broom phytoplasma within a *Catharanthus roseus* plant. Plant Pathol. J. 28, 191–196.

Li, Z-N., Liu, P., Zhang, L., Wu, Y-F., 2013. Detection and identification of the phytoplasma associated with China ixeris *(Ixeridium chinense)* fasciation. Botanical studies 54, 1–6.

Li, Z., Tang, Y., She, X., Yu, L., Lan, G., He, Z., 2019. First report of 16SrII-D phytoplasma associated with eggplant phyllody in China. J. Indian Dent. Assoc. 41, 339–344.

Lin, C-P., Chen, T.A., 1985. Monoclonal antibodies against the aster yellows agent. Science 227, 1233–1235.

Loi, N., Ermacora, P., Carraro, L., Osler, R., Chen, T.A., 2002. Production of monoclonal antibodies against apple proliferation phytoplasma and their use in serological detection. Eur. J. Plant Pathol. 108, 81–86.

Maejima, K., Oshima, K., Namba, S., 2014. Exploring the phytoplasmas, plant pathogenic bacteria. J. Gen. Plant Pathol. 80, 210–221.

References

Makarova, O., Contaldo, N., Paltrinieri, S., Bertaccini, A., Nyskjold, H., Nicolaisen, M., 2013. DNA barcoding for phytoplasma identification. In: Phytoplasma. Springer.

Manimekalai, R., Soumya, V., Sathish Kumar, R., Selvarajan, R., Reddy, K., Thomas, G., Sasikala, M., Rajeev, G., Baranwal, V., 2010. Molecular detection of 16SrXI group phytoplasma associated with root (wilt) disease of coconut *(Cocos nucifera)* in India. Plant Dis. 94, 636-636.

Marcone, C., Lee, I-M., Davis, R.E., Ragozzino, A., Seemüller, E., 2000. Classification of aster yellows-group phytoplasmas based on combined analyses of rRNA and *tuf* gene sequences. Int. J. Syst. Evol. Microbiol. 50, 1703−1713.

Marcone, C., Pierro, R., Tiwari, A., Rao, G.P., 2022. Major phytoplasma diseases of temperate fruit trees. Agrica 11 (1), 19−32.

Margaria, P., Palmano, S., 2013. Reverse transcription-PCR for phytoplasma detection utilizing crude sap extractions. In: Phytoplasma. Springer.

Martini, M., Botti, S., Marcone, C., Marzachi, C., Casati, P., Bianco, P., Benedetti, R., Bertaccini, A., 2002. Genetic variability among "flavescence dorée" phytoplasmas from different origins in Italy and France. Mol. Cell. Probes 16, 197−208.

Martini, M., Lee, I-M., Bottner, K., Zhao, Y., Botti, S., Bertaccini, A., Harrison, N., Carraro, L., Marcone, C., Khan, A., 2007. Ribosomal protein gene-based phylogeny for finer differentiation and classification of phytoplasmas. Int. J. Syst. Evol. Microbiol. 57, 2037−2051.

Martini, M., Lee, I-M., 2013. PCR and RFLP analyses based on the ribosomal protein operon. In: Phytoplasma. Springer.

Mehdi, A., Baranwal, V.K., Kochu Babu, M., Praveena, D., 2012. Sequence analysis of 16S rRNA and *secA* genes confirms the association of 16SrI-B subgroup phytoplasma with oil palm (*Elaeis guineensis* Jacq.) stunting disease in India. J. Phytopathol. 160, 6−12.

Mehle, N., Nikolić, P., Rupar, M., Boben, J., Ravnikar, M., Dermastia, M., 2013. Automated DNA extraction for large numbers of plant samples. In: Phytoplasma. Springer.

Mitrović, J., Kakizawa, S., Duduk, B., Oshima, K., Namba, S., Bertaccini, A., 2011. The *groEL* gene as an additional marker for finer differentiation of 'Candidatus Phytoplasma asteris'-related strains. Ann. Appl. Biol. 159, 41−48.

Musetti, R., DI Toppi, L.S., Ermacora, P., Favali, M., 2004. Recovery in apple trees infected with the apple proliferation phytoplasma: an ultrastructural and biochemical study. Phytopathology 94, 203−208.

Musetti, R., Favali, M., Pressacco, L., 2000. Histopathology and Polyphenol Content in Plants Infected by Phytoplasmas. Cytobios-Cambridge, pp. 133−148.

Musetti, R., Favali, M.A., 2004. Microscopy techniques applied to the study of phytoplasma diseases: traditional and innovative methods. Curr. Issues Multidisciplinary Microsc. Res. Educ. 2, 72−80.

Musetti, R., Loi, N., Carraro, L., Ermacora, P., 2002. Application of immunoelectron microscopy techniques in the diagnosis of phytoplasma diseases. Microsc. Res. Tech. 56, 462−464.

Nair, S., Manimekalai, R., Raj, P.G., Hegde, V., 2016. Loop mediated isothermal amplification (LAMP) assay for detection of coconut root wilt disease and arecanut yellow leaf disease phytoplasma. World J. Microbiol. Biotechnol. 32, 1−7.

Nair, S., Roshna, O., Soumya, V., Hegde, V., Kumar, M.S., Manimekalai, R., Thomas, G.V., 2014. Real-time PCR technique for detection of arecanut yellow leaf disease phytoplasma. Australas. Plant Pathol. 43, 527−529.

Namba, S., Yamashita, S., Doi, Y., Yora, K., 1981. Direct fluorescence detection method (DFD method) for diagnosing yellows-type virus diseases and mycoplasma diseases of plants. Japanese J. Phytopathol. 47, 258−263.

Nicolaisen, M., Bertaccini, A., 2007. An oligonucleotide microarray-based assay for identification of phytoplasma 16S ribosomal groups. Plant Pathol. 56, 332−336.

Nicolaisen, M., Nyskjold, H., Bertaccini, A., 2013. Microarrays for universal detection and identification of phytoplasmas. In: Phytoplasma. Springer.

Nipah, J., Jones, P., Dickinson, M., 2007. Detection of lethal yellowing phytoplasma in embryos from coconut palms infected with Cape St Paul wilt disease in Ghana. Plant Pathol. 56, 777–784.

Notomi, T., Okayama, H., Masubuchi, H., Yonekawa, T., Watanabe, K., Amino, N., Hase, T., 2000. Loop-mediated isothermal amplification of DNA. Nucleic Acids Res. 28, e63-e63.

Obura, E., Massiga, D., Wachira, F., Gurja, B., Khan, Z.R., 2011. Detection of phytoplasma by loop-mediated isothermal amplification of DNA (LAMP). J. Microbiol. Methods 84, 312–316.

Ong, S., Jonson, G.B., Calassanzio, M., Rin, S., Chou, C., Oi, T., Sato, I., Takemoto, D., Tanaka, T., Choi, I-R., 2021. Geographic distribution, genetic variability and biological properties of rice orange leaf phytoplasma in southeast Asia. Pathogens 10, 169.

Oropeza, C., Cordova, I., Chumba, A., Narváez, M., Sáenz, L., Ashburner, R., Harrison, N., 2011. Phytoplasma distribution in coconut palms affected by lethal yellowing disease. Ann. Appl. Biol. 159, 109–117.

Palmano, S., 2001. A comparison of different phytoplasma DNA extraction methods using competitive PCR. Phytopath. Mediterr. 40, 99–107.

Pasternak, T., Tietz, O., Rapp, K., Begheldo, M., Nitschke, R., Ruperti, B., Palme, K., 2015. Protocol: an improved and universal procedure for whole-mount immunolocalization in plants. Plant Methods 11, 1–10.

Poghosyan, A., Lebsky, V., Arce-Montoya, M., Landa, L., 2004. Possible phytoplasma disease in papaya (*Carica papaya* L.) from Baja California Sur: diagnosis by scanning electron microscopy. J. Phytopathol. 152, 376–380.

Purohit, S., Kg, R., Hc, A., 1978a. Light microscopic detection of mycoplasma-like organism (MLO) in sesamum phyllody. Curr. Science 47 (22), 866–867.

Quoc, N.B., Xuan, N.T.T., Nghiep, N.M., Phuong, N.D.N., Linh, T.B., Chau, N.N.B., Chuong, N.D.X., Nien, N.C., Dickinson, M., 2021a. Loop-mediated isothermal amplification (LAMP) assay for detection of sesame phyllody phytoplasmas in Vietnam. Folia Microbiol. 66, 273–283.

Quoc, N., Ntt, X., Ndn, P., Trang, H., Nnb, C., Duong, C., Dickinson, M., 2021b. Development of loop mediated isothermal amplification assays for the detection of sugarcane white leaf disease. Physiol. Mol. Plant Pathol. 113, 101595.

Quoc, N.B., Chau, N.N.B., Duong, C.A., 2020. Sugarcane white leaf and grassy shoot management for healthy seed production in Vietnam. In: Tiwari, A.K. (Ed.), Advances in Seed Production and Management. Springer, Singapore.

Raj, S., Khan, M., Snehi, S., Kumar, S., Mall, S., Rao, G., 2008. First report of phytoplasma 'Candidatus Phytoplasma asteris'(16SrI) from *Parthenium hysterophorus* L. showing symptoms of virescence and witches'broom in India. Australas. Plant Dis. Notes 3, 44–45.

Rao, G., 2018. Molecular characterization of phytoplasma associated with four important ornamental plant species in India and identification of natural potential spread sources. 3 Biotech 8, 1–12.

Rao, G.P., Madhupriya, T.V., Manimekalai, R., Tiwari, A.K., Yadav, A., 2017. A century progress of research on phytoplasma diseases in India. Phytopath. Mollic. 7, 1–38.

Russell, W., Newman, C., Williamson, D., 1975. A simple cytochemical technique for demonstration of DNA in cells infected with mycoplasmas and viruses. Nature 253, 461–462.

Schaper, U., Seemüller, E., 1982. Condition of the phloem and the persistence of mycoplasmalike organisms associated with apple proliferation and pear decline. Phytopathology 72, 736–742.

Schneider, B., Gibb, K.S., 1997. Sequence and RFLP analysis of the elongation factor Tu gene used in differentiation and classification of phytoplasmas. Microbiology 143, 3381–3389.

Shantha, P., Lakshmanan, M., 1984. Vector-borne MLOs of brinjal little leaf. Curr. Sci. 53 (5), 265–267.

Siddique, A., Guthrie, J., Walsh, K., White, D., Scott, P., 1998. Histopathology and within-plant distribution of the phytoplasma associated with Australian papaya dieback. Plant Dis. 82, 1112–1120.

Snehi, S., Parihar, S., Jain, B., 2021. First report of a jujube witches'broom phytoplasma (16SrV) strain associated with witches'broom and little leaf disease of *Solanum melongena* in India. New Dis. Reports 43, e12005.

Sugawara, K., Himeno, M., Keima, T., Kitazawa, Y., Maejima, K., Oshima, K., Namba, S., 2012. Rapid and reliable detection of phytoplasma by loop-mediated isothermal amplification targeting a housekeeping gene. J. Gen. Plant Pathol. 78, 389–397.

Sunpapao, A., 2014. Association of 'Candidatus Phytoplasma cynodontis' with the yellow leaf disease of ivy gourd in Thailand. Australas. Plant Dis. Notes 9, 1–3.

Tomlinson, J., 2013. In-field diagnostics using loop-mediated isothermal amplification. In: Phytoplasma. Springer.

Tomlinson, J., Boonham, N., Dickinson, M., 2010. Development and evaluation of a one-hour DNA extraction and loop-mediated isothermal amplification assay for rapid detection of phytoplasmas. Plant Pathol. 59, 465–471.

Valasevich, N., Schneider, B., 2017. Rapid detection of 'Candidatus Phytoplasma mali' by recombinase polymerase amplification assays. J. Phytopathol. 165 (11–12), 762–770.

Varma, A., Chenulu, V., Raychaudhuri, S., Prakash, N., Rao, P., 1969. Mycoplasma-like bodies in tissues infected with sandal spike and brinjal little leaf. Indian Phytopath. 22 (2), 289–291.

Vázquez-Euán, R., Harrison, N., Narvaez, M., Oropeza, C., 2011. Occurrence of a 16SrIV group phytoplasma not previously associated with palm species in Yucatan, Mexico. Plant Dis. 95, 256–262.

Vera, C., Milne, R., 1994. Immunosorbent electron microscopy and gold label antibody decoration of MLOs from crude preparations of infected plants and vector insects. Plant Pathol. 43, 190–199.

Viswanathan, R., 1997. Detection of phytoplasmas associated with grassy shoot disease of sugarcane by ELISA techniques/. J. Plant Dis. Protection 9–16.

Vu, N.T., Pardo, J.M., Alvarez, E., Le, H.H., Wyckhuys, K., Nguyen, K.L., Le, D.T., 2016. Extablishment of a loop-mediated isothermal amplification (LAMP assay for the detection of phytoplasma-associated cassava witches' broom disease. Appl. Biol. Chem. 59 (2), 151–156.

Wambua, L., Schneider, B., Okwaro, A., Wanga, J.O., Imali, O., Wambua, P.N., Agutu, L., Olds, C., Jones, C.S., Masiga, D., Midega, C., Khan, Z., Jores, J., Fischer, A., 2017. Development of field-applicable tests for rapid and sensitive detection of 'Candidatus Phytoplasma oryzae'. Mol. Cell. Probes. 35, 44–56.

Wang, Q., Valkonen, J., 2008. Efficient elimination of sweet potato little leaf phytoplasma from sweetpotato by cryotherapy of shoot tips. Plant Pathol. 57, 338–347.

Waters, H., Hunt, P., 1980. The *in vivo* three-dimensional form of a plant mycoplasma-like organism by the analysis of serial ultrathin sections. Microbiology 116, 111–131.

Wei, W., Davis, R.E., Lee, I-M., Zhao, Y., 2007. Computer-simulated RFLP analysis of 16S rRNA genes: identification of ten new phytoplasma groups. Int. J. Syst. Evol. Microbiol 57, 1855–1867.

Wei, W., Kakizawa, S., Jung, H-Y., Suzuki, S., Tanaka, M., Nishigawa, H., Miyata, S-I., Oshima, K., Ugaki, M., Hibi, T., 2004. An antibody against the SecA membrane protein of one phytoplasma reacts with those of phylogenetically different phytoplasmas. Phytopathology 94, 683–686.

Wei, W., Lee, I-M., Davis, R.E., Suo, X., Zhao, Y., 2008. Automated RFLP pattern comparison and similarity coefficient calculation for rapid delineation of new and distinct phytoplasma 16Sr subgroup lineages. Int. J. Syst. Evol. Microbiol. 58, 2368–2377.

Yu, S-S., Che, H-Y., Wang, S-J., Lin, C-L., Lin, M-X., Song, W-W., Tang, Q-H., Yan, W., Qin, W-Q., 2020. Rapid and efficient detection of 16SrI group Areca palm yellow leaf phytoplasma in China by loop-mediated isothermal amplification. Plant Pathol. J. 36, 459.

CHAPTER 2

Graft and vegetative transmission of phytoplasma-associated diseases in Asia and their management

Kadriye Caglayan[1], Elia Choueiri[2] and Govind Pratap Rao[3]

[1]*Plant Protection Department, Faculty of Agriculture, Hatay Mustafa Kemal University Antakya, Turkey;* [2]*Department of Plant Protection, Lebanese Agricultural Research Institute, Tal Amara, Zahlé, Lebanon;* [3]*Division of Plant Pathology, Indian Agricultural Research Institute, New Delhi, Delhi, India*

1. Introduction

Grafting is a vegetative plant propagation technique known and applied since ancient time. This practice was first applied 4000 years ago in China and Mesopotamia. A number of diverse grafting methods, linked to the characteristics of the scions and the rootstocks, can be employed to verify the presence of phytoplasmas in woody or herbaceous plants. Many plants carry phytoplasmas without symptoms, and the presence of these pathogens in the plants can be verified and confirmed by grafting the suspicious scions onto another highly susceptible variety that will display specific symptoms. There are several grafting methods (*e.g.,* bark or side grafting, whip, and bud grafting) with difference in the procedure and technical details including the types of scion and rootstock (Fig. 2.1) reported to transmit phytoplasmas efficiently (Zambon et al., 2017; Çağlayan et al., 2019; Akhtar et al., 2021). Sometimes one method is selected for particular plant species or scion–rootstock combination. Nevertheless, regardless of the method used, the principle involved remains the same; the vascular tissues of rootstock and scion have to be joined in a way they can continue to live and grow together as one grafted plant (Fig. 2.1). Graft inoculation is one of the important and widely used experimental means of transmitting phytoplasmas. Grafting may succeed in transmitting phytoplasmas where other methods fail, thus grafting can be used to differentiate plants infected with phytoplasma from plants showing phytoplasma-like symptoms due to nutritional deficiencies or other phenomena (Çağlayan et al., 2019). Although grafted plants have a long history of use in woody/fruit trees, grafting is also used in various herbaceous plants such as vegetables and flowers. Practicing grafting in herbaceous crops is an important tool in East Asia to manage various vulnerable issues to the intensive production.

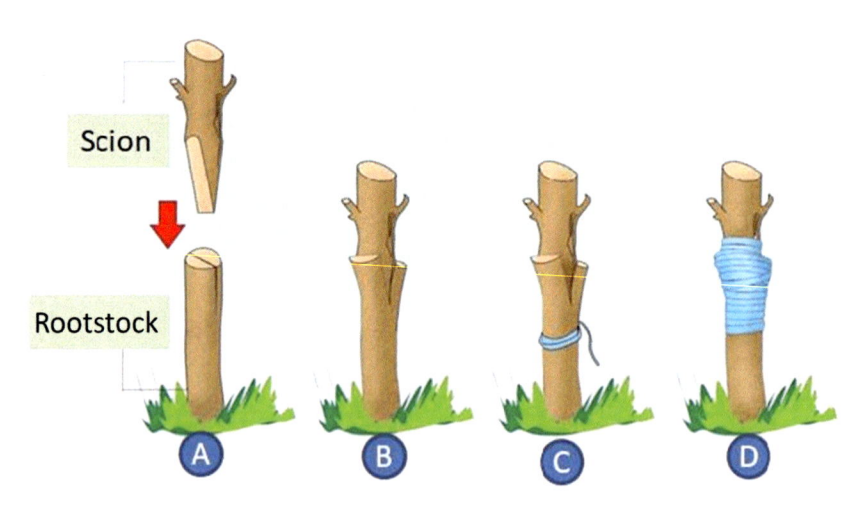

FIGURE 2.1
Process of grafting in fruit trees (A) Scion and rootstock, (B) graft union, (C) and (D) wrapping and tightening.

2. Graft transmission

In this chapter, an up-to-date information on graft transmission of phytoplasmas is described in woody, herbaceous and in vitro propagated plants along with suggested control measures.

2.1 Graft transmission in woody plants

In the woody plants, the vascular (main or wood) cambium is a thin cell layer located between bark and wood and is the important layer of actively growing cells of a tree that should receive the graft (rootstock) and must be put in contact with the same cell layer of the scion that is the plant part to be grafted (Barbosa et al., 2005; Bausher, 2013). A successful graft, at least for a few hours, is necessary for phytoplasma transmission in woody plants (Lee et al., 2000). Phytoplasma transmission is mostly reported very successful through grafting in horticulture crops especially in fruit crops and trees in Asia (Hemmati et al., 2021). '*Candidatus* Phytoplasma pyri', the agent of pear decline, was detected in the tree aerial parts of the pear trees in the Mediterranean areas, then in the central Europe orchards, and although it was shown that it is not transmissible at winter in central Europe, there are confirmations of its transmission by infected scions in the Mediterranean areas (Schaper and Seemüller, 1982; Garcia-Chapa et al., 2003; Yavuz et al., 2011). This variability of phytoplasma persistence according to the climatic conditions suggests the role of environments also in grafting success.

Using phytoplasma-infected planting material, regardless of the source of infection being either from rootstock or scion, the symptom manifestation may be delayed from a few months until several years, according to the plant species and the cultivar involved. Because of the latent period, apparently healthy grafted plants could carry phytoplasmas and this is also a problem when attempts are made to find out the source of infection. Although the phytoplasma detection molecular techniques are very sensitive, there are still difficulties due to seasonal variation of pathogen concentration, uneven distribution, and low concentrations of phytoplasma cells in the woody plants (Seemüller et al., 1984; Garcia-Chapa et al., 2003).

After the grafting from mother plants surviving a severe phytoplasma outbreak, the diverse apricot and Japanese plum varieties obtained were used to verify the phytoplasma graft transmission efficacy under experimental conditions. Three-year-old apricot plants were patch-grafted with apricot tissues infected by European stone fruit yellows (ESFY) and kept in screenhouse. Between one and six months after grafting, they were analyzed by polymerase chain reaction (PCR)/restriction fragment length polymorphism analyses and found positive for the phytoplasma presence. The phytoplasma transmission by patch grafting was demonstrated in apricot and Japanese plum varieties in Italy (Pastore et al., 2001).

Yavuz et al. (2011) and Çağlayan et al. (2014) reported the transmission efficiency of '*Ca.* P. pruni' and '*Ca.* P. pyri' by grafting when wild apricot cv. Zerdali and B29 were employed as rootstocks, which are the most common rootstocks of apricot and pear in Turkey. As inoculum source, phytoplasma-infected buds obtained from apricot cv. Tyrinthe and local pear cv. Deveci growing in open field were used. Fifty rootstocks from each cultivar were grafted by infected buds for each phytoplasma strain. All grafted plants were kept in screenhouse in field, and beside symptom observations, they were also tested every three months by nested PCR. The first positive results of PCR analysis for both phytoplasmas were obtained approximately one year after inoculation, and the transmission rates by grafting for '*Ca.* P. pyri' and '*Ca.* P. pruni' were 6% and 18%, respectively. All inoculated plants died two years after inoculation. These data confirm the high percentage of phytoplasma transmission when the grafting was performed in late summer with high titer of phytoplasmas in plants (Gazel et al., 2012). The correct choice of cultivar/rootstock combination may also affect the success rate of the graft transmission (Landi et al., 2010). Different pear cultivars and rootstock–scion combinations to pear decline phytoplasma were compared for their growth responses in experiments carried out under screenhouse in which each measurement was expressed as a proportion of the total average value of the variable observed in uninoculated control plants (Gazel et al., 2012; Çağlayan et al., 2022). The Turkish local cultivar Deveci was proven as most susceptible comparing to the other local cultivar Ankara and to the well-known pear cultivars, Williams and Santa Maria on Quince rootstock (BA29) than on *Prunus communis* seedlings. Most of the plants belonging to the Deveci x *P. communis* combinations died one year after inoculation.

Almond witches' broom (AlmWB) associated with '*Ca.* P. phoenicium' is an economically important and destructive phytoplasma strain in Lebanon where it was responsible for the death of more than 100,000 almond and peach trees (Choueiri et al., 2001; Abou-Jawdah et al., 2002; Verdin et al., 2003) and also serious decline in stone fruit trees in several provinces of Iran (Salehi et al., 2006, 2011, 2018). Experimental trials of grafting with infected almond shoots onto almond seedlings (*Prunus amygdalus*), plum (*Prunus mariana* GF8-1), and peach (*Prunus persica* GF305) (Fig. 2.2) have been carried out in greenhouse and symptoms appeared after a month (Verdin et al., 2003). However, apricot and plum scions grafted on almond trees infected with AlmWB (Fig. 2.2) did not develop any symptoms, and their growth remained symptomless for more than two years in the field (Tawidian et al., 2017; E. Choueiri, unpublished data) and also tested negative in PCR assay for phytoplasma presence. Additionally, shoots developed from Farclo apricot trees grafted onto AlmWB-infected trees in the field showed clear symptoms after two months from grafting, but after three months, they were recovered and remained symptomless for about 2.5 years (Tawidian et al., 2017).

FIGURE 2.2

(A) Inoculation by grafting: experimental transmission of infected almond shoots by '*Candidatus* phytoplasma phoenicium' on peach GF 305 seedling, (B) plum scions grafted in the field on almond trees infected with almond witches' broom: no symptoms in plum branches and witches' broom in the main trunk of infected almond tree.

Courtesy: Elia Choueiri, LARI, Lebanon.

However, in Iran, '*Ca.* P. phoenicium' from almond-infected scions was transmitted to peach trees and developed symptoms of phytoplasma in graft transmission trials (Salehi et al., 2006). In addition, symptomatic shoots of a wild almond tree (*Prunus scoparia*) with severe witches' brooms associated with the presence of '*Ca.* P. phoenicium' were considered a source of AlmWB transmission in different Iranian provinces and used for disease transmission testing. Graft transmission experiments applying side-grafted scions (twigs with two to three leaves) taken from the infected *P. scoparia* trees on two-year-old seedlings of bitter almond (*Prunus dulcis* var. *amara*), peach (*P. persica* cv. Alberta), nectarine (*P. persica* var. *nucipersica*), and apricot (*Prunus armeniaca* cv. Asefi) were carried out and maintained in an insect-free greenhouse. The *P. scoparia* witches' broom agent from different locations was successfully transferred to all inoculated seedlings of peach nectarine, apricot, and bitter almond. All grafted plants tested positive in PCR assay. Same symptoms in inoculated plants as those produced in naturally infected plants such as witches' broom, yellowing, and spindly shoots were produced with a disease latent period ranging from five months in peach to eight months in apricot seedlings (Salehi et al., 2015). Orange (*Citrus sinensis*) buds infected by 16SrIX group phytoplasma ('*Ca.* P. phoenicium') in Kerman province of Iran were grafted on healthy Valencia and a local orange cultivars, and typical phytoplasma symptoms were observed eight months after inoculation and the results were confirmed by nested PCR analysis (Abbasi et al., 2019).

Graft transmission of apricot yellows associated with '*Ca.* P. phoenicium' was successfully attempted by side grafting of healthy seedlings of bitter almond (*P. amygdalus*) and apricot (*P. armeniaca*: cultivars Nouri, Talkh, Asefi, Tokhme Morghee, and Shekarpareh) in Iran (Salehi et al., 2018). '*Ca.* P. aurantifolia'-related strain showing ESFY symptoms in apricot trees was successfully transmitted to healthy GF-677 (peach × almond) trees through side grafting with symptomatic apricot scions in Iran. The symptoms were expressed after 21 months from inoculation in all the grafted symptomatic GF-677 trees which also tested phytoplasma-positive in PCR assays (Rasoulpour et al., 2019).

In China, jujube witches' broom (JWB) phytoplasma was reported to be graft transmissible in jujube tree (*Zizyphus jujube* cv. Inermis Rehd.) (Kim, 1965; La and Woo, 1980). JWB disease was also found in Korea and Japan (Tsai et al., 1988; Lee, 1988; Ohashi et al., 1996), and the graft transmission risk is very high in these countries too. *Sophora japonica* trees widely planted as decorative ornamental varieties along the streets in China are grafted trees expressing leaf yellowing coloration in normal foliage branches and look naturally beautiful along the streets and were found phytoplasma infected (Duduk et al., 2010; Bertaccini et al., 2014). But these types of practices should be avoided because they would increase the chances of transmission of phytoplasmas strains through insect vectors to other important agricultural/horticultural crops. Petrovic et al. (2000) used the tissue culture technique to allow the identification of increasing grapevine yellows (GY) phytoplasma concentration in the tissues of infected grapevine to improve its detection. The risk of GY dissemination by propagation material is very much linked to its graft transmissibility during the vegetative propagation.

The periwinkle [*Catharanthus roseus* (G) Don] is a common ornamental plant in its ability to form also distant interspecific grafts even though woody buds were used as phytoplasma inoculum source (Çağlayan et al., 2019). Successful graft transmission was achieved with apple proliferation phytoplasma from infected scions to periwinkle in Iran (Aldaghi et al., 2007). '*Ca*. P. asteris' was also successfully transmitted through wedge grafting from the periwinkle plants showing phytoplasma symptoms and produced typical phyllody and virescence symptoms within 45 days (Kumar, 2010). The grapevine phytoplasma agents have been also transmitted from grapevine to periwinkle via grafting in some Asian countries (Tanne and Orenstein, 1997).

2.2 Graft transmission in herbaceous plants

In the early 20th century, the grafting practices were introduced in vegetables and ornamentals. There are a number of phytoplasma strains reported as graft transmissible in different vegetable crops. Eggplant infected by little leaf phytoplasma disease could be successfully transmitted by wedge grafting from eggplant to eggplant (Kumar and Rao, 2017). Phytoplasmas associated with chrysanthemum yellows, *Crotalaria saltiana* phyllody, strawberry green petal, sweet potato little leaf, and potato witches' broom maintained in periwinkle have been also successfully graft transmitted in periwinkle. Typical disease symptoms were observed after six to eight weeks in grafted plants, and nested PCR and *secA* gene amplification and sequencing demonstrated that the transmission of the phytoplasma was successful (Kawicha et al., 2012). Toria phyllody phytoplasma (16SrIX group strain) was graft transmitted to healthy toria plants as well as to other brassicaceous plants in India. This procedure allowed to induce phytoplasma symptoms in yellow and brown sarson, toria, *Eruca sativa*, and *Brassica napus* plants after two months from grafting (Azadvar et al., 2011).

The healthy sugarcane or periwinkle seedling's grafting using scions from sugarcane white leaf diseased plants (16SrXI group phytoplasmas) was also able to successfully transmit the phytoplasma which is further confirmed by PCR assay in both graft recipients, sugarcane, and periwinkle plants in Thailand (Wongkaew and Fletcher, 2004). Damam (2012) reported that sesame phyllody phytoplasma can be successfully transmitted from infected to healthy sesame and produced typical phyllody symptoms within 25–35 days by side patch grafting. Recently, different transmission assays were employed through grafting (wedge, patch, leaf disc, and plug) to see the efficacy of transmission of sesame phyllody phytoplasma, 16SrII-D strain in India. In wedge grafting, the success rate of '*Ca.* P. australasia' (16SrII-D) transmission was higher in sesame-to-sesame plants with an efficiency of 80% in comparison with sesame to periwinkle grafting (77.77%) (Fig. 2.3). The leaf disc grafting showed 75% successful transmission of 16SrII-D subgroup phytoplasma strain in sesame-to-sesame leaf-disc grafted plants. However, the transmission efficiency was recorded low (33.33%) in sesame to periwinkle grafted plants with symptoms of little leaf and leaf chlorosis (Fig. 2.4). The success of patch grafting was observed minimal (<25%) in transmission from sesame to sesame and sesame to periwinkle plants. The plug grafting method showed moderate transmission efficiency of less than 10% in both sesame and periwinkle grafted plants that were confirmed by nested PCR assays (Fig. 2.5). These results clearly indicated that wedge grafting was proved successful in transmission of sesame phyllody phytoplasma in India (Ranebennur et al., 2022). The results observed in patch and plug grafting were not much encouraging with transmission efficiency of <10% which was in agreement with earlier findings of Pastore et al. (2001) who also concluded that though the patch grafting is an efficient method of transmission of '*Ca.* P. prunorum' to apricot and Japanese plum trees in orchards, it is not efficient for experimental transmission when used on young apricot and pear plants and hence these grafting practices would not be recommended. In plug grafting, the survival of inoculated plants (sesame or periwinkle) was the major issue, and the transmission success rate was quite low as compared to other grafting methods.

Recently, a method for successful transmission for the citrus greening bacterium ('*Candidatus* Liberibacter asiaticus') in citrus plants has been proposed, which involves inserting the leaf disc of a

FIGURE 2.3

Wedge grafting procedure (A) wedge-shaped cut at the base of infected sesame scion, (B) wedge placed in the slit of healthy sesame plants inserting phytoplasma-infected sesame scion on to healthy periwinkle plants and (C) wrapping of graft union.

FIGURE 2.4

Leaf disk grafting in periwinkle and sesame (A) and (C) circular leaf disc taken out from healthy sesame and periwinkle leaves (B) and (D) leaf of healthy sesame and periwinkle seedlings grafted with infected leaf disc and adhesive tapes was put above and down the leaf disc grafted union.

diseased leaf into the leaf of a healthy plant (Zambon et al., 2017). This method was used for screening of resistance in citrus germplasms against "huanglongbing" disease. The leaf disc graft transmission was also found successful in transmission of 16SrII-D strain of phytoplasma in sesame associated with sesame phyllody disease in India (Ranebennur et al., 2022). The leaf disc method allows the grafting of single leaf of young seedlings which overcomes the disadvantage of the conventional grafting practices that needs the mature bark and girth of the stem which takes a significant time to reach to the required stage.

Big bud disease associated with phytoplasmas is an emerging threat to tomato production all over Asia (Kumari et al., 2019). The development of resistant varieties would be an effective approach to manage this problem. It requires an appropriate screening technique for indexing tomato germplasm for resistance. Recently, Akhtar et al. (2021) demonstrated a simple and efficient chip graft inoculation assay (CGIA) for transmission of a tomato big bud associated phytoplasma enclosed in the 16SrII-D subgroup. This technique was proved useful in screening tomato germplasms against tomato leaf curl New Delhi virus and tomato big bud phytoplasma in Pakistan. CGIA success rate and phytoplasma transmission was 100% since all the grafts survived, and phytoplasma was detected in these plants using nested PCR. In addition to phytoplasma transmission, CGIA can also be used for better understanding the plant—phytoplasma interactions and biology (Akhtar et al., 2021). This technique involves creating a slit below the tip on the main stem of the tomato seedlings. Small pieces of (about 2.5 to 4 cm long) of tender stem from a phytoplasma-infected source plant were cut, and the bark was removed from two opposite sides of stem expose to the cambium layer. This chip was then inserted into the cut slit of the test plant and tightened with parafilm wrapping. The major advantage of the CGIA

FIGURE 2.5

Plug grafting method (A) plug removed from the healthy sesame plant, (B) replaced with the plug of infected sesame plant, (C) wrapped the plug with the tape, (D) plug grafted sesame plant, and (E) plug grafted periwinkle plant.

Source Ranebennur, H. et al. 2022.

over other grafting methods is that it allows one infected plant to provide a lot of chips (scions) (Akhtar et al., 2019). The CGIA, therefore, offers a promising tool for efficient assessment of resistance against phytoplasma infection across a range of *Solanum* germplasm to develop high-yielding and resistant genotypes/hybrids.

Among the phytoplasma diseases of vegetable crops, alfalfa witches' broom was identified in Fars Province of Iran using graft transmission (Salehi and Izadpanah, 1993). On the other hand, tomato witches' broom associated with a phytoplasma of 16SrII-D subgroup showing stunting, proliferation of the axillary buds, and witches' broom was also transmitted to eggplant by graft inoculation developing witches' broom, flower virescence, yellowing, and phyllody symptoms (Salehi et al., 2014). In addition, a 16SrVI-A strain in tomato fields was transmitted from infected tomato to healthy tomato, eggplant, and periwinkle using grafting and dodder (Salehi et al., 2017). Carrot witches' broom reported in carrot fields of Yazd province was transmitted to healthy periwinkle seedlings developing later phytoplasma symptoms by grafting and dodder (Salehi et al., 2016).

2.3 Transmission by micropropagation

In vitro micropropagation is another useful method which overcomes the limitations of other vegetative propagation methods (Bertaccini et al., 1992; Jarausch et al., 1996; Monteuuis, 2012). The in vitro technique may also employed for the elimination of some phytoplasma strains from diseased plants.

Although the in vitro micropropagation has been used to eliminate pathogens from diseased plants, a few reports indicate micropropagation as a screening or indexing method for the phytoplasma resistance. Mostly in vitro propagation has been attempted in European countries with success rate of over 90% for apple proliferation and ESFY phytoplasmas (Jarausch et al., 1994, 1996). These results demonstrated that this method is a valid tool to carry out a screening for the preliminary assessment of phytoplasma resistance in *Malus* and *Prunus* genotypes. Its advantages compared to usual procedures are because that it is a method rapid, easy, requiring little space, not affected with environmental conditions and/or by the presence of other pathogens. But these techniques are little attempted in Asian countries.

By micropropagation, it is possible to maintain the phytoplasma strain transmitted to periwinkle but also in other naturally infected plant species. The phytoplasma detection in these periwinkle shoots micropropagated was detected positive for up to 30 years (Bertaccini et al., 2012; A. Bertaccini, personal communication).

2.4 Dodder transmission

The thread-like grown vines of dodder (*Cuscuta reflexa*, family: Convolvulaceae) are a very important source of vegetative transmission of phytoplasma strains in different plant species. Several reports are available for successful phytoplasma transmission in periwinkle, toria, sesame, brinjal, coconut, and areca nut through using dodders in different Asian countries (Azadvar et al., 2011; Rao et al., 2015; Kumar and Rao 2017; Ranebennur et al., 2022). In the process, the grown dodder threads were wrapped on phytoplasma-positive symptomatic plants (tested positive for phytoplasma strain). After 15 days, the established dodder strands on infected plants were collected and again wrapped to healthy plants in separate pots under insect-proof conditions (Fig. 2.6). In most of the studies, phytoplasmas are transmitted in two weeks' time. The vegetative transmission studies through dodder in India revealed that '*Ca.* P. australasia' sesame strain could be efficiently transmitted through dodder from infected sesame to healthy sesame (93.33%) and periwinkle plants (92.85%) (Ranebennur et al., 2022). These methods would be helpful in maintaining the phytoplasma strain for further characterization study and can also be suitably used further for indexing sesame germplasm accessions for resistance. Similar results were reported earlier in the transmission of sesame phyllody phytoplasma (16SrI and 6SrII-D) from diseased to healthy sesame plant by dodder (Sertkaya, 2007; Rao et al., 2015; Gogoi et al., 2017). Subsequently, phytoplasmas associated with different plants also proved to be transmitted by *Cuscuta* spp.

The two species of dodder, *Cuscuta europia* and *C. campestris* could successfully transmit different phytoplasma strains worldwide, ESFY, pear decline, *Picris echioides* yellows, rubus stunt, cotton

FIGURE 2.6

Dodder transmission (A) dodder established on infected sesame plant, (B) infected dodder wrapped on healthy sesame plant, (C) infected dodder wrapped on periwinkle plants, (D) dodder reestablished on sesame plants and (E) dodder reestablished on a healthy periwinkle plants.

Source Ranebennur, H. et al. 2022.

phyllody, sesame phyllody, toria phyllody, brinjal little leaf, coconut root wilt, areca nut leaf yellows, and safflower phyllody (Marcone et al., 1997; Salehi et al., 2009; Azadvar et al., 2011). Dodder can be very well used for transmission of phytoplasma strains associated with different plant species for maintaining phytoplasma strains under glasshouse conditions for biological studies and molecular characterization.

3. Management

The use of phytoplasma-free clean planting material is one of the major tools in the management of phytoplasma diseases (Roddee et al., 2018). Establishing of new plantations with certified pathogen-free planting materials is of primary importance for a successful phytoplasma prevention and control

(Bianco et al., 2018). Adequate sampling and using of highly sensitive PCR and recently developed isothermal assays would be useful to ensure that nuclear/mother stocks are free from phytoplasma infections.

In addition, if the vegetative materials would transfer to the new area, then heating or cooling treatment, dipping in hot water, and use of radiation would decrease the chance of new disease introduction to other countries. Heat therapy, hot water treatments, in vitro meristem tip cultures, and tetracycline in combination with in vitro techniques are suggested to be useful in producing clean propagation materials in most of horticultural crops and are efficient in checking further spread of the disease in the establishment of new orchards and/or in cultivation practices (Singh et al., 2007; Tiwari et al., 2011; Linck et al., 2019).

Tissue culture and thermotherapy are techniques recommended for eliminating the phytoplasmas from the diseased plants. For the eradication of '*Ca*. P. prunorum' in diseased *Prunus* plants, in vitro thermotherapy and meristem tip culture have been applied (Laimer, 2003; Laimer and Bertaccini, 2008). Several tissue culture techniques coupled with thermotherapy have been suggested for the production of phytoplasma-free fruit trees plantlets (Laimer and Bertaccini 2019). Stem-cutting culture combined with thermotherapy was also reported to be an effective technique against phytoplasmas. Subsequently shoot tip and stem-cutting cultures associated with heat treatment and shoot tip micrografting were found to be suitable for '*Ca*. P. phoenicium' elimination from regenerated shootlets in almonds (Chalak et al., 2005). Tissue culture coupled with heat therapy or hot water treatment was applied to eliminate grapevine "bois noir" phytoplasma and producing 100% sanitized shoots in Lebanon (Chalak et al., 2013).

Meristem tip culture is also used to produce phytoplasma-free grapevine and stone fruit nursery stock in Turkey. The selection of certified seeds or planting material is one of the best options for the management of witches' broom disease of acid lime in Oman and Iran (Al-Sadi et al., 2012; Siampour et al., 2019). Another effective method for pathogen eradication from infected tissue is cryotherapy of shoot tips (Wang and Valkonen, 2008). This method was effective to eliminate several phytoplasma strains and some other graft-transmissible pathogens from different crop plants. Sweet potato little leaf phytoplasma has been successfully eradicated from a sweet potato line using cryotherapy of shoot tip (Wang and Valkonen, 2008, 2009). But this technique is not prophylactic, and when plants are exposed to nature, the insect vectors again makes them infected.

Cultural control is commonly practiced for the management of various emerging phytoplasma diseases. Several cultural and conventional practices have an indirect effect on graft-transmitted phytoplasma disease spread. Implementation of appropriate prophylactic measures also supports the disease management. Removal of diseased plants or wild hosts from natural habitat as inoculum reservoirs contributes to the eradication of the disease from the plantations, especially in areas where the disease has limited distribution (Duduk et al., 2018). In Lebanon, the eradication of almond trees infected with '*Ca*. P. phoenicium' has been implemented to reduce the *foci* of infection (Molino Lova et al., 2014). Some growers practiced to remove the infected trees and replace them with other species or cultivars (Al-Sadi et al., 2012). In Oman, growers usually remove severely infected acid lime trees that are affected by witches' broom disease. The regular removal of symptomatic branches is also reported to reduce phytoplasma titers and vector attraction to the symptomatic branches (Al-Subhi et al., 2021). In addition, this practice also helps citrus trees to get rid of the symptomatic and nonproductive branches, since they do not produce any fruits or bear only small nonmarketable fruits (Hemmati et al., 2021).

The scientific efforts on phytoplasma resistance have been ended up with the identification of some resistant genotypes and progenies. The genotype "Xingguang" (clone of the cultivar Junzao) crossed and released in 2005 in China (Liu et al., 2004; Zhao et al., 2009) has very high resistance to JWB phytoplasma, one of the most destructive disease of Chinese jujube (*Z. jujuba* Mill.) also transmitted by grafting (Liu et al., 2014).

The new reports of outbreaks of phytoplasma diseases in certain Asian countries and the exchange of plant propagation material of unknown sanitary status pose a high risk for the establishment and spread of phytoplasma diseases in Asia; this requires the following actions: (1) rehabilitation of laboratories that implement sanitation and micropropagation procedures for producing phytoplasma-free propagation materials; (2) developing self-control on seedlings quality, development and modernization of the nursery sector, and the circulation of certified plant propagation material; (3) strengthening of phytoplasmas diagnosis; (4) establishing controlled greenhouse for dodder transmission procedures and grafting of suspected plant samples on susceptible woody indicators, in addition to the laboratories of serological and molecular detection assays; (5) field surveys and continuous monitoring; and (6) inspection at the ports of entry in all Asian countries.

4. Conclusions and perspectives

Phytoplasmas systemically colonize their host tissues can spread by vegetative propagation methods such as grafting, cuttings, micropropagation, and other methods and should be avoided in genetic crossing. Grafting branches or buds from diseased plants on healthy plants is one of the oldest transmission ways for the phytoplasmas. Natural grafting is also possible by roots, and this is a transmission that mostly take place in forest, tropical environments, and for some fruit trees. In quarantine or certification programs, despite the wide adoption of serological and molecular methods for the detection of graft-transmissible diseases, the use of biological indexing on susceptible woody indicators should be extensively adopted and considered one of the methods of choice for the detection of uncharacterized graft-transmissible or undefined agents including phytoplasmas. Therefore, phytoplasma graft and other vegetative transmission procedures are of relevance in both the disease management and the phytosanitary certification. The phytoplasma transmission in woody plant is not easy because of low titer of phytoplasmas and their seasonal title variation. Some problems can be overcomed by the phytoplasma transmission to periwinkle (*C. roseus*), commonly employed as experimental plant to propagate the strains since it is harboring almost all the studied phytoplasmas.

References

Abbasi, A., Hasanzadeh, N., Ghayeb Zamharir, M., Tohidfar, M., 2019. Identification of a group 16SrIX '*Candidatus* Phytoplasma phoenicium' phytoplasma associated with sweet orange exhibiting decline symptoms in Iran. Australas. Plant Dis. Notes 14, 11.

Abou-Jawdah, Y., Karakashian, A., Sobh, H., Martini, M., Lee, I-M., 2002. An epidemic of almond witches' broom in Lebanon: classification and phylogenetic relationship of the associated phytoplasma. Plant Dis. 86, 477–484.

Akhtar, K.P., Akram, A., Ullah, N., Saleem, M.Y., Saeed, M., 2019. Evaluation of *Solanum* species for resistance to tomato leaf curl New Delhi virus using chip grafting assay. Sci. Hortic. 256, 108646.

References

Akhtar, K., Ullah, N., Saleem, M., 2021. Validation of chip grafting inoculation assay to assess the resistance of *Solanum* species against phytoplasma. Plant Genet. Resour. Charact. Util. 19 (2), 178–182.

Al-Sadi, A.M., Al-Moqbali, H.S., Al-Yahyai, R.A., Al-Said, F.A., 2012. AFLP data suggest a potential role for the low genetic diversity of acid lime (*Citrus aurantifolia* Swingle) in Oman in the outbreak of witches' broom disease of lime. Euphytica 188, 285–297.

Al-Subhi, A.M., Al-Sadi, A.M., Al-Yahyai, R.A., Chen, Y., Mathers, T., Orlovskis, Z., Moro, G., Mugford, S., Al-Hashmi, K.S., Hogenhout, S.A., 2021. Witches' broom disease of lime contributes to phytoplasma epidemics and attracts insect vectors. Plant Dis. 105 (9), 2637–2648.

Aldaghi, M., Massart, S., Steyer, S., Lateur, M., Jijakli, M.H., 2007. Study on diverse grafting techniques for their capability in rapid and efficient transmission of apple proliferation disease to different host plants. Bull. Insectol. 60, 381–382.

Azadvar, M., Baranwal, V.K., Yadava, D.K., 2011. Transmission and detection of toria [*Brassica rapa* L. subsp. dichotoma (Roxb.)] phyllody phytoplasma and identification of a potential vector. J. Gen. Plant Pathol. 77, 194–200.

Barbosa, C.J., Pina, J.A., Pérez-Panadés, J., Bernad, L., Serra, P., Navarro, L., Duran-Vila, N., 2005. Mechanical transmission of citrus viroids. Plant Dis. 89, 749–754.

Bausher, M.G., 2013. Serial transmission of plant viruses by cutting implements during grafting. Hort. Sci. 48, 37–39.

Bertaccini, A., Davis, R.E., Lee, I-M., 1992. *In vitro* micropropagation for maintenance of mycoplasmalike organisms in infected plant tissues. Hortic. Sci. 27, 1041–1043.

Bertaccini, A., Paltrinieri, S., Martini, M., Tedeschi, M., Contaldo, N., 2012. Micropropagation and maintenance of phytoplasmas in tissue culture. In: Dickinson, M., Hodgetts, J. (Eds.), Phytoplasma Methods and Protocols, Methods in Molecular Biology. Springer, New York, p. 421.

Bertaccini, A., Duduk, B., Paltrinieri, S., Contaldo, N., 2014. Phytoplasmas and phytoplasma diseases: a severe threat to agriculture. Am. J. Plant Sci. 5, 1763–1788.

Bianco, P.A., Romanazzi, G., Mori, N., Myrie, W., Bertaccini, A., 2018. Integrated management of phytoplasma diseases. In: Bertaccini, A., Weintraub, P.G., Rao, G.P., Mori, N. (Eds.), Phytoplasmas: Plant Pathogenic Bacteria—II. Springer Nature Singapore, pp. 237–257.

Çağlayan, K., Gazel, M., Ulubaş Serçe, Ç., Kaya, K., Can Cengiz, F., 2014. Türkiye'de bazı meyve ağaçlarında saptanan fitoplazmalar ve olası vektörleri. V Turkish Plant Protection Congress, p. 190.

Çağlayan, K., Gazel, M., Škorić, D., 2019. Transmission of phytoplasmas by agronomic practices. In: Bertaccini, A., Weintraub, P.G., Rao, G.P., Mori, N. (Eds.), Phytoplasmas: Plant Pathogenic Bacteria-II. Springer, Singapore, pp. 149–163.

Çağlayan, K., Gazel, M., Serçe, Ç.U., Kaya, K., 2022. Assessment of susceptibility of different rootstock/variety combinations of pear to 'Candidatus Phytoplasma pyri' and experimental transmission studies by *Cacopsylla pyri*. Eur. J. Plant Pathol. 163 (4), 1–9.

Chalak, L., Elbitar, A., Rizk, R., Choueiri, E., Salar, P., Bové, J.M., 2005. Attempts to eliminate 'Candidatus Phytoplasma phoenicium' from infected Lebanese almond varieties by tissue culture techniques combined or not with thermotherapy. Eur. J. Plant Pathol. 112, 85–89.

Chalak, L., Elbitar, A., Mourad, N., Mortada, C., Choueiri, E., 2013. Elimination of grapevine bois noir phytoplasma by tissue culture coupled or not with heat therapy or hot water treatment. Adv. Crop Sci. Technol. 1, 107.

Choueiri, E., Jreijiri, F., Issa, S., Verdin, E., Bové, J-M., Garnier, M., 2001. First report of a phytoplasma disease of almond (*Prunus amygdalus*) in Lebanon. Plant Dis. 85, 802.

Damam, S., 2012. Studies on Phytoplasma with Special Reference to Sesame Phyllody. M.Sc. thesis. University of Agricultural Science, Dharwad, Inida, p. 86.

Duduk, B., Stepanovic, J., Yadav, A., Rao, G.P., 2018. Phytoplasmas in weeds and wild plants. In: Rao, G.P., Bertaccini, A., Fiorem, N., Liefting, L.W. (Eds.), Phytoplasmas: Plant Pathogenic Bacteria—I. Springer, Singapore, pp. 313—345.

Duduk, B., Tian, J.B., Contaldo, C., Fan, X.P., Paltrinieri, S., Chen, Q.F., Zhao, Q.F., Bertaccini, A., 2010. Occurrence of phytoplasmas related to "stolbur" and to 'Candidatus Phytoplasma japonicum' in woody host plants in China. J. Phytopath. 158 (2), 100—104.

Garcia-Chapa, M., Medina, V., Viruel, M.A., Lavina, A., Battle, A., 2003. Seasonal detection of pear decline phytoplasma by nested PCR in different pear cultivars. Plant Pathol. 52, 513—520.

Gazel, M., Serce, C.U., Yavuz, S., Gültekin, H., Caglayan, K., 2012. Susceptibility of some apricot and pear cultivars on various rootstocks to 'Candidatus Phytoplasma pruni' and 'Ca. P. pyri'. Petria 22, 183.

Gogoi, S.H., Kalita, M.K., Nath, P.D., 2017. Biological characterization of sesamum phyllody disease in Assam, India. Int. J. Curr. Microbiol. Appl. Sci. 6, 1862—1875.

Hemmati, C., Nikooei, M., Al-Sadi, A.M., 2021. Five decades of research on phytoplasma-induced witches' broom diseases. CAB Rev. 16, 1—16.

Jarausch, W., Lansac, M., Dosba, F., 1994. Micropropagation for maintenance of mycoplasma-like organisms in infected *Prunus marianna* GF 8-1. Acta Hortic. 359, 169—176.

Jarausch, W., Lansac, M., Dosba, F., 1996. Long-term maintenance of non-culturable apple proliferation phytoplasmas in their micropropagated natural host plant. Plant Pathol. 45, 778—786.

Kawicha, P., Hodgetts, M., Dickinson, M., 2012. A simple method for phytoplasmas transmission by grafting. Petria 22, 206.

Kim, C.J., 1965. Witches' broom of jujube tree (*Zizyphus jujube* Mill. var. inermis Rehd.). Transmission by grafting. Korean J. Microbiol. 3, 1—6.

Kumar, S., 2010. Studies on Phytoplasma Disease of Periwinkle [*Catharanthus Roseus* (L.) G. Don.]. M.Sc. thesis. University of Agricultural Science, Dharwad, India, p. 89.

Kumar, M., Rao, G.P., 2017. Molecular characterization, vector identification and sources of phytoplasmas associated with brinjal little leaf disease in India. 3 Biotech 7 (1), 7.

Kumari, S., Nagendran, K., Rai, A.B., Singh, B., Rao, G.P., Bertaccini, A., 2019. Global status of phytoplasma diseases in vegetable crops. Front. Microbiol. 10. https://doi.org/10.3389/fmicb.2019.01349.

La, Y., Woo, K., 1980. Transmission of jujube witches' broom mycoplasma by the leafhopper *Hishimonus sellatus* Uhler. J. Kor. For. Soc. 48, 29—39.

Laimer, M., 2003. Detection and elimination of viruses and phytoplasmas from pome and stone fruit trees. Hortic. Rev. 28, 187—236.

Laimer, M., Bertaccini, A., 2008. European stone fruit yellows. In: Harrison, N.A., Rao, G.P., Marcone, C. (Eds.), Characterization, Diagnosis and Management of Phytoplasmas. Studium Press, Texas, USA, pp. 73—92, 2008.

Laimer, M., Bertaccini, A., 2019. Phytoplasma elimination from perennial horticultural crops. In: Bertaccini, A., Weintraub, P.G., Rao, G.P., Mori, N. (Eds.), Phytoplasma: Plant Pathogenic Bacteria II: Transmission and Management of Phytoplasma-Associated Diseases. Springer, Singapore, pp. 185—206.

Landi, F., Prandini, A., Paltrinieri, S., Missere, D., Bertaccini, A., 2010. Assessment of susceptibility to European stone fruit yellows phytoplasma of new plum variety and of five rootstock/plum variety combinations. Julius-Kühn-Archiv 427, 378—382.

Lee, J.T., 1988. Investigation on jujube diseases and their severities of incidence. Res. Rept. RDA 31, 155—161.

Lee, I-M., Davis, R.E., Gundersen-Rindal, D.E., 2000. Phytoplasma: phytopathogenic mollicutes. Annu. Rev. Microbiol. 54, 221—255.

Linck, H., Lankes, C., Krüger, E., Reineke, A., 2019. Elimination of phytoplasmas in *Rubus* mother plants by tissue culture coupled with heat therapy. Plant Dis. 103 (6), 1252—1255.

Liu, M.J., Zhou, J.Y., Zhao, J., 2004. Screening of Chinese jujube germplasm with high resistance to witches' broom disease. Acta Hortic. 663, 575—580.

Liu, Z., Wang, Y., Xiao, J., Zhao, J., Liu, M., 2014. Identification of genes associated with phytoplasma resistance through suppressive subtraction hybridization in Chinese jujube. Physiol. Mol. Plant Pathol. 86, 43–48.

Marcone, C., Ragozzino, A., Seemüller, E., 1997. Dodder transmission of alder yellows phytoplasma to the experimental host *Catharanthus roseus* (periwinkle). Eur. J. For. Res. 27 (6), 347–350.

Molino Lova, M., Abou Jawdah, Y., Choueiri, E., Beyrouthy, M., Fakhr, R., Bianco, P.A., Alma, A., Sobh, H., Jawahri, M., Mortada, C., Najjar, P., Casati, P., Quaglino, F., Picciau, L., Tedeschi, R., Khalil, S., Maacaroun, R., Makhfoud, C., Haydar, L., Al Achi, R., 2014. Almond witches' broom phytoplasma: disease monitoring and preliminary control measures in Lebanon. In: Bertaccini, A. (Ed.), Phytoplasmas and Phytoplasma Disease Management: How to Reduce Their Economic Impact. Chapter 2. COST Action FA0807 Integrated Management of Phytoplasma Epidemics in Different Crop Systems, pp. 71–75.

Monteuuis, O., 2012. *In vitro* grafting of woody species. Propag. Ornam. Plants 12, 11–24.

Ohashi, A., Nohira, T., Yamaguchi, K., Kusunoki, M., Shiomi, T., 1996. Jujube (*Zizyphus jujuba*) witches' broom caused by phytoplasma in Gifu prefecture. Trans. Forest Soc. Japan 107, 309–310.

Pastore, M., Piccirillo, P., Tian, J., Simeone, A.M., Paltrinieri, S., Bertaccini, A., 2001. Transmission by patch grafting of ESFY phytoplasma to apricot (*Prunus armeniaca* L.) and Japanese plum *(Prunus salicina* Lindl). Acta Hortic. 550, 339–344.

Petrovic, N., Jeraj, N., Ravnikar, M., 2000. The use of tissue culture for improved detection of phytoplasma in grapevines. 13th ICVG Conference, Adelaide, Australia, 12–17. March, pp. 119–120.

Ranebennur, H., Rawat, K., Rao, A., Kumari, P., Chalam, V.C., Meshram, N., Rao, G.P., 2022. Transmission efficiency of a 'Candidatus Phytoplasma australasia' related strain (16SrII-D) associated with sesame phyllody by dodder, grafting and leafhoppers. Eur. J. Plant Pathol. 164 (8), 193–208.

Rao, G.P., Madhupriya, V.T., Manimekalai, R., Tiwari, A., Yadav, A., 2015. A century progress of research on phytoplasma diseases in India. Phytopath. Moll. 7, 1–38.

Rasoulpour, R., Salehi, M., Bertaccini, A., 2019. Association of a 'Candidatus Phytoplasma aurantifolia'-related strain with apricot showing European stone fruit yellows symptoms in Iran. 3 Biotech 9, 65.

Roddee, J., Kobori, Y., Hanboonsong, Y., 2018. Multiplication and distribution of sugarcane white leaf phytoplasma transmitted by the leafhopper, *Matsumuratettix hiroglyphicus* (Matsumura) (Hemiptera: Cicadellidae), in infected sugarcane. Sugar Tech 20, 445–453.

Salehi, M., Izadpanah, K., 1993. A comparative study of transmission, host range and symptomatology of sesame phyllody and alfalfa witches' broom. Iran. J. Plant Pathol. 29, 157–158.

Salehi, M., Izadpanah, K., Heydarnejad, J., 2006. Characterization of a new almond witches' broom phytoplasma in Iran. J. Phytopathol. 154, 386–391.

Salehi, M., Izadpanah, K., Siampour, M., Firouz, R., Salehi, E., 2009. Molecular characterization and transmission of safflower phyllody phytoplasma in Iran. J. Plant Pathol. 91, 453–458.

Salehi, M., Haghshenas, F., Khanchezar, A., Esmailzadeh-Hosseini, S.A., 2011. Association of 'Candidatus Phytoplasma phoenicium' with GF-677 witches' broom in Iran. Bull. Insectol. 64 (Suppl.), S113–S114.

Salehi, E., Salehi, M., Taghavi, S., Izadpanah, K., 2014. A 16SrII-D phytoplasma strain associated with tomato witches' broom in Bushehr province, Iran. J. Crop Prot. 3, 377–388.

Salehi, M., Salehi, E., Abbasian, M., Izadpanah, K., 2015. Wild almond (*Prunus scoparia*), a potential source of almond witches' broom phytoplasma in Iran. J. Plant Pathol. 97, 377–381.

Salehi, M., Esmailzadeh Hosseini, S.A., Salehi, E., Bertaccini, A., 2016. Molecular and biological characterization of a 16SrII phytoplasma associated with carrot witches' broom in Iran. J. Plant Pathol. 98, 83–90.

Salehi, E., Salehi, M., Masoumi, M., 2017. Biological and molecular characterization of the phytoplasma associated with tomato big bud disease in Zanjan province, Iran. Iran. J. Plant Pathol. 52, 415–427.

Salehi, M., Salehi, E., Siampour, M., Quaglino, F., Bianco, P.A., 2018. Apricot yellows associated with 'Candidatus Phytoplasma phoenicium' in Iran. Phytopath. Mediterr. 57, 269–283.

Schaper, U., Seemüller, E., 1982. Condition of the phloem and the persistence of mycoplasmalike organisms associated with apple proliferation and pear decline. Phytopathology 72, 736–742.

Seemüller, E., Kunze, L., Schaper, U., 1984. Colonization behaviour of MLO and symptom expression of proliferation-diseased apple trees and decline-diseased pear trees over a period of several years. J. Plant Dis. Prot. 91, 525–532.

Sertkaya, G., Martini, M., Musetti, R., Osler, R., 2007. Detection and molecular characterization of phytoplasmas infecting sesame and solanaceous crops in Turkey. Bull. Insectol. 60, 141–142.

Siampour, M., Izadpanah, K., Salehi, M., Afsharifar, A., 2019. Occurrence and distribution of phytoplasma diseases in Iran. In: Sustainable Management of Phytoplasma Diseases in Crops Grown in the Tropical Belt. Springer, pp. 47–86, 2019.

Singh, P.K., Akram, M., Vajpeyi, M., Srivastava, R.L., Kumar, K., Naresh, R., 2007. Screening and development of resistant sesame varieties against phytoplasma. Bull. Insectol. 60, 303–304.

Tanne, E., Orenstein, S., 1997. Identification and typing of grapevine phytoplasma amplified by graft transmission to periwinkle. Vitis 36 (1), 35–38.

Tawidian, P., Jawhari, M., Hana Sobh, H., Bianco, P.A., Abou-Jawdah, Y., 2017. The potential of grafting with selected stone fruit varieties for management of almond witches' broom. Phytopath. Mediterr. 56, 458–469.

Tiwari, A.K., Tripathi, S., Lal, M., Sharma, M.L., Chiemsombat, P., 2011. Elimination of sugarcane grassy shoot disease through apical meristem culture. Arch. Phytopath. Plant Protect. 44 (20), 1942–1948.

Tsai, J.H., Chen, X.Y., Shen, C.Y., Jin, K.X., 1988. Mycoplasmas and fastidious vascular prokaryotes associated with tree disease in China. In: Hiruki, C.J. (Ed.), Tree Mycoplasmas and Mycoplasma Diseases. Alberta Press, Canada, pp. 69–96.

Verdin, E., Salar, P., Danet, J-L., Choueiri, E., Jreijiri, F., El Zammar, S., Gélie, B., Bové, J-M., Garnier, M., 2003. 'Candidatus Phytoplasma phoenicium' sp. nov., a novel phytoplasma associated with an emerging lethal disease of almond trees in Lebanon and Iran. Int. J. Syst. Evol. Microbiol. 53, 833–838.

Wang, Q.C., Valkonen, J.P.T., 2008. Efficient elimination of sweetpotato little leaf phytoplasma from sweetpotato by cryotherapy of shoot tips. Plant Pathol. 57, 338–347.

Wang, Q.C., Valkonen, J.P.T., 2009. Cryotherapy of shoot tips: novel pathogen eradication method. Trends Plant Sci. 14, 119–122.

Wongkaew, P., Fletcher, J., 2004. Sugarcane white leaf phytoplasma in tissue culture: long-term maintenance, transmission, and oxytetracycline remission. Plant Cell Rep. 23, 426–434.

Yavuz, Ş., Gültekin, H., Gazel, M., Ulubaş Serçe, Ç., Çağlayan, K., 2011. Studies on the Transmission Efficiency of European Stone Fruit Yellows and Pear Decline Phytoplasmas by Grafting. IV Turkish Plant Protection Congress, Kahramanmaraş, p. 322.

Tabay Zambon, F., Plant, K., Etxeberria, E., 2017. Leaf-disc grafting for the transmission of 'Candidatus Liberibacter asiaticus' in citrus (Citrus sinensis; Rutaceae) seedlings. Appli. Plant Sci. 5 (1), 1600085.

Zhao, J., Liu, M.J., Liu, X.Y., Zhao, Z.H., 2009. Identification of resistant cultivar for jujube witches' broom disease and development of management strategies. Acta Hortic. 840, 409–412.

CHAPTER 3

Transmission of lime witches' broom

Mohammad Mehdi Faghihi[1], Abdolnabi Bagheri[2], Mohammad Salehi[1] and Majid Siampour[3]

[1]*Plant Protection Research Department, Fars Agricultural and Natural Resources Research and Education Centre, AREEO, Zarghan, Iran;* [2]*Plant Protection Research Department, Hormozgan Agricultural and Natural Resources Research and Education Centre, AREEO, Bandar Abbas, Iran;* [3]*Department of Plant Protection, College of Agriculture, Shahrekord University, Shahrekord, Iran*

1. Introduction

Phytoplasmas are a taxon of *Mollicutes* that are associated with diseases in hundreds plant species in the world. Being diverse, phytoplasmas were able to adapt to different ecological niches in various plant and insect species. Based on the restriction fragment length polymorphism (RFLP) analysis of the 16S rRNA gene, phytoplasmas have been so far divided into 33 16Sr groups. To date, diverse ribosomal groups of phytoplasmas were reported from Iran. Witches' broom disease of lime (WBDL) has been considered as the most economically important phytoplasma disease in Iran. The disease is associated with 'Candidatus Phytoplasma aurantifolia' (a 16SrII-B subgroup strain). Small fruited acid lime (*Citrus aurantifolia*) is the most important citrus species naturally infected by the WBDL phytoplasma (Zreik et al., 1995). The disease was so far reported from Iran, Oman, and the United Arab Emirates (UAE) (Garnier et al., 1991; Salehi et al., 1997). In Iran, it was primarily found in a small area in southeast of the country in 1997 (Salehi et al., 1997). Since then, WBDL has spread to many areas, devastating Mexican lime plantations in the major lime growing provinces of Hormozgan, Sistan and Baluchestan, Kerman, and Fars (Salehi et al., 2008). In the field, the phytoplasma is spread by the leafhopper *Hishimonus phycitis* (Salehi et al., 2007; Bagheri et al., 2009). In nature, the WBDL phytoplasma has a narrow host range; however, the pathogen has been experimentally transmitted to several species of citrus by graft inoculation, and to periwinkle and a number of solanaceous plants such as tomato, eggplant, and tobacco by dodder or graft inoculation (Salehi et al., 2000, 2002, 2005).

2. Vector transmission
2.1 *Hishimonus phycitis*

H. phycitis is a well-defined species, occurring in tropical and subtropical Asian countries from Iran to Malaysia (Fig. 3.1). It is a polyphagous insect feeding on *Citrus* spp. and *Solanum melongena* either as a leafhopper extracting host nutrients through feeding or as a vector of WBDL phytoplasma, which limits production of *C. aurantifolia*, and in India as a vector of brinjal little leaf phytoplasma impacting

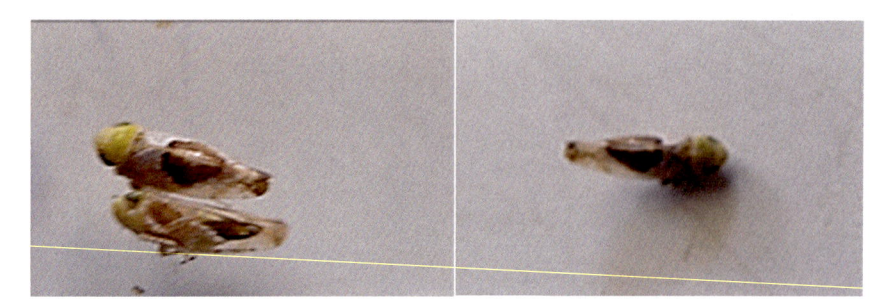

FIGURE 3.1

Hishimonus phycitis, the leafhopper vector of WBDL phytoplasma.

S. melongena yields. Although the disease spread rapidly in many citrus-growing regions, all attempts to find the vector had been failed for years. At the same time, the disease was successfully transmitted to several citrus species by grafting and to a number of herbaceous hosts by dodder. As the leafhopper *H. phycitis* (Distant) (Hemiptera: Cicadellidae) was the most common phloem-feeding species associated with lime trees in the citrus-growing areas, many efforts were made to establish the transmission of this disease by *H. phycitis* (Salehi et al., 2007; Bagheri et al., 2009; Hemmati et al., 2020). Accordingly, Salehi et al. (2007) could successfully transmit the WBDL to bakraee (*Citrus reticulata* hybrid) seedlings by feral individuals of *H. phycitis* collected from WBDL-infected lime orchards. In addition, transmission of the disease to mature healthy lime plants by *H. phycitis* was confirmed by Bagheri et al. (2009). Hemmati et al. (2020) also reported infection of lime seedling using feral *H. phycitis* individuals collected from infected lime orchards.

2.1.1 Host range

Identifying herbaceous and nonherbaceous host plants of *H. phycitis* helps to reduce the infection rate of WBDL through decreasing the population density of the insect vector by removing weeds or spraying them with insecticides. In a study, Abbaszadeh et al. (2010) verified the host range of *H. phycitis* in southern Iran and found that it was able to complete its life cycle on sweet orange (*Citrus sinensis*), grapefruit (*Citrus paradise*), Mexican lime (*C. aurantifolia*), Mediterranean sweet lemon (*Citrus limetta*), mandarin (*Citrus reticulate*), lemon (*Citrus limon*), sour orange (*Citrus aurantium*), rough lemon (*Citrus jambhiri*), volkamer lemon (*Citrus volkameriana*), and *Ziziphus* (*Ziziphus spina-christi*). Although many of these species were not reported as natural host of the *H. phycitis* and WBDL phytoplasma, they should be considered as a potential reservoir for WBDL phytoplasma especially when infected lime trees were eradicated in the region.

2.1.2 Host preference

In a study conducted in 2010, the density of *H. phycitis* was investigated on Mexican lime, (symptomatic and asymptomatic trees), sweet orange, and mandarin. The results showed that the density of *H. phycitis* on Mexican lime (both symptomatic and asymptomatic trees) was significantly higher than on the other hosts species mentioned above. In addition, by comparing symptomatic and asymptomatic trees, it was revealed that the number of insects on the symptomatic branches was significantly higher than in those with no symptoms (Bagheri and Ameri, 2010).

2.2 Diaphorina citri

One of the serious pests in citriculture is the Asian citrus psyllid (ACP), *Diaphorina citri* Kuwayama (Hemiptera: Liviidae), vector of the phloem-limited bacteria '*Candidatus* Liberibacter asiaticus' and '*Ca*. L. americanus', the agents of the Asian and American forms of the "huanglongbing" (HLB) disease. The ACP was not found in many WBDL-infected lime orchards in Iran and Oman in the early years of WBDL occurrence, but it was later introduced to these citrus-growing areas. During the recent decade, several attempts were made to transmit WBDL phytoplasma by ACP. In one instance, Queiroz et al. (2016) reported that *D. citri* was also able to transmit WBDL phytoplasma to Mexican lime seedlings. However, they reported that the rate of WBDL transmission by *D. citri* was lower than that by *H. phycitis*. Furthermore, it was shown that *D. citri* individuals fed on hosts infected by WBDL phytoplasma had higher reproduction rates than those fed on healthy hosts. On the other hand, attempts made by Siampour et al. (unpublished data) to transmit WBDL phytoplasma to the lime seedlings by *D. citri* were not successful.

Siampour et al. (2006) detected WBDL phytoplasma within the body of *Recilia schmidtgeni* and *Idioscopus clypealis* leafhoppers collected on witches' broom−affected lime groves. These insects, however, were only considered as carriers of the WBDL phytoplasma.

2.2.1 Management of the insect vector

To reach a proper control of the vector, it is needed to acquire enough information regarding population fluctuation *of H. phycitis* more specially to find dates on which the vector has the highest density and highest number of infectious individuals. To this end, Bagheri et al. (2010a) studied the population fluctuation of *H. phycitis* in two WBDL-infected orchards in southern Iran. They also determined the number of phytoplasma-infected individuals over two consecutive years at monthly intervals. They found that *H. phycitis* has two main peaks per year, including an autumn-winter peak (known as the large peak of the vector, takes about two to three months) and a small peak in spring (about one month). In the latter case, due to the presence of vegetative flushing and temperate conditions, *H. phycitis* may play a much more effective role in the disease transmission. However, from late May, the insect population gradually decreases so that in summer it experiences the lowest population density. From early autumn, the population gradually increases so that it reaches the highest possible density in late autumn and then decreases. In late winter and early spring, the population density increases again and experiences a small peak. The results of the number of phytoplasma-infected individuals were approximately similar to what revealed for the population fluctuations, and the peaks of this curve corresponded perfectly to the two peaks of the population fluctuation. By this information, it is revealed that the vector needs to be controlled by appropriate insecticides two times per year (once after harvesting in November and another before flowering in January). Lime seedlings can be a potential source for the disease transmission. Therefore, lime nurseries should be located in areas where *H. phycitis* is less populated. However, in cases where there is no choice to produce seedlings in/or near infected orchards, the spraying should be repeated at intervals as necessary. In many Mexican lime orchards in southern Iran, the systemic insecticides such as Imidacloprid and Acetamiprid, are applied by drench application, causing the insecticides be less degraded by photolysis, and lime trees have a longer protection from the infestation by sucking insects.

3. Seed transmission

Several studies have focused on the detection of WBDL phytoplasma in various parts of the lime seed. El-Kharbotly et al. (2000) were the first to suggest the possibility of seed transmission of WBDL phytoplasma. Later, Khan (2005) used the polymerase chain reaction (PCR) assay to study seed transmission of WBDL phytoplasma in six-month-old seedlings. Based on this report, the majority of seed from infected plants in Oman yielded infected seedlings. The author also reported the presence of phytoplasma-like structures in seed tissues by electron microscopy. Such a high rate of seed transmission is unusual for a phloem-limited organism. Further investigation in affected citrus groves of southern Iran fails to support seed transmissibility of this phytoplasma (Faghihi et al., 2011). Lime trees are mainly propagated through seed. In addition, lime seedlings are used as rootstock for many citrus species and cultivars. Therefore, the issue of seed transmissibility of 'Ca. P. aurantifolia' is important because it relates directly to the implementation of control strategies. In a study by Faghihi et al. (2011), claims of seed transmission of WBDL phytoplasma were examined. Fruits were collected from infected trees in Minab (Hormozgan province) and from symptomless trees in noninfected areas. Lime seed from symptomless and witches' broom–affected trees were sown in separate beds in an insect-proof screenhouse, and the resulting seedlings were examined for phytoplasma presence. During two years of experiment period, repeated PCR tests on the seedlings did not reveal the presence of phytoplasma DNA. Likewise, symptoms of the disease were not observed on these seedlings after two years. PCR assays detected the phytoplasma in coats of some seed from infected trees; however, no excised embryos were positive for the phytoplasma. The evidence show that WBDL was not seed transmissible and it reduced the concern associated with the routine practice of lime propagation by seed.

4. Graft transmission and experimental hosts

Based on the disease symptoms and PCR assays and other molecular analyses, several citrus and some herbaceous plants were reported as natural and experimental hosts of WBDL phytoplasmas. In the UAE, not only lime but also citron, sweet limetta, and Indian Palestine sweet lime showed severe symptoms of the WBDL (El Shereiqi and Gassouma, 1993). Moghal et al. (1998) reported the list of hosts infected by the WBDL phytoplasma in Oman that included several citrus species such as *C. aurantifolia, C. limetta, Citrus medica, C. jambhiri*, and *C. limon*. Different surveys in Iran confirmed natural infection of lime, thornless lime, bakraee (*C. reticulata* hybrid), Valencia sweet orange (*C. sinensis*), Minneola tangelo, Orlando tangelo (cross between Duncan grapefruit and Dancy mandarin), Kinnow (a cross of *Citrus nobilis* Lou. and *Citrus deliciosa*) (Salehi et al., 2005b, 2008), red blush and Duncan cultivars of grape fruit (*Citrus paradisi*) (Salehi et al., 2008; Bagheri et al., 2010b; Najafi and Azadwar 2016), limequat (Faghihi et al., 2017), and citron (*C. medica*) (Azadvar et al. 2015) with WBDL phytoplasmas. Graft inoculation experiments confirmed that several other citrus species are also susceptible to 'Ca. P. aurantifolia'. In Oman, the disease agent was graft-transmitted from lime to lime as well as to lemon, rough lemon (*C. jambhiri*), trifoliate orange (*Poncirus trifoliata*), Troyer citrange (*Poncirus trifoliate* x *C. sinensis*), *C. excelsa, C. ichangensis, C. karna,* Alemow *(C. macrophylla),* Kaffir lime (*C. hystrix*), rough lemon (*C. jambhiri*), Etrog citron (*C. medica*), Rangpur lime (*C. limonia*), Meyer lemon (*C. meyeri*), and mandarin lime (*C. limonia*)

(Garnier et al., 1991; Bové et al., 1993; Bové et al., 1996). Graft inoculation trials in Iran showed that the following citrus plants were experimental hosts of WBDL phytoplasma: pear-shaped (probably a lemon hybrid), cucumber-shaped lime (a lime hybrid), sour orange (*C. aurantium*), Taiwanica (*C. taiwanica*), Lisbon lemon (*C. limon*), Cleopatra mandarin (*C. reshni*), volkamer lemon, Rough lemon, Rangpur lime, Alemow, and citron (Salehi et al., 2004, 2005, 2008; Hassanzadeh et al., 2019). Graft transmission of WBDL revealed that some of lime biotypes (Hassanzadeh et al., 2019) and lime hybrids (Rezazadeh et al., 2019) were resistant to '*Ca*. P. aurantifolia'. Based on dodder and graft inoculation and subsequent PCR assay, periwinkle and a number of solanaceous species including eggplant, ornamental eggplant, Jimson weed, tobacco, nightshade, and tomato are reported as experimental herbaceous hosts of WBDL phytoplasmas (Salehi et al. 2000, 2002).

5. Conclusion and perspectives

WBDL is a destructive disease associated with phytoplasma strains of the subgroup 16SrII-B. The disease has so far reported from Iran, the UAE, and Oman. After a long survey, *H. phycitis* leafhopper was identified as the main insect vector of WBDL phytoplasmas. Despite WBDL phytoplasma is highly virulent and pathogenic to lime (and to a few other citrus hosts), the leafhopper vector *H. phycitis* apparently benefits from infection by WBDL phytoplasma. In this regard, a sharp rise was found in the population density of *H. phycitis* on the lime branches expressing witches' broom symptoms. Due to the dual benefits, a symbiosis association could be proposed between WBDL phytoplasma and *H. phycitis* insect vector. The narrow feeding habitat of *H. phycitis* has been determinant for the natural host range of WBDL phytoplasma. In this regard, the WBDL management in the affected areas could be achieved through monitoring and control of *H. phycitis* population and immediate uprooting of the infected lime (or other citrus) trees. Also, the transportation of infected materials should be restricted from the affected areas. Additionally, quarantine programs should be applied to prevent the entry of *H. phycitis* from affected areas to unaffected or protected areas.

References

Abbaszadeh, G., Bagheri, A., Faghihi, M.M., Askari, M., 2010. Host range of *Hishimonus phycitis*, vector of WBDL in lime orchards of Iran. 19th Iranian Plant Protection Congress, p. 579.

Azadvar, M., Ranjbar, S., Najafinia, M., Baranwal, V.K., 2015. First report of natural infection of citron (*Citrus medica* L.) by '*Candidatus* Phytoplasma aurantifolia' in Iran. J. Agric. Biotechnol. 6, 15–22.

Bagheri, A.N., Salehi, M., Faghihi, M.M., Samavi, S., Sadeghi, A., 2009. Transmission of '*Candidatus* Phytoplasma aurantifolia' to Mexican lime by the leafhopper *Hishimonus phycitis* in Iran. J. Plant Pathol. 91 (4), S105.

Bagheri, A., Ameri, A., 2010. Population dynamism of *Hishimonus phycitis* (Cicadellidae), vector of WBDL on lime, lemon, sweet orange and mandarin. 19th Iranian Plant Protection Congress, p. 529.

Bagheri, A.N., Faghihi, M.M., Salehi, M., Siampour, M., Samavi, S., 2010a. Population fluctuations of *Hishimonous phycitis*, the vector of witches' broom disease of lime. In: Hormozgan Province. 19th Iranian Plant Protection Congress, p. 630.

Bagheri, A., Faghihi, M., Salehi, M., Khanchezar, A., 2010b. First report of natural infection of grapefruit trees to lime witches' broom phytoplasma. 19th Iranian Plant Protection Congress, p. 409.

Bové, J-M., Navarro, L., Bonnet, P., Zreik, L., Garnier, M., 1996. Reaction of *Citrus* cultivars to graft inoculation of phytoplasma aurantifolia-infected lime shoots. Proc. 13th Conf. IOCV. 249–251.

Bové, J-M., Zreik, L., Danet, J-L., Bonfils, J., Mjeni, A., Garnier, M., 1993. Witches' broom disease of lime trees: monoclonal antibody and DNA probes for the detection of the associated MLO and the identification of a possible vector. In: International Organization of Citrus Virologists Conference Proceedings, vol 12. eScholarship, University of California Riverside, pp. 342–348.

El-Kharbotly, A., Al-Shanfari, A., Al-Subhi, A., 2000. Molecular evidence for the presence of the 'Ca. Phytoplasma aurantifolia' in lime seeds and transmission to seedlings. In: Proc. Int. Soc. Citricult. Orlando, FL, pp. 97–98.

El Shereiqi, R.K., Gassouma, S., 1993. Witches' broom disease of lime in the United Arab Emirates. In: Proc. 12th Conf. IOCV., IOCV, Riverside, pp. 453–454.

Faghihi, M., Bagheri, A., Askari Seyahooei, M., Pezhman, A., Faraji, G., 2017. First report of a '*Candidatus* Phytoplasma aurantifolia'-related strain associated with witches' broom disease of limequat in Iran. New Dis. Rep. 35, 24.

Faghihi, M.M., Bagheri, A.N., Bahrami, H.R., Hasanzadeh, H., Rezazadeh, R., Siampour, M., Samavi, S., Salehi, M., Izadpanah, K., 2011. Witches' broom disease of lime affects seed germination and seedling growth but is not seed transmissible. Plant Dis. 95, 419–422.

Garnier, M., Zreik, L., Bové, J-M., 1991. Witches' broom, a lethal mycoplasmal disease of lime trees in the Sultanate of Oman and the United Arab Emirates. Plant Dis. 75, 546–551.

Hassanzadeh, H., Bahrami, H.R., Faghihi, M.M., Bagheri, A., 2019. Reaction of some commercial citrus species and Iranian lime biotypes to witches' broom disease of lime. Crop Protect. 122, 23–29.

Hemmati, C., Askari Seyahooei, M., Nikooei, M., Modarees Najafabadi, S.S., Goodarzi, A., Amiri Mazraie, M., Faghihi, M.M., 2020. Vector transmission of lime witches' broom phytoplasma to Mexican lime seedlings under greenhouse condition. J. Crop. Prot. 9 (2), 209–215.

Khan, I.A., 2005. Management and control of witches' broom disease of lime. In: International Tropical Fruits Workshop (Citrus and Mango).

Moghal, M.S., Zidghali, A.D., Moustaffa, S.S., 1998. Natural host range and reaction of *Citrus* species to witches' broom disease of lime in Oman. Proc. IPM Conf. SQU 143–154.

Najafian, M., Azadvar, M., 2016. Witches' broom disease of lime and its management. Ind. Phytopathol. 69, 330–332.

Queiroz, R.B., Donkersley, P., Silva, F.N., Al-Mahmmoli, I.H., Al-Sadi, A.M., Carvalho, C.M., Elliot, S.L., 2016. Invasive mutualisms between a plant pathogen and insect vectors in the Middle East and Brazil. R. Soc. Open Sci. 3 (12), 160557.

Rezazadeh, N., Asadi Abkenar, A., Rouhibakhsh, A., 2019. Reaction assessment of a number of citrus hybrids to '*Candidatus* Phytoplasma aurantifolia'. Iran. J. Plant Pathol. 54 (3), 173–183.

Salehi, M., Izadpanah, K., Siampour, M., Bagheri, A., Faghihi, M.M., 2007. Transmission of '*Candidatus* Phytoplasma aurantifolia' to Bakraee (*Citrus reticulata* hybrid) by feral *Hishimonus phycitis* leafhoppers in Iran. Plant Dis. 91 (4), 466.

Salehi, M., Heydarnejad, J., Izadpanah, K., 2005. Molecular characterization and grouping of 35 phytoplasmas from central and southern provinces in Iran. Iran. J. Plant Pathol. 41 (1), 62–65.

Salehi, M., Izadpanah, K., Rahimian, H., 1997. Witches' broom disease of lime in Sistan-Baluchistan. Iran. J. Plant Pathol. 33, 3–4.

Salehi, M., Izadpanah, K., Taghizadeh, M., 2002. Witches' broom disease of lime in Iran: new distribution areas, experimental herbaceous hosts and transmission trials. In: Proc. 15th Conf. IOCV, Riverside, CA, pp. 293–296.

Salehi, M., Nejat, N., Tvakoli, A.R., Izadpanah, K., 2005b. Reaction of citrus cultivars to '*Candidatus* Phytoplasma aurantifolia' in Iran. Iran. J. Plant Pathol. 41, 147–149.

Salehi, M., Nejat, N., Tavakkoli, A.R., Karampour, F., Izadpanah, K., 2004. Reaction of citrus species to phytoplasmal agent of witches' broom disease of lime. In: 16th Plant Prot. Cong. Iran. Tabriz. Univ. Tabriz, Iran, p. 334.

Salehi, M., Izadpanah, K., Taghizadeh, M., 2000. Herbaceous host range of lime witches' broom phytoplasma in Iran. J. Plant Pathol. 36, 343–353.

Salehi, M., Firooz, R., Taghizadeh, M., 2008. Witches' broom disease of lime: new hosts records. In: Proc. 18th Plant Prot. Cong. Iran. Isfahan. Univ. Bu-Ali Sina, Hamedan, Iran, p. 415.

Siampour, M., Izadpanah, K., Afsharifar, A.R., Salehi, M., Taghizadeh, M., 2006. Detection of phytoplasma in insects collected in witches' broom affected lime groves. Iran. J. Plant Pathol. 42 (1), 139–158.

Zreik, L., Carle, P., Bové, J-M., Garnier, M., 1995. Characterization of the mycoplasma like organism associated with witches' broom disease of lime and proposition of a '*Candidatus*' taxon for the organism, '*Candidatus* phytoplasma aurantifolia'. Int. J. Syst. Evol. Microbiol. 45 (3), 449–453.

CHAPTER 4

Major insect vectors of phytoplasma diseases in Asia

Chamran Hemmati[1,2], Mehrnoosh Nikooei[1,2] and Abdullah Mohammed Al-Sadi[3]

[1]Department of Agriculture, Minab Higher Education Center, University of Hormozgan, Bandar Abbas, Iran; [2]Plant Protection Research Group, University of Hormozgan, Bandar Abbas, Iran; [3]Department of Plant Sciences, College of Agriculture and Marine Sciences, Sultan Qaboos University, Muscat, Oman

1. Introduction

Phytoplasmas, previously known as mycoplasma-like organism (MLO), are wall-less, pleomorphic, and Gram-negative bacteria associated with more than a 1000 diseases in economically important crops. They inhabit phloem sieve tubes and are transmitted to other plants by phloem-sucking insects. Followed by initial evidence regarding the possible growth of phytoplasmas in cell-free media (Contaldo et al., 2016), forecasted biological features and metabolisms of phytoplasmas were provided. Up to the present time, 33 ribosomal groups and more than 120 subgroups have been identified based on restriction fragment length polymorphism of 16Sr gene and 49 'Candidatus Phytoplasma' species have been proposed (Hemmati et al., 2021a; Marcone et al., 2022). It should be noted that the identified insect vectors are much less than the described phytoplasmas (Alma et al., 2019).

The Hemiptera are a large and diverse order of exopterygote insects that occur in all zoogeographic regions of the world. There are more than 50,000 species in about 100 families. The Hemiptera are divided into three suborders: Heteroptera (true bugs), Sternorrhyncha (scale insects, aphids, whiteflies, and psyllids), and Auchenorrhyncha (leafhoppers, planthoppers, cicadas, treehoppers, and spittlebugs) (Szwedo, 2016). Since phytoplasmas are phloem-limited, only phloem-feeding insects can potentially acquire and transmit the pathogen. Four specific characteristics suggest them be vectors of phytoplasmas including being hemimetabolous (nymphs and adults can transmit the pathogen), feeding specifically and selectively on certain plant tissues, having a propagative and persistent relationship with phytoplasmas, and finally having transovarial transmission for some phytoplasmas (Weintraub and Beanland, 2006). Within the groups of phloem-feeding insects, only a small number, primarily in three taxonomic groups, have been confirmed as vectors of phytoplasmas; Cicadellidae, Fulgoromorpha (in which four families of vector species are found), and two genera in the Psyllidae (Dietrich, 2009). Most Auchenorrhyncha feeds on phloem tissue, but two superfamilies (Cicadoidea: cicadas; Cercopoidea: spittlebugs) and a subfamily of the Cicadellidae (Cicadellinae) feed on both phloem and xylem tissues. In addition, the majority of species in the leafhopper subfamily Typhlocybinae feeds by removing the cell contents from mesophyll cells (Weintraub and Beanland, 2006).

2. Insect acquisition and transmission of phytoplasmas

Once insects are feeding on the phloem of infected plants, the phytoplasma restricted in the phloem can passively enter the vector body. Acquisition access period (AAP) is defined as the necessary time when a sufficient phytoplasma titer can be acquired by insects. The acquisition may occur by an insect within a few minutes; however, it is normally within an hour and the extended the AAP, the greater the chance of acquisition. This time also depends on the titer of the phytoplasma in the plants (Alma et al., 2015). Although there is no evidence regarding the effect of phytoplasma titer and fluctuations on AAP, Wei et al. (2004) stated that phytoplasma titer can increase sixfolds weekly in onion yellows phytoplasma.

The time taken by an insect to vector the initially acquired phytoplasma is known as the latent period (LP) or incubation period. The LP can be changed on the basis of temperature and ranges from a few to 80 days. When phytoplasmas enter the vector body, they should finally enter the salivary gland to be able to be transmitted. They should pass through the midgut cell and enter to hemocoel (Alma et al., 2018). To pass from midgut cell, they pass intracellularly through the epithelial cells and replicate within a vesicle or they can pass to midgut cells and enter the hemocoel through the basement membrane. Phytoplasma circulating in the hemolymph may enter other organs including fat body, reproductive organs, and Malpighian tubes into which they may reproduce although this is not necessary for transmission. Reproducing and staying in such organs show the long-lasting evolution between the insect, host plant, and pathogen (Weintarube and Beanland, 2006). Multiplication of phytoplasma in insect vector body is also related to the vector species. For example, Bosco et al. (2014) studied the titer of chrysanthemum yellows phytoplasma in three vectors after acquisition. They revealed that this phytoplasma multiplied faster in *Macrosteles quadripunctulatus* than *Euscelis incisus* and *E. variegatus*. In addition, the LP in *M. quadripunctulatus* was shorter (18 d) than in two other species (30 d).

For successful transmission, phytoplasmas should enter into specific cells of the salivary glands and should reach to a high number in the posterior acinar cells of the salivary gland (Lefol et al., 1994). However, if the phytoplasmas could not enter or exit the tissue, the insect species cannot transmit the phytoplasmas to other host plants. There are three barriers in the salivary gland that phytoplasmas should pass through before they can be ejected with the saliva: the basal lamina, the basal plasmalemma, and apical plasmalemma (Chuche and Thiéry, 2014). Leafhoppers that acquire phytoplasmas and are not able to transmit them to healthy plants may not be able to pass through these three barriers in salivary glands (Weintraub and Benaland, 2006). Such evidence was confirmed by Pagliari et al. (2019) who studied the multiplication of the phytoplasma in transmitting and nontransmitting *E. variegatus*. They showed that there were phytoplasmas in nontransmitting leafhoppers, suggesting that the salivary glands and midgut cell play a crucial role in transmission.

3. How the phytoplasma insect vectors can be identified?

Classically, once new phytoplasma diseases occurred, the disease epidemiology should be defined through the identification of the insects active in fields/orchards or near the diseased plants. To do so, the researcher should monitor the plants for a long time, and the most known way to do monitoring is using sticky traps, vacuum sampling by D-Vac, and sweeping net, of which the two latter are used to capture live insects in the fields (Weintraub and Orenstein, 2004). To yield live insects on trees, shaking the branches can provide the falling insects. Traps are also used in phytoplasma vectors' identification survey, as malaise trap can capture insects that are not acquired in other traps. Malaise trap can provide information about the movement pattern of the insects and movement pattern between crops and forest vegetation (Pilikington et al., 2004).

After collection, they can be tested in terms of phytoplasma presence by molecular methods. Correlation between disease incidence and species abundance can provide clues to identify vector species. Finally, transmission experiments make available the most substantial evidence of the vector ability of tested candidate vector species. For some species, a sucrose-based feeding medium has been developed to overcome unsuccessfull cage transmission trials. In this method, potential vectors were collected and held in a small tube in which sucrose was covered by a parafilm (Bosco and Tedeschi, 2013). After a few days, the medium can be tested for phytoplasma presence by polymerase chain reaction (PCR). Being positive shows that the phytoplasma is present in the salivary glands and able to transmit to healthy plants (Tanne et al., 2001).

Much research has been done on the identification, biology, and ecology of insect vectors, Weintraub and Beanland (2006) collected all the published papers and reviewed all aspects of insect vectors and the relationship between pathogen and vectors. Research on this subject continued, and many papers have been published in recent years. However, there is no update data regarding the advances on insect vectors of phytoplasmas in Asian countries. In this chapter, an overview of the recent developments concerning insect vector identification (Table 4.1), the relationship between phytoplasmas and their vectors, and host plants as well as the most recent control strategy applied in Asian countries has been provided.

Table 4.1 List of the insect vectors and putative vectors of phytoplasma diseases in Asia.

Location	Disease name	Phytoplasma ribosomal group	Insect vector	Family	References
Iran	*Zinnia elegans* witches' broom	16SrII-D	*Austroagallia sinuata*	Cicadellidae	Hemmati and Nikooei (2019)
Iran	Almond witches' broom	16SrIX-C	*Frutioidea bisignata*	Cicadellidae	Salehi et al. (2006); Zirak et al. (2010)
Iran, Oman, UAE	Lime witches' broom	16SrII-B	*Hishimonus phycitis*	Cicadellidae	Bagheri et al. (2009)
Iran	Cabbage yellowing	16SrVI-A	*Neoaliturus haematoceps*	Cicadellidae	Salehi et al. (2007)
Iran	Cucumber phyllody	16SrII-M	*Orosius albicinctus*	Cicadellidae	Salehi et al. (2015)
Iran	Squash phyllody	16SrII-D	*O. albicinctus*	Cicadellidae	Salehi et al. (2015)
Iran	Beet witches' broom	16SrII-E	*O. albicinctus*	Cicadellidae	Mirzaie et al. (2007)
Iran	Lettuce phyllody	16SrIX-C	*Neoaliturus fenetratus*	Cicadellidae	Salehi et al. (2007c)
Iran	Carrot witches' broom	16SrII-C	*O. albicinctus*	Cicadellidae	Salehi et al. (2016a)

Continued

Table 4.1 List of the insect vectors and putative vectors of phytoplasma diseases in Asia.—cont'd

Location	Disease name	Phytoplasma ribosomal group	Insect vector	Family	References
Iran	Faba bean phyllody	16SrII-D	*O. albicinctus*	Cicadellidae	Salehi et al. (2016b)
Iran	*Petunia violacea* witches' broom	16SrII-D	*O. albicinctus*	Cicadellidae	Hemmati et al. (2019a)
Iran	Sesame phyllody	16SrII-D 16SrVI-A 16SrIX-C	*Neoaliturus haematoceps*	Cicadellidae	Salehi et al. (2017b)
Iran	Sesame phyllody	16SrII-D	*O. albicinctus*	Cicadellidae	Salehi et al. (2017b)
Iran	Rapeseed phyllody	16SrI-B	*N. haematoceps*	Cicadellidae	Salehi et al. (2011)
Iran	Bermuda grass white leaves	16SrXIV-A and 16SrXIV-B	*Exitianus capicola*	Cicadellidae	Salehi et al. (2009)
Iran	*Aerva javanica* leaf roll	16SrII-D	*Austroagallia sinuata*	Cicadellidae	Hemmati et al. (2019b)
Iran	Alfalfa witches' broom	16SrII-D	*O. albicinctus*	Cicadellidae	Salehi et al. (2011)
Lebanon	Almond witches' broom	16SrIX-B	*Tachycixius viperinus*	Cixiidae	Abou-Jawdah et al. (2003)
Lebanon	Almond witches' broom	16SrIX-B	*Asymmetrasca decedens*	Cicadellidae	Abou-Jawdah (2011)
Israel	Carrot yellows	16SrI 16SrIII 16SrV	*N. haematoceps* *N. fenestratus*	Cicadellidae	Weintraub and Orenstein (2004)
Israel	*Limonium* hybrids yellows	16SrV	*Circulifer oreintalis*	Cicadellidae	Weintraub et al. (2004)
		16SrIX	*E. capicola*	Cicadellidae	
Turkey	Sesame phyllody	16SrVI-A	*O. albicinctus*	Cicadellidae	Ikten et al. (2014)
India	Sesame phyllody	16SrII-D	*Hishimonus phycitis*	Cicadellidae	Un Nabi et al. (2015)
India	Kerala wilt		*Stephanitis typical*	Tingidae	Mathen et al. (1990)
India	Sandal spike	16SrI-B	*Coelidia indica*	Cicadellidae	Rangaswami and Griffith (1941)
India	Grassy shoot of sugarcane	16SrXI	*Deltocephalus vulgaris* *Exitianus indicus* *Maiestas portico* *Cofana unimaculata*	Cicadellidae	Srivastava et al. (2006); Rao et al. (2014); Tiwari et al. (2016), (2017)

Table 4.1 List of the insect vectors and putative vectors of phytoplasma diseases in Asia.—cont'd

Location	Disease name	Phytoplasma ribosomal group	Insect vector	Family	References
China	Jujube witches' broom	16SrV-B	*Hishimonoides chinensis*	Cicadellidae	Tsai et al. (1988)
China	Mulberry dwarf	16SrI-B	*Hishimonoides sellatiformis*	Cicadellidae	Ishijima and Ishiie (1981)
China	Mulberry dwarf	16SrI-B	*H. sellatus*	Cicadellidae	Ishijima and Ishiie (1981)
China	Jujube witches' broom	16SrV-B	*Hishimonus sellatus*	Cicadellidae	Kusunoki et al. (2002)
Japan	*Cryptotania japonica* witches' broom	16SrI	*H. sellatus*	Cicadellidae	Nishimura et al. (2004)
Japan	Potato purple top; Gentian witches' broom; Tsuwabuki witches' broom	16SrIII	*Scleroracus flavopictus*	Cicadellidae	Okuda (1997)
Southeast Asia	Coconut root wilt	16SrIV	*Proutista moesta*	Derbidae	Ponnamma and Solomon (1998)
East Asia	Paulownia witches' broom	16SrI-D	*Halyomorpha halys*	Pentatomidae	Hiruki et al. (1997)
Asia	Rice yellow dwarf	16SrXI-A	*Nephotettix virescens*	Cicadellidae	Chancellor and Cook (1995)

4. Major insect vectors of phytoplasmas
4.1 *Hishimonus* spp.

The genus *Hishimonus* was erected by Ishihara (1953) with type species *Thammotettix sellata* Uhler (1896), originally designated as *Acocephalus disguttus* by Walker (Niranjana, 2019).

4.1.1 *Hishimonus phycitis* (Distant, 1908)

Taxonomy and distribution: *Hishimonus phycitis* was described by Distant (1908), as having the coloration that varies considerably including the development of the median spot. This species is closely related to *Hishimonus sellatus* but differs principally in the absence of the concavity on the lateral margin of the shafts of the aedeagus (Uhler, 1896) (Fig. 4.1). *H. phycitis* belongs to the Deltocephalinae of the Cicadellidae in the suborder Auchenorrhyncha (Da Graça et al., 2007).

The distribution of *H. phycitis* is covering a wide area in Asian countries including Iran, Oman, the United Arab Emirates, India, China, Macau, Malaysia, Pakistan, the Philippines, Sri Lanka, Taiwan,

FIGURE 4.1

Dorsal and lateral views of *Hishimonus phycitis*.

and Thailand (EPPO, 2020). It was also reported from the Netherlands and Slovenia in Europe (EPPO, 2020).

4.1.1.1 History of its identification as insect vector

This species is regarded in the Middle East more significantly as a vector of witches' broom disease of lime (WBDL) phytoplasma, which limits the production of *Citrus aurantifolia* (Bagheri et al., 2009), and in India as a vector of brinjal little leaf (BLL) phytoplasma (Kumar and Rao, 2017) and sesame phyllody phytoplasma (Un Nabi et al., 2015) impacting *Solanum melongena* and *Sesamum indicum* yields.

WBDL is considered the most devastating disease of Mexican lime and has been reported from Oman, the United Arab Emirates, and Iran (Garnier et al., 1991). The phytoplasma associated with this disease is 'Candidatus Phytoplasma aurantifolia', and is vectored by the leafhopper, *H. phycitis* (Salehi et al., 2007a; Bagheri et al., 2009). Efforts to find the vector of lime witches' broom disease dated back to 1993, when Bové and colleagues collected all insects feeding on weeds and trees in severely infected orchards. The associated phytoplasma was detected in only *H. phycitis*, although several insects have been collected from the orchards. Siampour et al. (2006) conducted the same research and collected all insects in infected citrus orchards in southern Iran. They observed that the associated phytoplasma was detected in some sap-feeding insects including *H. phycitis, Recilia schmidtgeni, Idioscopus clypealis,* and *Diaphorina citri*. However, further research confirmed the associated phytoplasma in *H. phycitis* and *D. citri* body 40 and 30 days after collection, respectively. Although the associated phytoplasma could be detected in saliva, salivary glands, and head of *H. phycitis*, the transmission trials by feral collected insects to lime seedlings had failed in laboratory conditions. One year after, Salehi et al. (2007a) could confirm the vector status of *H. phycitis* for '*Ca.* P. aurantifolia' associated with Bakraee seedlings. They collected *H. phycitis* leafhoppers from infected orchards and released them on Bakraee seedlings (five individuals per plant at the two-leaf stage), and after six months from inoculation, the 30% of seedlings showed symptoms of phytoplasma infection including bud proliferation, chlorosis, and stunting. The presence of phytoplasmas in challenged plants was also confirmed by PCR assays. In 2009, the vector status of *H. phycitis* was completely confirmed by Bagheri et al. (2009) who collected the feral insects from the infected orchards and released them on 15-year-old Mexican lime trees. The severe phytoplasma disease symptoms appeared after six months from inoculation, and the phytoplasma infection was also confirmed by PCR assays with specific primers. Recently, Hemmati et al. (2020) reported the transmission of the phytoplasma to the one-year Mexican lime seedlings by feral-infected insects under semifield conditions.

Studies on the identification of BLL phytoplasma's vector dated back to 1969 where Bindra and Singh confirmed the presence of phytoplasmas in this species collected from the infected fields. Later, Azadvar and Baranwal (2012) also confirmed the presence of phytoplasmas in the collected leafhoppers by PCR assays. In 2017, Kumar and colleague confirmed the vector status of *H. phycitis* by transmission trials. They collected all stages of *H. phycitis* from disease-free fields and then released on phytoplasma-free brinjal plants under laboratory conditions. Then, the leafhoppers were transferred to diseased plants (positive in PCR assays) to acquire phytoplasma for three days and then transferred to healthy plants for seven days as AAP. The plants were monitored daily for 60 days after inoculation. After this time, 40%–60% of brinjal plants showed typical little leaf disease symptoms and also tested positive for the presence of phytoplasma in PCR assays. By these results, Kumar and Rao (2017) confirmed that *H. phycitis* is a vector of BLL phytoplasma strain belonging to the 16SrVI-D subgroup in India.

H. phycitis is also known as the vector of sesame phyllody disease which was associated with 16SrI-B in India (Un Nabi et al., 2015). They collected leafhoppers from infected sesame fields and observed that only *H. phycitis* was positive in PCR assays. Then, they conducted transmission trials with this species based on above-mentioned assays. Finally, they could confirm that *H. phycitis* was the vector of this disease in India. It should be noted that *Orosius albicinctus* and *O. argentatus* had been considered as sesame phyllody disease vectors in India since 1995 (Vasudeva and Sahambi, 1995; Vasudeva, 1995). Very recently, Tiwari et al. (2021) recorded infestation of *H. phycitis* near *Catharanthus roseus* plants in rainy session, and plants were with the symptoms of witches' broom and phyllody disease. The phytoplasma association in both the insect and *C.roseus* was confirmed as 16SrI-B subgroup phytoplasma.

4.1.1.2 Biology

The biology of *H. phycitis* in Iran was analyzed by Salehi et al. (2017a) who revealed that the peak of population was observed in February/March and the population decreased gradually during the warmer months from May to October. In addition, the high population densities of *H. phycitis* in brinjal fields were observed from July to October followed by the high incidence of BLL disease symptoms in Delhi, India (Kumar and Rao, 2017). Un Nabi et al. (2015) demonstrated that the *H. phycitis* leafhopper population was found to increase from July to September and then decreased from October onwards in sesame fields at Kushinagar and Delhi, India. Biological parameters and life table history were analyzed in phytoplasma infected and noninfected *H. phycitis* (Hemmati et al., 2021b). *H. phycitis* harboring phytoplasmas resulted in having more fecundity compared to uninfected insects. Moreover, the nymph development rate also increased. The finite rate of population increases and net reproductive rate were both higher among phytoplasma-infected *H. phycitis* indicating that the overall infected population fitness was improved as confirmed also by the higher number of offsprings produced. The results showed that female longevity was lower than male infected counterparts suggesting that a direct result of the physiological costs associated with the multiplication of the bacteria within *H. phycitis* female. The results could be used in the management of the disease as the infected trees may play a key role in the spread of the disease. The infected trees are considered as an initial inoculum source and the presence of the phytoplasmas resulted in more offspring and fecundity of the vector. This can lead to a higher transmission rate and population size in infected areas. So, the elimination of the infected trees is the most relevant measure to be taken in the disease management programs (Hemmati et al., 2021b).

Al-Subhi et al. (2020) revealed that witches' brooms were more attractant to *H. phycitis* in comparison with healthy limes. They observed that defense genes that have a role in plant defense responses to phytoplasma and insects were more downregulated in witches' brooms trees compared to the healthy ones. Based on the results, it can be concluded that witches' broom—affected parts of the trees contribute to the witches' broom of lime epidemics by supporting higher phytoplasma titers and attracting insect vectors. The same trend had been also observed in infected brinjal plants in which the infected parts attracted more *H. phycitis* leafhoppers than asymptomatic parts (Srinivasan and Chelliah, 1980). The infected parts that had more moisture, total carbohydrates, sugars, and organic acids than healthy ones can be a vital role in attracting leafhoppers.

4.1.2 *Hishimonus sellatus* (Uhler, 1896)

The species is classified in the Deltocephalinae subfamily of the Cicadellidae family. It is distributed in East Asia including China, South Korea, Malaysia, Indonesia, the Philippines, Sri Lanka, and Japan (EPPO, 2020). The species was first known as the vector of mulberry dwarf disease in Japan, where Ishijiam and Ishiie (1981) observed the MLO (later identified as a 16SrI-B subgroup) in salivary glands of *H. sellatus*. No transmission trials were done at that time. Several years later, Nishimura et al. (2004) revealed that *H. sellatus* can be the vector of *Cryptotania japonica* witches' broom and onion yellows phytoplasmas in Japan. Kusunoki et al. (2002) revealed that *H. sellatus* could transmit two phylogenetically distinct phytoplasma groups. They showed that the leafhopper was able to vector Rhus yellow (16SrI) and jujube witches' broom (JWB) (16SrV-B) phytoplasmas to *Rhus javanica* and *Ziziphus jujube*, respectively, in Japan (Kusunoki et al., 2002).

4.1.3 Management
4.1.3.1 Chemical control
Control of phytoplasma disease relies on controlling insect vectors and it is mostly achieved by spraying insecticides. Some systemic insecticides such as Dimethoate, Chlorpyrifos, Eforia, Pymethrozin, and imidacloprid are used against *H. phycitis* in infected orchards and can successfully decrease the populations (C. Hemmati, unpublished).

4.1.3.2 Biological controls
Information about natural enemies of the insect vectors is rare and limited to a primary study conducted by Bagheri et al. (2017) who revealed that two species of predatory spiders namely *Plexippus iranus* Longunov and *Thyene imperialis*, belonging to Salticidae family, voraciously fed *H. phycitis*. In addition, the eggs deposited on the infested leaves of the host tree were collected and maintained inside a Petri dish until parasitoid adult emergence. The parasitoid was then identified as *Polynema* sp. belonging to the Mymaridae family.

4.1.3.3 New control approaches

Accurate information from the insect vector populations including the genetic structure of the vector in infested areas can be a part of lime witches' broom disease management (Bové et al., 2000). Knowledge of the population genetic structure of *H. phycitis* may pave the way for understanding different aspects of the biology and ecology of this insect vector. In addition, it may shed light on gene flow among the populations of this insect which may be used in vector-based management strategies of WBDL. The DNA-based molecular markers have been widely used as a tool to assess genetic diversity in a number of insect species (Behura, 2006). Shabaniet al. (2013) analyzed the genetic structure of *H. phycitis* population using mitochondrial cytochrome *c* oxidase I and nine microsatellite DNA marker. They tested seven populations (one from Oman and six from Iran) with not enough variation. Shabani et al. (2013) revealed that *H. phycitis* population in North Oman has become separate and distinct from those distributed in Iran. The former study could not separate various WBDL vector populations which may be stemming from their inability to discriminate of low genetic differences. Hemmati et al. (2018) used six inter-simple sequence repeat (ISSR) primers to investigate the genetic structure of *H. phycitis* collected from 13 geographical localities (Kerman, Hormozgan, Sistan and Baluchestan, and Fars provinces) in Iran. Based on the ISSR results, there was no significant divergence among all populations except Forg (Fars province) and Qale'e Qazi (Hormozgan province) populations which diverged from the others. The Forg and Qale'e Qazi populations were found to be completely different and showed the highest and significant genetic distance compared with the other populations. The existence of genetically distinct *H. phycitis* populations could have important consequences for managing this vector in the regions where lime trees are grown. For instance, *H. phycitis* populations could differ in terms of their ability to transmit '*Ca*. P. aurantifolia' in different regions (Fars and Hormozgan) which would affect the status of *H. phycitis* as a phytoplasma vector in those regions. Furthermore, preferences for host plants, reproductive behavior, and natural enemies of this leafhopper vector might vary in different regions and consequently might affect its pest status and efficiency of controlling programs (Hemmati et al., 2018).

Phytoplasmas are not the only microorganisms living in the vectors' body, and they are involved in complex microbiomes enclosing bacteria, viruses, and fungi, in turn, influencing such associations. Insect symbionts including bacteria, fungi (yeast and yeast-like), and viruses are known effectors of ecology, life history, and development of their hosts. Similar to all sap feeders, phytoplasma vectors depend on bacterial symbionts supplementing their unbalanced diet (Zchori-Fein and Bourtzis, 2011). Insect microbial symbionts accompanied by supplying the host with nutrients offer protection against pathogens or other stress factors while manipulating the reproduction. Such multifaceted interaction could be used for developing microbe-based control approaches against phytoplasma-borne diseases (Dale and Moran, 2006).

Numerous symbiotic organisms could be used for control. Symbiotic control is known as a new biological control technique for plant diseases. In this technique, symbiotic microorganisms are isolated, genetically modified, and then reintroduced to express an antipathogenic agent in the insect vector (Wangkeeree et al., 2012). The identification of insect-associated microorganisms is the first step toward symbiotic control of insect pests. To find the bacterial symbionts of *H. phycitis*, 13 populations of *H. phycitis* were collected from WBDL-contaminated lime orchards from southern Iran. Sequencing and phylogenetic analysis of the 16S rRNA and *wsp* genes uncovered two obligate endosymbionts, '*Ca*. Sulcia muelleri' and '*Ca*. Nasuia deltocephalinicola', both of which exhibited 100% infection frequencies. Five facultative endosymbionts, *Wolbachia*, *Arsenophorus*, *Pantoea*,

Diplorickettsia, and *Spiroplasma* exhibited 70%, 90%, 57%, 48%, and 92% infection frequencies, respectively. *Wolbachia* was detected in all tested populations (Hemmati et al., 2021). This bacterium is currently being used to control dengue fever by inducing abnormal reproduction (Ruang-Areerate and Kittayapong, 2006). The level of gene flow observed in *H. phycitis* populations (Hemmati et al., 2018) would assist to introduce a specific gene or a transgenic population to a region and spreading these traits between populations. The technique has been proposed to control "flavescence dorée" phytoplasma vectored by the leafhopper *Scaphoideus titanus* Ball by cross-colonizing it with the specific bacterium *Asaia* (Crottiet al., 2009). Taking into accounts the above facts, it is possible to employ these bacteria to reach a new method to control this vector as well as WBDL.

The evolutionary and ecological success of the insects might rely on yeast and yeast-like symbiont (YLS) (Douglas, 2011). Yeast-like relationships with insects are recognized as obligate and common symbiotic relationship. A novel biological control technique for vector-borne diseases is paratransgenesis. A requirement for the progress of the strategies for the symbiotic control of insect vectors is to identify and characterize the insect-associated YLSs. YLSs living in *H. phycitis* were investigated in insects collected from 13 localities of citrus orchards distributed in southern Iran (Kerman, Hormozgan, Sistan and Baluchestan, and Fars provinces). Results revealed that this vector harbored two YLSs namely YLS of *H. phycitis* and *Candida pimensis* with a similarity of 98%–99% to those reported from the other cicadellids (Hemmati et al., 2017). These complex symbionts (bacterial and fungal symbionts) population living in *H. phycitis* provide a research road to find a novel way for WBDL control.

4.2 *Orosius albicinctus* Distant, 1918

Taxonomy and distribution: This species is classified in Cicadellidae family, Deltocephalinae subfamily, and Opsiini tribe. The leafhopper subfamily Deltocephalinae is the largest and is divided into 38 tribes with approximately 6700 valid species. Opsiini is one of the most significant tribes for their ability to transmit some plant pathogens, and the genus *Orosius* is one of the most important vectors of both phytoplasma and virus (Zahniser and Dietricj, 2013).

The genus *Orosius* was described by Distant (1918) with *O. albicinctus* as type species. This genus can be distinguished by creamy with irregular transverse brown spots and margins on vortex, pronotum, mesoscutellum, forewing, and legs and by having a very long and narrowed spine of the anal tube (Fig. 4.2). Its aedegaus is a long, triangle with well-diverging arms (Fletcher et al., 2017).

It is distributed in most Asian countries including Iran, Iraq, Palestine, Israel, Saudi Arabia, Oman, Turkey, India, and Pakistan (El-Sonbati et al., 2019). Due to that this species can transmit diverse phytoplasmas to diverse plant species, it will be described based on the host plants.

4.2.1 Vector status

Cucumber and squash phyllody: *O. albicinctus* is known as a vector of several phytoplasma diseases in Asian countries. Salehi et al. (2015) reported that this species was vector of cucumber and squash phyllody associated in Iran with the 16SrII-D and 16SrII-M subgroups, respectively. During a survey, they collected eight leafhopper species from the infected fields of which *O. albicinctus* was positive in PCR assays. Then, they collected *O. albicinctus* from the infected fields and released on 15 young healthy cucumber and squash plants to test its transmission ability. After passing four weeks from inoculation, typical symptoms of phytoplasma diseases had been observed on 7/15 squash and 9/15

FIGURE 4.2

Dorsal and lateral view of *Orosius albicinctus*.

cucumber plants. They also tested the host range of this phytoplasma strain by *O. albicinctus* transmission trials. They stated that *O. albicinctus* could also transmit the associated phytoplasma to periwinkle, carrot, sunflower, eggplant, tomato, parsley, alfalfa, and pot marigold plants.

Sesame phyllody: phyllody is one of the most important diseases of sesame and causes significant economic losses in Iran (associated with 16SrII-D, 16SrVI-A, and 16SrIX-C), India (16SrI, 16SrII, 16SrVI, and 16SrIX), Turkey (16SrII-D and 16SrIX-C), and Pakistan (16SrII-D) (Esmailzadeh Hosseini et al., 2007; Ikten et al., 2014; Un Nabi et al., 2015; Salehi et al., 2017b). Based on reports, *O. albicinctus* is the vector of all the associated phytoplasmas. The first report of *O. albicinctus* as the vector of sesame phyllody dated back to 1995 where Vasudeva and Sahambi (1995) demonstrated that this species is the vector of the agent of sesame phyllody in India; however, it was thought the causal agent was a virus. In 2007, Esmaeilzadeh Hosseini et al. confirmed the vector status of *O. albicinctus* for 16SrII phytoplasma associated with sesame phyllody in Iran. They collected the leafhoppers from the healthy fields and released them on healthy sesame plants. After four generations, they transferred leafhoppers to graft-infected sesame plants, and after 30 days, they were released on healthy plants. After two months, the typical symptoms were observed on challenged plants; however, no symptoms were observed on the plants fed by leafhoppers reared on healthy ones. The vector status of *O. albicinctus* was also reported for sesame phyllody phytoplasma in Pakistan in 2009 by conducting the same trials (Akhtar et al., 2009). In Turkey, Ikten and colleagues also confirmed that *O. albicinctus* was the vector of the disease in 2014; however, they didn't demonstrate which strain (16SrII or 16SrIX) could be transmitted. In Syria, this species is considered a putative vector of sesame phyllody (Khabbaz et al., 2013).

Alfalfa witches' broom: this disease, associated with diverse phytoplasmas groups (16SrI, 16SrII, 16SrVI, and 16SrXII), is one of the most destructive diseases in Asian countries especially in Iran,

Saudi Arabia, and Oman. It has been confirmed that *O. albicinctus* is the vector of 16SrII phytoplasmas in Iran (Esmailzadeh Hosseini et al., 2011). In other countries, its status has not been confirmed; however, some other species such as *Empoasca decipiens* (Paoli) and *Cicadulina bipunctata* in Saudi Arabia (Al-Saleh et al., 2014) and *Austroagallia avicula* (Rebaust) and *Epoasca* sp. in Oman (Khan et al., 2002) are considered as putative vectors for the phytoplasma associated with this disease.

Carrot witches' broom: the disease was reported from some countries including Iran, Israel, India, and Saudi Arabia (Hemmati et al., 2021a). Salehi et al. (2016a) confirmed that *O. albicinctus* is the vector of carrot witches' broom (associated with 16SrII) by transmission trials in Iran. The field-collected insects were released on healthy carrots and alfalfa plants. The plants were monitored, and after two months, the disease symptoms had been observed on the challenged plants. No reports are available on the vector status of *O. albicinctus* for carrot witches' broom in other countries; however, in Israel, two other leafhopper species namely *Circulifer haematoceps* and *Neoaliturus fenestratus* were reported as potential vectors (Weintraub and Orenstein, 2004).

Faba bean phyllody: the disease has been reported from Iran, Saudi Arabia, and Oman associated with the 16SrII-D subgroup phytoplasmas (Hemmati et al., 2021a). The only insect vector identified to date is *O. albicinctus* leafhopper that has been shown to vector faba bean phyllody phytoplasma in Iran (Salehi et al., 2016b).

Petunia witches' broom: the disease with witches' broom, little leaves, and stunting symptoms has been reported from Iran. Hemmati et al. (2019a) revealed that the associated phytoplasma can be vectored by *O. albicinctus* to healthy *Petunia* under laboratory conditions.

***Limonium* spp**: symptoms like those induced by phytoplasma including leaf yellowing, stunting, phyllody, and virescence have been reported on *Limonium* sp. in Israel (Weintraub et al., 2004). To find the vector of the associated phytoplasma, leafhoppers were collected and screened for phytoplasma presence by PCR assays. The associated phytoplasma has been detected in *O. albicinctus* body, and then transmission trials have been conducted by feral leafhoppers. The results confirmed that *O. albicinctus* is the natural vector of this disease in Israel (Weintraub et al., 2004).

Purple top-roll, marginal flavescence, and witches' broom disease of potatoes: the disease has been reported from India in 1974, and transmission trials confirmed that *O. albicinctus* is the vector of the disease (Nagaich et al., 1974).

4.2.2 Management
4.2.2.1 Chemical control
Sarnaik et al. (1986) revealed that three sprays of fenvalerate 0.01% gave the most effective control in the sesame field in India. In addition, Reddy et al. (2019) tested some insecticides against *O. albicinctus* in sesame fields in India. They revealed that chemical treatments with pymetrozine showed the highest percent age of reduction of the nymphal population (91.5%) followed by imidaclopride + ethiprole (80%), acephate (73%), and dinotefuran (70%). The lowest percent age of reduction was recorded in flonicamid (39.3%), thiamethoxam (seed treatment) (25.6%), and imidaclopride (seed treatment) (12.7%). There is no information about the other control ways of *O. albicinctus*.

4.3 *Neoaliturus* spp.
4.3.1 *Neoaliturus haematoceps* (Mulsant and Rey, 1855)
This species was first described as *Jassus haematoceps* Mulsant and Rey (1855), then was classified in *Circulifer* genus Zachvatkin, 1935. The new checklist, published by Pakarpour Rayeni et al. (2015), has classified it as *Neoaliturus haematoceps* (Fig. 4.3).

4. Major insect vectors of phytoplasmas   **57**

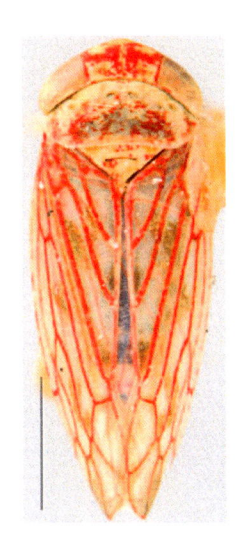

FIGURE 4.3

Dorsal and lateral views of *Neoaliturus haematoceps*.

It is widely distributed in Iran, Iraq, Israel, Saudi Arabia, Syria, Afghanistan, Jordan, Lebanon, Armenia, Azerbaijan, Georgia, Turkmenia, Kirghizia, Kazakhstan, and Turkey (EPPO, 2020). The species an important insect vector of plant pathogens in some Asian countries that can transmit phytoplasmas and other bacteria like *Spiroplasma citri*, the causal agent of stubborn disease in Iran (Omidi et al., 2011).

Cabbage yellow: the disease, associated with the 16SrVI group, was first reported from Iran in 2007 (Salehi et al., 2007b). During a survey, they collected the leafhoppers from the infected fields and released them on healthy cabbage under laboratory conditions. After two months from inoculation, typical symptoms of phytoplasma disease had been observed. In addition, they released feral leafhoppers on other plant species including cauliflower, rape, and periwinkle which showed the phytoplasma disease symptoms. According to the results, the vector ability of *N. haematoceps* for 16SrVI phytoplasma was confirmed (Salehi et al., 2007b).

Rapeseed phyllody: phytoplasma infection of the rapeseed was described in Iran in 2011. The phytoplasma associated with the diseases was characterized as the 16SrI group. Salehi et al. (2011) revealed that *N. haematoceps* was the vector of the diseases in Iran. The feral leafhoppers were collected from the infected fields and released on rapeseed, periwinkle, sesame, stock, mustard, radish, and rocket plants. In addition, leafhoppers were caught from the healthy fields and reared on healthy rapeseed in the laboratory. Then, they transferred the leafhoppers to experimentally infected rapeseed plants for phytoplasma acquisition. Then, the leafhoppers were transferred on healthy rapeseed plants for three months. All the challenged plants (exposed to naturally collected leafhoppers and experimentally infected leafhoppers) showed typical symptoms of phytoplasma diseases.

Sesame phyllody: sesame phyllody is one of the most important diseases of sesame that can lose in yield near 100%. It has been reported from Iran, Turkey, India, Israel, Thailand, Pakistan, Oman, Iraq, and Myanmar (Martini et al., 2018). In Turkey, this disease has been reported to be associated with 16SrII-d and 16SrIX-C phytoplasma groups (Ikten et al., 2014). It has been confirmed that the associated phytoplasma can be vectored to healthy sesame plants by *N. haematoceps* in Turkey (kersting, 1993). Salehi and Izadpanah (1992) also reported that *N. haematoceps* was the vector of sesame phyllody in Iran.

Limonium latifolium witches' broom: this disease has been reported from Israel with an infection rate of 60%. Weintraub et al. (2004) reported that the associated phytoplasmas (16SrII, 16SrV, and 16SrIX) can be transmitted to the healthy *L. latifolium* by *N. haematoceps*, however, they didn't specify which strain can be vectored by the leafhopper.

4.3.2 Neoaliturus tenellus

Limonium latifolium witches' broom: it has been reported that *Neoaliturus tenellus* can also be the vector of *L. latifolium* witches' broom in Israel (Weintraub et al., 2004).

4.3.3 Neoaliturus fenestratus

This species had been described as *Jassus fenestrarus*. The species is widely distributed in Armenia, India, Iraq, Israel, Japan, Saudi Arabia, Syria, and Turkey (Zahniser, 2007).

Lettuce and wild lettuce phyllody: the disease associated with 16SrIX phytoplasma was first reported by Salehi et al. (2007c) in Iran. It has been confirmed that *N. fenestratus* was the vector of lettuce and wild lettuce phyllody phytoplasma (Salehi et al., 2007c). In addition, the feral leafhopper collected from infected lettuce fields could transmit the phytoplasma to wild lettuce, sowthistle, and periwinkle under laboratory conditions.

5. Some minor phytoplasma insect vectors in Asia

This section will focus on the vectors that may transmit one or a few phytoplasmas to plants or diseases that are of less importance.

Almond witches' broom (AlmWB) is known as the most destructive phytoplasma disease of stone fruits in Lebanon and Iran (Abou-Jawdah et al., 2003). Studies conducted by Abou-Jawdah et al. (2011) revealed that polyphagous leafhopper *Asymmetrasca decedens* is the vector of 16SrIX-B in Lebanon (Abou-Jawdah et al., 2011). Further transmission trials studies revealed that *Tachycixius viperinus* Dlabola can transmit the 16SrIX-B phytoplasma to almonds in Lebanon (Tedeschi et al., 2015). The vector of AlmWB disease has not been identified in Iran; however, two leafhoppers *Zygina flammigera* and *Frutioidea bisignata* were collected from infected almond orchards showing that they can be the potential vector of the 16SrIX-C subgroup in Iran (Salehi et al., 2006; Zirak et al., 2010).

Sandalwood (*Santalum album*) is the host of 16SrI-B phytoplasma which induces symptoms including spike-like, crowded leaves, and phyllody. The associated phytoplasma is transmitted to healthy sandals by *Coelidia indica* Walker classified in the Coelidiinae subfamily of the Cicadellidae family (Rao et al., 2017).

Sugarcane grassy shoot (SCGS) disease, associated with 16SrXI phytoplasma, is one of the most important diseases of sugarcane in India (Rao et al., 2017). Srivastava et al. (2006) demonstrated that *Deltocephalus vulgaris* (Cicadellidae: Deltocephalinae), as a natural vector of SCGS, moreover *Exitianus indicus* (Cicadellidae: Deltocephalinae) and *Pyrilla perpusilla* (Lophopidae) were reported as putative insect vectors of the diseases in India (Rao et al., 2014; Tiwari et al., 2016). Recently, two additional leafhoppers *Maiestas portico* (Cicadellidae: Deltocephalinae) and *Cofana unimaculata* (Cicadellidae: Cicadellinae) were also reported as vectors of SCGS phytoplasma (Tiwari et al., 2017).

JWB disease, associated with 16SrV-B, destroyed the sustainable jujube industry in China, Japan, and Korea for several years. It has been confirmed that JWB phytoplasma is transmitted by *Hishimonoides chinensis* Anufriev (Cicadellidae: Deltocephalinae) (Tsai et al., 1988).

Mulberry dwarf disease, associated with 16SrI-B phytoplasma, is one of the most important diseases of mulberry reported from China. *Hishimonoides sellatiformis* Ishihara (Cicadellidae: Deltocephalinae) is known as the vector of mulberry dwarf phytoplasma (Ishijima and Ishiie, 1981). It has been confirmed that the associated phytoplasma has transovarial transmission.

'*Ca.* P. oryae' (16SrXI-A) is associated with the rice yellow dwarf disease was reported from China, India, Iran, Bangladesh, Japan, Malaysia, the Philippines, Sri Lanka, Taiwan, and Thailand (Jung et al., 2003). *Nephotettix virescens* Distant (Cicadellidae: Deltocephalinae), known as green paddy leafhopper, can vector the associated phytoplasma to healthy plants (Chancellor and Cook, 1995). The vector is widely distributed in India, Malaysia, Brunei, Laos, Taiwan, Thailand, Vietnam, the Philippines, Indonesia, Bangladesh, Azerbaijan, and Cambodia (EPPO, 2020).

Scleroracus flavopictus (Ishihara, 1935) (Cicadellidae: Deltocephalinae) is known as the vector of associated phytoplasma with potato purple top gentian witches' broom and tsuwabuki witches' broom associated with 16SrIII in Japan. The species is only reported from Japan (Okuda, 1997).

Austroagallia sinuatea (Mulsant and Rey, 1855) has been confirmed as the vector of *Aerva javanica* leaf roll phytoplasma and *Zinnia elegans* phyllody phytoplasma associated with 16SrII-D subgroup in Iran (Hemmati and Nikooei, 2019; Hemmati et al., 2019b).

Bermuda grass white leaves disease is the most important phytoplasma disease associated with 16SrXIV-A and 16SrXIV-B subgroups. Salehi et al. (2009) reported that the associated phytoplasma was vectored by *Exitianus capicola* in Iran (Salehi et al., 2009).

Proutista moesta (Westwood) (Derbidae) can transmit coconut Kerala root wilt phytoplasma. It is also known as the vector of grassy shoot disease which is one of the most important diseases of sugarcane in India (Ponnamma and Solomon, 1998).

True bugs, classified in Heteroptera suborder can also be the vector of phytoplasmas. *Hyalomorpha halys* Stål (Hom: Pentatomidae) is known as the vector of paulownia witches' broom disease reported from East Asian countries (Hiruki et al., 1997). In addition, *Stephanitis typica* (Distant) (Hom.: Tingidae) distributed in most Asian countries is known as the vector of coconut root wilt disease associated with 16SrXI phytoplasma (Mathen et al., 1990).

6. General management

Control of phytoplasma disease relies on controlling insect vectors and is mostly achieved by spraying insecticides. Another way to control such diseases is the eradication of host plants that play the role as alternative or reservoirs of phytoplasmas. Habitat management could reduce vector incidence resulting in the reduction of phytoplasma diseases. Genetic modification of plants that can produce defensive compounds against insect vectors or phytoplasmas can be an effective way. Insect symbionts (fungi, bacteria, and viruses) are recognized effectors of ecology, life history, and the development of their hosts. Similar to all sap feeders, phytoplasma vectors, indeed leafhoppers, psyllids, and planthoppers depend on bacterial symbionts supplementing their unbalanced diet (Zchori-Fein and Bourtzis, 2011). Such multifaceted interaction could be used for developing microbe-based control approaches against phytoplasma-borne diseases through the overall idea of microbial resource management defined as the resolution for practical problems through the management of complex microbial systems and their related metabolic abilities (Crotti et al., 2012) to endophytes and insect symbionts. Most recently, RNA-based gene silencing has been proposed for controlling phytoplasma disease. In such a method, the genes involved in phytoplasma attachment in the vector body are silenced and therefore interfere with phytoplasma transmission. It is, moreover, necessary to understand the taxonomy and classification of phytoplasma insect vector for a concrete management strategy.

7. Conclusion

There is much to be learned from collecting and conducting research on phytoplasma vectors. In disease management, early detection of the disease and vector can help to implement the best approach and preventing the spread of the disease. Most phytoplasma vectors reported in Asia are members of the Cicadellidae. Focus on vector species usually occurs when economic crops are affected, which may skew understanding of the true range of vector species. Vector studies must be expanded to include a variety of plant phloem feeders, including the Membracidae. The primary means of controlling phytoplasma vectors is by insecticides; however, increasing pressure to find less toxic and more biologically based techniques to control, or at least manage, insect vectors necessitates an even greater reliance on solid understandings of the biology of insect vectors from the cellular to the ecological level. The choice of specific strategy depends on the biology of plants and insect vectors: migration and transovarial transmission are successful for inoculating annual crops, whereas phytoplasma diseases of tree crops are easily maintained in plants along years and they often follow a closed epidemiological cycle. In addition, the degree of specialization influences the efficiency of insect vectors, monophagous species being more active than polyphagous ones. That is why phytoplasmas transmitted by generalist species may share more than one vector. The pathogen-vector-plant-habitat complex must therefore be considered as a single issue when dealing with the epidemics of phytoplasma disease.

References

Abou-Jawdah, Y., Dakhil, H., El-Mehtar, S., Lee, I.M., 2003. Almond witches' broom phytoplasma: a potential threat to almond, peach, and nectarine. J. Indian Dent. Assoc. 25 (1), 28–32.

Abou-Jawdah, Y., Dakhil, H., Molino-Lova, M.M., Sobh, H., Nehme, M., Fakhr-Hammad, E.A., Alma, A., Samsatly, J., Jawhari, M., Abdul-Nour, H., Bianco, P.A., 2011. Preliminary survey of potential vectors of 'Candidatus Phytoplasma phoenicium' in Lebanon and probability of occurrence of apricot chlorotic leaf roll (ACLR) phytoplasma. Bull. Insectol. 64 (Suppl.), S123–S124.

Akhtar, K.P., Sarwar, G., Dickinson, M., Ahmad, M., Haq, M.A., Hameed, S., Iqbal, M.J., 2009. Sesame phyllody disease: its symptomatology, etiology, and transmission in Pakistan. Turk. J. Agric. For. 33 (5), 477–486.

Al-Saleh, M.A., Amer, M.A., Al-Shahwan, I.M., Abdalla, O.A., Damiri, B.V., 2014. Detection and molecular characterization of alfalfa witches' broom phytoplasma and its leafhopper vector in Riyadh region of Saudi Arabia. Int. J. Agric. Biol. 16 (2), 300–306.

Al-Subhi, A.M., Al-Sadi, A.M., Al-Yahyai, R., Chen, Y., Mathers, T., Orlovskis, Z., Moro, G., Mugford, S., Al-Hashmi, K., Hogenhout, S., 2020. Witches' broom disease of lime contributes to phytoplasma epidemics and attracts insect vectors. Plant Dis. 104, 2637–2648.

References

Alma, A., Tedeschi, R., Lessio, F., Picciau, L., Gonella, E., Ferracini, C., 2015. Insect vectors of plant pathogenic Mollicutes in the Euro-Mediterranean region. Phytopath. Moll. 5 (2), 53–73.

Alma, A., Lessio, F., Gonella, E., Picciau, L., Mandrioli, M., Tota, F., 2018. New insights in phytoplasma-vector interaction: acquisition and inoculation of "flavescence dorée" phytoplasma by *Scaphoideus titanus* adults in a short window of time. Ann. Appl. Biol. 173 (1), 55–62.

Alma, A., Lessio, F., Nickel, H., 2019. Insects as phytoplasma vectors: ecological and epidemiological aspects. In: Bertaccini, A., Weintraub, P.W., Rao, G.P., Mori, N. (Eds.), Phytoplasmas: Plant Pathogenic Bacteria-II. Springer, Singapore, pp. 1–25.

Azadvar, M., Baranwal, V.K., 2012. Multilocus sequence analysis of phytoplasma associated with brinjal little leaf disease and its detection in Hishimonusphycitis in India. Phytopath. Moll. 2 (1), 15–21.

Bagheri, A.N., Salehi, M., Faghihi, M.M., Samavi, S., Sadeghi, A., 2009. Transmission of '*Candidatus* Phytoplasma aurantifolia' to Mexican lime by the leafhopper *Hishimonus phycitis* in Iran. J. Plant Pathol. 91 (4), S105.

Bagheri, A., Hemmati, C., Askari Seyahooei, M., Modarres Najafabadi, S.S., Nikooei, M., 2017. Preliminary study on natural enemies of *Hishimonus phycitis* (Distant, 1908) (Hemi., Cicadellidae), the vector of lime witches' broom phytoplasma in Hormozgan province. In: 2nd Iranian International Congress of Entomology. 2-4 Sep. University of Tehran, Karaj, Iran.

Behura, S.K., 2006. Molecular marker systems in insects: current trends and future avenues. Mol. Ecol. 15 (11), 3087–3113.

Bosco, D., Tedeschi, R., 2013. Insect vector transmission assays. In: Phytoplasma. Humana Press, Totowa, NJ, pp. 73–85.

Bosco, D., Galetto, L., Leoncini, P., Saracco, P., Raccah, B., Marzachì, C., 2014. Interrelationships between '*Candidatus* Phytoplasma asteris' and its leafhopper vectors (Homoptera: Cicadellidae). J. Econ. Entomol. 100 (5), 1504–1511.

Bové, J-M., Garnier, M., 2000. Witches' broom disease of lime. Arab J. Plant Protect. 18 (2), 148–152.

Chancellor, T.C.B., Cook, A.G., 1995. The bionomics and population dynamics of *Nephotettix* spp. (Hemiptera: Cicadellidae) in South and Southeast Asia with particular reference to the incidence of rice tungro virus disease. Trop. Sci. 35 (2), 200–216.

Chuche, J., Thiéry, D., 2014. Biology and ecology of the "flavescence dorée" vector *Scaphoideus titanus*: a review. Agron. Sustain. Dev. 34 (2), 381–403.

Contaldo, N., Satta, E., Zambon, Y., Paltrinieri, S., Bertaccini, A., 2016. Development and evaluation of different complex media for phytoplasma isolation and growth. J. Microbiol. Methods 127, 105–110.

Crotti, E., Damiani, C., Pajoro, M., Gonella, E., Rizzi, A., Ricci, I., Negri, I., Scuppa, P., Rossi, P., Ballarini, P., Raddadi, N., 2009. *Asaia*, a versatile acetic acid bacterial symbiont, capable of cross-colonizing insects of phylogenetically distant genera and orders. Environ. Microbiol. 11 (12), 3252–3264.

Crotti, E., Balloi, A., Hamdi, C., Sansonno, L., Marzorati, M., Gonella, E., Favia, G., Cherif, A., Bandi, C., Alma, A., Daffonchio, D., 2012. Microbial symbionts: a resource for the management of insect-related problems. Microb. Biotechnol. 5 (3), 307–317.

Da Graça, J.V., Sétamou, M., Skaria, M., French, J.V., 2007. Arthropod vectors of exotic citrus diseases: a risk assessment for the Texas citrus industry. Subtrop. Plant. Sci. 59, 64–74.

Dale, C., Moran, N.A., 2006. Molecular interactions between bacterial symbionts and their hosts. Cell 126 (3), 453–465.

Dietrich, C.H., 2009. Auchenorrhyncha:(cicadas, spittlebugs, leafhoppers, treehoppers, and planthoppers). In: Encyclopedia of Insects. Academic Press, pp. 56–64.

Douglas, A.E., 2011. Lessons from studying insect symbioses. Cell Host Microbe 10 (4), 359–367.

El-Sonbati, S.A., Wilson, M.R., Dhafer, H.M.A., 2019. Revision of the leafhopper genus *Orosius* Distant, 1918 (Hemiptera: Cicadellidae: Deltocephalinae: Opsiini) in the Arabian peninsula with the description of a new species. Zootaxa 4565 (1) zootaxa.4565.1.2.

EPPO, 2020. EPPO Global database. In: EPPO Global Database. EPPO, Paris, France. https://gd.eppo.int.

Esmailzadeh Hosseini, S.A., Salehi, M., Mirzaie, A., 2011. Alternate hosts of alfalfa witches' broom phytoplasma and winter hosts of its vector *Orosius albicinctus* in Yazd-Iran. Bull. Insectol. 64 (Suppl.), S247–S248.

Esmailzadeh-Hosseini, S.A., Mirzaie, A., Jafari-Nodooshan, A., Rahimian, H., 2007. The first report of transmission of a phytoplasma associated with sesame phyllody by *Orosius albicinctus* in Iran. Australas. Plant Dis. Notes 2 (1), 33–34.

Fletcher, M., Löcker, H., Mitchell, A., Gopurenko, D., 2017. A revision of the genus *Orosius* Distant (Hemiptera: Cicadellidae) based on male genitalia and DNA barcoding. Austral. Entomology 56 (2), 198–217.

Garnier, M., Zreik, L., Bové, J-M., 1991. Witches' broom, a lethal mycoplasmal disease of lime trees in the sultanate of Oman and the United Arab Emirates. Plant Dis. 75 (6), 546–551.

Hemmati, C., Nikooei, M., 2019. *Austroagallia sinuata* transmission of '*Candidatus* Phytoplasma aurantifolia' to *Zinnia elegans*. J. Plant Pathol. 101 (4), 1223.

Hemmati, C., Moharramipour, S., Askari Siahooei, M., Bagheri, A., Mehrabadi, M., 2017. Identification of yeast and yeast-like symbionts associated with *Hishimonus phycitis* (Hemiptera: Cicadellidae), the insect vector of lime witches' broom phytoplasma. J. Crop Protect 6 (4), 439–446.

Hemmati, C., Moharramipour, S., Askari Seyahooei, M., Bagheri, A., Mehrabadi, M., 2018. Population genetic structure of *Hishimonus phycitis* (Hem.: Cicadellidae), vector of lime witches' broom phytoplasma. J. Agric. Sci. Technol. 20 (5), 999–1012.

Hemmati, C., Nikooei, M., Bertaccini, A., 2019a. Identification, occurrence, incidence and transmission of phytoplasma associated with *Petunia violacea* witches' broom in Iran. J. Phytopathol. 167 (10), 547–552.

Hemmati, C., Nikooei, M., Bertaccini, A., 2019b. Identification and transmission of phytoplasmas and their impact on essential oil composition in *Aerva javanica*. 3 Biotech 9 (8), 1–7.

Hemmati, C., Askari Seyahooei, M., Nikooei, M., Modarees Najafabadi, S.S., Goodarzi, A., Amiri Mazraie, M., Faghihi, M.M., 2020. Vector transmission of lime witches' broom phytoplasma to Mexican lime seedlings under greenhouse condition. J. Crop Protect. 9 (2), 209–215.

Hemmati, C., Nikooei, M., Al-Subhi, A.M., Al-Sadi, A.M., 2021a. History and current status of phytoplasma diseases in the Middle East. Biology 10, 226.

Hemmati, C., Nikooei, M., Al-Sadi, A.M., 2021b. '*Candidatus* Phytoplasma aurantifolia' increased the fitness of *Hishimonus phycitis*; the vector of lime witches' broom disease. Crop Protect. 142, 105532.

Hiruki, C., 1997. Paulownia witches' broom disease important in East Asia. International Symposium on Urban Tree Health 496, 63–68.

Ikten, C., Catal, M., Yol, E., Ustun, R., Furat, S., Toker, C., Uzun, B., 2014. Molecular identification, characterization and transmission of phytoplasmas associated with sesame phyllody in Turkey. Eur. J. Plant Pathol. 139 (1), 217–229.

Ishihara, T., 1953. Tentative Check List of the Superfamily Cicadelloidea of Japan (Homoptera).

Ishijima, T., Ishiie, T., 1981. Mulberry dwarf: first tree mycoplasma disease. In: Mycoplasma Diseases of Trees and Shrubs. Academic Press, pp. 147–184.

Jung, H-Y., Sawayanagi, T., Wongkaew, P., Kakizawa, S., Nishigawa, H., Wei, W., Oshima, K., Miyata, S.I., Ugaki, M., Hibi, T., Namba, S., 2003. '*Candidatus* Phytoplasma oryzae', a novel phytoplasma taxon associated with rice yellow dwarf disease. Int. J. System.Evol. Microbiol. 53 (6), 1925–1929.

Kersting, U., 1993. Symptomatology, etiology and transmission of sesame phyllody in Turkey. J. Turk. Phytopathol. 22 (2), 47–54.

Khabbaz, S.E., Alnabhan, M., Arafeh, M., 2013. First report of the natural occurrence of phyllody disease of sesame in Syria. Int. J. Sci. Res 2, 1–3.

Khan, A.J., Botti, S., Al-Subhi, A.M., Gundersen-Rindal, D.E., Bertaccini, A.F., 2002. Molecular identification of a new phytoplasma associated with alfalfa witches' broom in Oman. Phytopathology 92, 1038–1047.

Kumar, M., Rao, G.P., 2017. Molecular characterization, vector identification and sources of phytoplasmas associated with brinjal little leaf disease in India. 3 Biotech 7 (1), 7.

Kusunoki, M., Shiomi, T., Kobayashi, M., Okudaira, T., Ohashi, A., Nohira, T., 2002. A leafhopper (*Hishimonus sellatus*) transmits phylogenetically distant phytoplasmas: rhus yellows and *Hovenia* witches' broom phytoplasma. J. Gen. Plant Pathol. 68 (2), 147–154.

Lefol, C., Lherminier, J., Boudon-Padieu, E., Larrue, J., Louis, C., Caudwell, A., 1994. Propagation of "flavescence dorée" MLO (mycoplasma-like organism) in the leafhopper vector *Euscelidius variegatus* Kbm. J. Invertebr. Pathol. 63 (3), 285–293.

Marcone, C., Pierro, R., Tiwari, A.K., Rao, G.P., 2002. Major phytoplasma diseases of temperate fruit trees. Agrica 11 (1), 20–31.

Martini, M., Delić, D., Liefting, L., Montano, H., 2018. Phytoplasmas infecting vegetable, pulse and oil crops. In: Phytoplasmas: Plant Pathogenic Bacteria-I. Springer, Singapore, pp. 31–65.

Mathen, K., Rajan, P., Nair, C.R., Sasikala, M., Gunasekharan, M., Govindankutty, M.P., Solomon, J.J., 1990. Transmission of root (wilt) disease to coconut seedlings through *Stephanitis typica* (Distant) (Heteroptera: Tingidae). Trop. Agri. 67 (1), 69–73.

Mirzaie, A., Esmailzadeh-Hosseini, S.A., Jafari-Nodoshan, A., Rahimian, H., 2007. Molecular characterization and potential insect vector of a phytoplasma associated with garden beet witches' broom in Yazd, Iran. J. Phytopathol. 155 (4), 198–203.

Mulsant, M.E., Rey, C., 1855. Description of some new or little known Hemiptera-Homoptera. Ann. Linnean Soci. Lyon 2 (2), 197–249, 426.

Nagaich, B.B., Puri, B.K., Sinha, R.C., Dhingra, M.K., Bhardwaj, V.P., 1974. Mycoplasma-like organisms in plants affected with purple top-roll, marginal flavescence and witches' broom diseases of potatoes. J. Phytopathol. 81 (3), 273–279.

Niranjana, G., 2019. Study on Systematics of Leafhopper Fauna at Different Altitudes (Doctoral dissertation, Division of Entomology Icar-Indian Agricultural Research Institute New Delhi).

Nishimura, N., Nakajima, S., Kawakita, H., Sato, M., Namba, S., Fujisawa, I., Tsuchizaki, T., 2004. Transmission of *Cryptotaenia japonica* witches' broom and onion yellows by *Hishimonoides sellatiformis*. Japan. J. Phytopathol. 70 (1), 22–25.

Okuda, S., Prince, J.P., Davis, R.E., Dally, E.L., Lee, I-M., Mogen, B., Kato, S., 1997. Two groups of phytoplasmas from Japan distinguished on the basis of amplification and restriction analysis of 16S rDNA. Plant Dis. 81 (3), 301–305.

Omidi, M., Hosseini-Pour, A., Rahimian, H., Massumi, H., Saillard, C., 2011. Identification of *Circulifer haematoceps* (Hemiptera: Cicadellidae) as vector of *Spiroplasma citri* in the Kerman province of Iran. J. Plant Pathol. 167–172.

Pagliari, L., Chuche, J., Bosco, D., Thiéry, D., 2019. Phytoplasma transmission: insect rearing and infection protocols. In: Phytoplasmas. Humana Press, New York, NY, pp. 21–36.

Pakarpour Rayeni, F., Nozari, J., Seraj, A.A., 2015. A checklist of Iranian Deltocephalinae (Hemiptera: Cicadellidae). Iran. J. Animal Biosyst. 11 (2), 121–148.

Pilkington, L.J., Gurr, G.M., Fletcher, M.J., Nikandrow, A., Elliott, E., 2004. Vector status of three leafhopper species for Australian lucerne yellows phytoplasma. Aust. J. Entomol. 43 (4), 366–373.

Ponnamma, K.N., Solomon, J.J., 1998. Yellow leaf disease of areca palms-critical studies on transmission with the vector *Proutista moesta* Westwood (Homoptera; Derbidae). J. of Plant. Crops 26, 75–76.

Rangaswami, S., Griffith, A.L., 1941. Demonstration of *Jassus indicus* (Walk) as a vector of the spike disease of sandal (*Santalum album*, Linn.). Indian For. 67, 387–394.

Rao, G.P., Madhupriya, Tiwari, A.K., Kumar, S., Baranwal, V.K., 2014. Identification of sugarcane grassy shoot-associated phytoplasma and one of its putative vectors in India. Phytoparasitica 42, 349–354.

Rao, G.P., Madhupriya, T.V., Manimekalai, R., Tiwari, A.K., Yadav, A., 2017. A century progress of research on phytoplasma diseases in India. Phytopath. Moll. 7 (1), 1–38.

Reddy, T.V., Prasad, K.H., Chalam, M., Viswanath, K., 2019. Management of leafhopper (*Orosius albicintus*) of sesamum with certain insecticides. Andhra Pradesh J Agri. Sci. 5 (3), 181–186.

Ruang-Areerate, T., Kittayapong, P., 2006. *Wolbachia* transinfection in *Aedes aegypti*: a potential gene driver of dengue vectors. Proc. Natl. Acad. Sci. U. S. A. 103 (33), 12534–12539.

Salehi, M., Izadpanah, K., 1992. Etiology and transmission of sesame phyllody in Iran. J. Phytopathol. 135 (1), 37–47.

Salehi, M., Izadpanah, K., Heydarnejad, J., 2006. Characterization of a new almond witches' broom phytoplasma in Iran. J. Phytopathol. 154 (7-8), 386–391.

Salehi, M., Izadpanah, K., Siampour, M., Bagheri, A., Faghihi, S.M., 2007a. Transmission of '*Candidatus* Phytoplasma aurantifolia' to Bakraee (*Citrus reticulata* hybrid) by feral *Hishimonus phycitis* leafhoppers in Iran. Plant Dis. 91 (4), 466.

Salehi, M., Izadpanah, K., Siampour, M., 2007b. Characterization of a phytoplasma associated with cabbage yellows in Iran. Plant Dis. 91 (5), 625–630.

Salehi, M., Izadpanah, K., Nejat, N., Siampour, M., 2007c. Partial characterization of phytoplasmas associated with lettuce and wild lettuce phyllodies in Iran. Plant Pathol. 56 (4), 669–676.

Salehi, M., Izadpanah, K., Siampour, M., Taghizadeh, M., 2009. Molecular characterization and transmission of Bermuda grass white leaf phytoplasma in Iran. J. Plant Pathol. 655–661.

Salehi, M., Izadpanah, K., Siampour, M., 2011. Occurrence, molecular characterization and vector transmission of a phytoplasma associated with rapeseed phyllody in Iran. J. Phytopathol. 159 (2), 100–105.

Salehi, M., Siampour, M., Esmailzadeh Hosseini, S.A., Bertaccini, A., 2015. Characterization and vector identification of phytoplasmas associated with cucumber and squash phyllody in Iran. Bull. Insectol. 68 (2), 311–319.

Salehi, M., Hosseini, S.E., Salehi, E., Bertaccini, A., 2016a. Molecular and biological characterization of a 16SrII phytoplasma associated with carrot witches' broom in Iran. J. Plant Pathol. 83–90.

Salehi, M., Rasoulpour, R., Izadpanah, K., 2016b. Molecular characterization, vector identification and partial host range determination of phytoplasmas associated with faba bean phyllody in Iran. Crop Protect. 89, 12–20.

Salehi, M., Bagheri, A., Faghihi, M.M., Izadpanah, K., 2017a. Study of partial biological and behavioral traits of *Hishimonus phycitis*, vector of lime witches' broom, for management of the disease. Iran. J. Plant Pathol. 53 (1), 75–96.

Salehi, M., Hosseini, S.E., Salehi, E., Bertaccini, A., 2017b. Genetic diversity and vector transmission of phytoplasmas associated with sesame phyllody in Iran. Folia Microbiol. 62 (2), 99–109.

Sarnaik, D.N., Ghode, R.N., Peshkar, L.N., Satpute, U.S., 1986. Insecticidal control of sesamum jassids *Orosius albicinctus* (Distant). PKV Res. J. 10 (1), 41–43.

Shabani, M., Bertheau, C., Zeinalabedini, M., Sarafrazi, A., Mardi, M., Naraghi, S.M., Rahimian, H., Shojaee, M., 2013. Population genetic structure and ecological niche modelling of the leafhopper *Hishimonus phycitis*. J. Pest. Sci. 86 (2), 173–183.

Siampour, M., Izadpanah, K., Afsharifar, A.R., Salehi, M., Taghizadeh, M., 2006. Detection of phytoplasma in insects collected in witches' broom affected lime groves. Iran. J. Plant Pathol. 42 (1).

Srinivasan, K., Chelliah, S., 1980. The mechanism of preference of the leafhopper vector, *Hishimonus phycitis* (Distant) for eggplants infected with little leaf disease. Proc. Indian Nat. Sci. Aca. B 46 (6), 786–796.

Srivastava, S., Singh, V., Gupta, P.S., Sinha, O.K., Baitha, A., 2006. Nested PCR assay for detection of sugarcane grassy shoot phytoplasma in the leafhopper vector *Deltocephalus vulgaris*: a first report. Plant Pathol. 55 (1), 25–28.

Szwedo, J., 2016. The unity, diversity and conformity of bugs (Hemiptera) through time. Earth Environment. Sci. Transact. Royal Soci. Edinburgh 107 (2−3), 109−128.
Tanne, E., Boudon-Padieu, E., Clair, D., Davidovich, M., Melamed, S., Klein, M., 2001. Detection of phytoplasma by polymerase chain reaction of insect feeding medium and its use in determining vectoring ability. Phytopathology 91 (8), 741−746.
Tedeschi, R., Picciau, L., Quaglino, F., Abou-Jawdah, Y., Molino Lova, M., Jawhari, M., Casati, P., Cominetti, A., Choueiri, E., Abdul-Nour, H., Bianco, P.A., 2015. A cixiid survey for natural potential vectors of '*Candidatus* Phytoplasma phoenicium' in Lebanon and preliminary transmission trials. Ann. Appl. Biol. 166 (3), 372−388.
Tiwari, A.K., Madhupriya, S.K.P., Pandey, B.S., Rao, G.P., 2016. Detection of sugarcane grassy shoot phytoplasma (16SrXI-B subgroup) in *Pyrilla perpusilla* Walker in Uttar Pradesh, India. Phytopath Mollicut. 6 (1), 56−59.
Tiwari, A.K., Kumar, S., Mall, S., Jadon, V., Rao, G.P., 2017. New efficient natural leafhopper vectors of sugarcane grassy shoot phytoplasma in India. Sugar Tech 19 (2), 191−197.
Tiwari, N.N., Jain, R.K., Tiwari, A.K., 2021. *Hishimonas phycitis*: a possible vector of 16SrI-B subgroup phytoplasma. Agrica 10 (1), 72−75.
Tsai, J.H., Chen, Z., Shen, C.Y., Jin, K.X., 1988. Mycoplasmas and fastidious vascular prokaryotes associated with tree diseases in China. Tree Mycoplasm. Mycoplasma Dis. 69.
Uhler, P.R., 1896. Summary of the Hemiptera of Japan, Presented to the United States National Museum by Professor Mitzukuri, vol 19. US Government Printing Office.
Un Nabi, S., Dubey, D.K., Rao, G.P., Baranwal, V.K., Sharma, P., 2015. Molecular characterization of '*Candidatus* Phytoplasma asteris' subgroup I-B associated with sesame phyllody disease and identification of its natural vector and weed reservoir in India. Australas. Plant Pathol. 44 (3), 289−297.
Vasudeva, R.S., 1955. Phytopathological News from India, vol. 1. CommonwPhytopathol News, p. 420.
Vasudeva, R.S., Sahambi, H.S., 1955. Phyllody in sesamum (*Sesamum orientale* L.). Indian Phytopathol. 8, 124−129.
Wangkeeree, J., Miller, T.A., Hanboonsong, Y., 2012. Candidates for symbiotic control of sugarcane white leaf disease. Appl. Environ. Microbiol. 78 (19), 6804−6811.
Wei, W., Kakizawa, S., Suzuki, S., Jung, H.Y., Nishigawa, H., Miyata, S.I., Oshima, K., Ugaki, M., Hibi, T., Namba, S., 2004. In planta dynamic analysis of onion yellows phytoplasma using localized inoculation by insect transmission. Phytopathology 94 (3), 244−250.
Weintraub, P.G., Beanland, L., 2006. Insect vectors of phytoplasmas. Annu. Rev. Entomol. 51, 91−111.
Weintraub, P.G., Orenstein, S., 2004. Potential leafhopper vectors of phytoplasma in carrots. Int. J. Trop. Insect Sci. 24 (3), 228−235.
Weintraub, P.G., Pivonia, S., Rosner, A., Gera, A., 2004. A new disease in *Limonium latifolium* hybrids. II. Investigating insect vectors. Hortscience 39 (5), 1060−1061.
Zahniser, J.N., 2007. An Online Interactive Key and Searchable Database of Deltocephalinae (Hemiptera: Cicadellidae).
Zahniser, J.N., Dietrich, C., 2013. A review of the tribes of Deltocephalinae (Hemiptera: Auchenorrhyncha: Cicadellidae). European J. Taxon. (45).
Zchori-Fein, E., Bourtzis, K. (Eds.), 2011. Manipulative Tenants: Bacteria Associated with Arthropods. CRC press.
Zirak, L., Bahar, M., Ahoonmanesh, A., 2010. Characterization of phytoplasmas related to '*Candidatus* Phytoplasma asteris' and peanut WB group associated with sweet cherry diseases in Iran. J. Phytopathol. 158 (1), 63−65.

Further reading

Bindra, O.S., Singh, B., 1969. Biology and bionomics of *Hishimonus phycitis* (Distant) 1, A Jassid vector of little-leaf disease of brinjal (*Solanum melongena* L.). Indian J. Agric. Sci. 39 (9), 912.

Bove, J.M., Zreik, L., Danet, J.L., Bonfils, J., Mjeni, A.M.M., Garnier, M., 1993. Witches' broom disease of lime trees: monoclonal antibody and DNA probes for the detection of the associated MLO and the identification of a possible vector. In: International Organization of Citrus Virologists Conference Proceedings (1957-2010), vol. 12. No. 12.

Gera, A., Maslenin, L., Rosner, A., Zeidan, M., Pivonia, S., Weintraub, P.G., 2004. A new disease in *Limonium* hybrids. I. Molecular identification. Hortscience 39 (5), 1056–1059.

Gera, A., Maslenin, L., Rosner, A., Zeidan, M., Weintraub, P.G., 2006. Phytoplasma diseases in ornamental crops in Israel. Acta Hortic. 108–202.

Ishiie, T., 1970. Mycoplasma-like organism, causal agent of mulberry dwarf disease. JARQ (Jpn. Agric. Res. Q.) 5 (3), 48–53.

Queiroz, R.B., Donkersley, P., Silva, F.N., Al-Mahmmoli, I.H., Al-Sadi, A.M., Carvalho, C.M., Elliot, S.L., 2016. Invasive mutualisms between a plant pathogen and insect vectors in the Middle East and Brazil. Royal Soc. Open Sci. 3 (12), 160557.

Rao, G.P., Nabi, S.U., 2015. Overview on a century progress in research on sesame phyllody disease. Phytopath. Moll. 5 (2), 74–83.

Tanaka, M., Osada, S., Matsuda, I., 2000. Transmission of rhus (*Rhus javanica* L.) yellows by *Hishimonus sellatus* and host range of the causal phytoplasma. J. Gen. Plant Pathol. 66 (4), 323–326.

CHAPTER 5

Genomic studies on Asian phytoplasmas

Ching-Ting Huang[1], Shen-Chian Pei[1,2] and Chih-Horng Kuo[1]

[1]*Institute of Plant and Microbial Biology, Academia Sinica, Taipei, Taiwan;* [2]*Department of Plant Pathology and Microbiology, National Taiwan University, Taipei, Taiwan*

1. Background

Phytoplasmas are plant-pathogenic bacteria that belong to the class Mollicutes and can infect more than 300 plant species through insect transmission (Hogenhout et al., 2008; Lee et al., 2000; Namba, 2019). Infected plants often show dramatic changes in their development and morphology. Typical symptoms include stunting, dwarfism, virescence (greening of flowers), phyllody (abnormal development of floral parts into leaf-like tissues), and witches' broom (proliferation of stems and leaves). Due to these symptoms, plant diseases associated with phytoplasmas have resulted in considerable agricultural losses across the globe. In Asia, phytoplasma diseases affecting herbaceous and woody plants have been reported in at least 15 countries (Hoat et al., 2015; Kumari et al., 2019). Some notable examples include the jujube production in China that suffers up to 80% losses (Zhao et al., 2019) and the sugarcane production in India that suffers 5%–20% losses (Anuradha et al., 2019) due to phytoplasma infections.

To alleviate the agricultural impact by phytoplasmas, a better understanding of their biology is critical. However, despite decades of effort, the culture of phytoplasmas has remained elusive, which impedes phytoplasma research in several critical ways. Phytoplasmas are assigned to a '*Candidatus* (*Ca.*)' genus (IRPCM, 2004). Although 49 '*Ca.* Phytoplasma' species have been described (Bertaccini and Lee, 2018; Bertaccini et al., 2022) (Fig. 5.1), many more phytoplasmas are described based solely on the hosts and associated symptoms, which make it difficult to ascertain how the existing biological diversity is organized and whether the same biological entity is related to certain phytoplasma diseases or not. Finally, the unavailability of genetic tools to investigate gene functions represents a formidable challenge for in-depth characterization of their pathogenicity mechanisms.

Fortunately, the advancements in genomics provided culture-independent approaches to resolve the aforementioned issues. Over the past decade, several reviews were published to summarize the key findings (Hogenhout and Seruga Music, 2009; Kube et al., 2012, 2019; Oshima et al., 2013; Sugio and Hogenhout, 2012). In this chapter, we provide further updates and emphasize the phytoplasma genomics research in Asia.

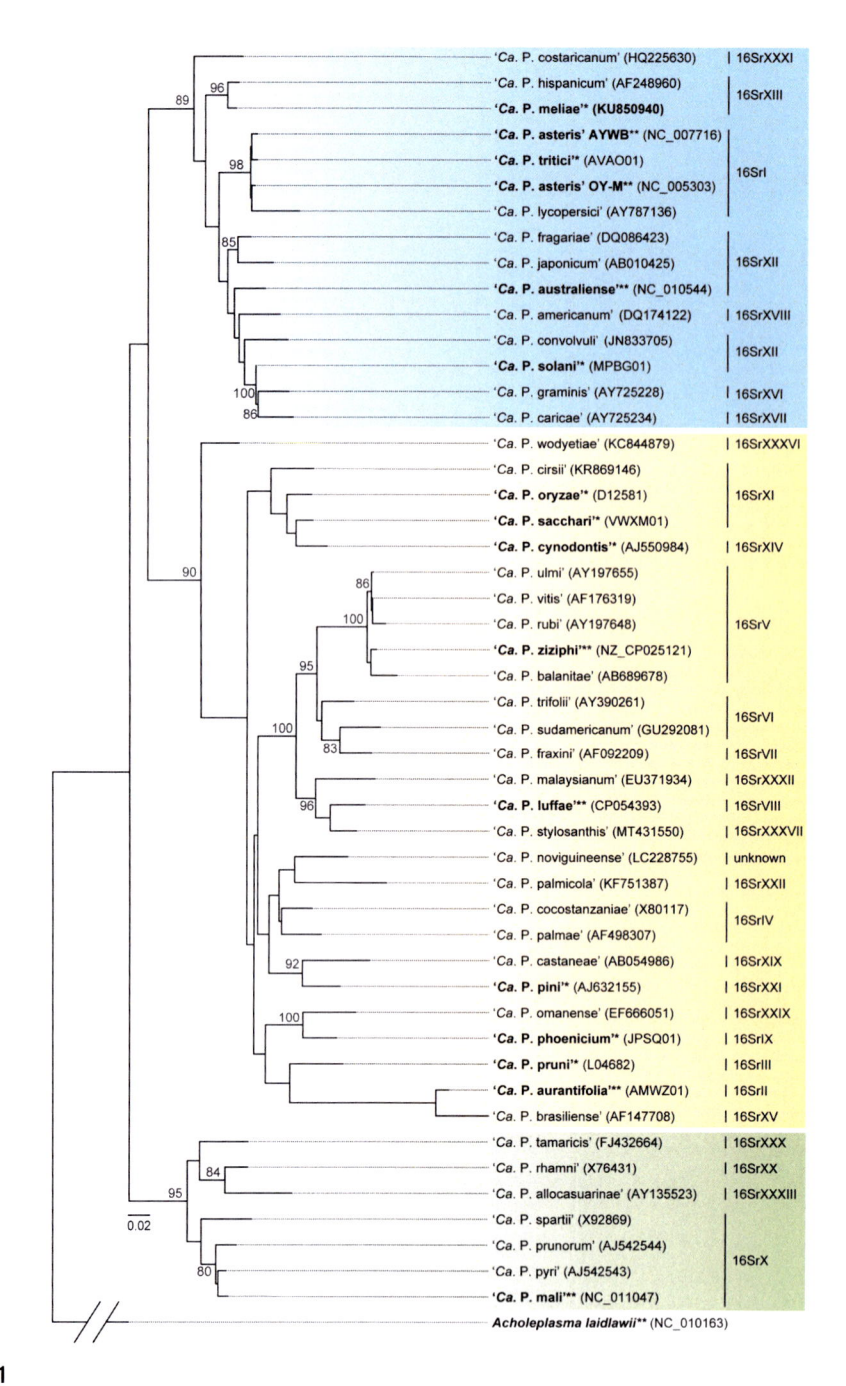

FIGURE 5.1

Maximum likelihood phylogeny of representative phytoplasmas based on the 16S rRNA gene sequences. Sequence accession numbers are provided in parentheses. Assignments of the 16S rRNA gene RFLP (16Sr) groups are labeled to the right according to available information (Bertaccini and Lee, 2018; IRPCM, 2004). Strains with genome sequences available are highlighted in bold (*, draft; **, complete). The three major clusters of phytoplasmas are indicated by colored backgrounds (I, blue; II, green; III, yellow). *Acholeplasma laidlawii* is included as the outgroup. The procedure for phylogenetic inference is based on that described previously (Cho et al., 2019b). Numbers next to internal branches indicate the level of bootstrap support based on 1,000 resampling; only values ≥80% are shown.

2. Historical perspective

The history of phytoplasma genomics research in Asia can be traced back to the 1990s. A Japanese group, headed by Shigetou Namba at the University of Tokyo, started the first genome sequencing projects in 1994 to work on two '*Ca*. Phytoplasma asteris' strains associated with onion yellows (Namba, 2019). As recently reviewed (Kube et al., 2019), several technical challenges have to be overcome in these projects. These include the enrichment of phytoplasma DNA from infected hosts, construction of DNA libraries for whole-genome shotgun sequencing based on the Sanger sequencing method, and data processing of sequences with low G + C content bias and abundant repeats. It took nearly a decade of effort for this team to publish the draft genome assembly of '*Ca*. P. asteris' OY-W in 2002 (Oshima et al., 2002) and later the complete genome sequence of '*Ca*. P. asteris' OY-M in 2004 (Oshima et al., 2004). Following these pioneering works, several other genome sequences were published for phytoplasmas collected elsewhere in the world based on similar methods. These include '*Ca*. P. asteris' AY-WB from lettuce plants in the United States of America that was published in 2006 (Bai et al., 2006) and two others published in 2008, which are '*Ca*. P. australiense' PAa from a cotton bush in Australia (Tran-Nguyen et al., 2008) and '*Ca*. P. mali' AT from apple trees in Germany (Kube et al., 2008). Also, a second strain of '*Ca*. P. australiense', NZSb11 associated with the strawberry lethal yellows in Australia, was published in 2013 (Andersen et al., 2013). These early works provided the first look into the genome organization and gene content of phytoplasmas. However, the laborious processes and high cost associated with these Sanger sequencing-based methods hampered further sequencing of more diverse phytoplasmas.

At the beginning in the late 2000s, the advent of next-generation sequencing (NGS) technologies, particularly the Illumina platform, as well as various bioinformatics tools, provided solutions to the aforementioned challenges. The substantial increase in sequencing throughput makes the enrichment for phytoplasma DNA by using pulsed-field gel electrophoresis or cesium chloride density-gradient centrifugation an optional step. Instead, the DNA samples prepared from infected plants or insects could be sequenced directly and processed as metagenomes. With the phytoplasma genome sequences from those pioneering studies (Andersen et al., 2013; Bai et al., 2006; Kube et al., 2008; Oshima et al., 2002, 2004; Tran-Nguyen et al., 2008), phytoplasma contigs could be extracted from the metagenomes based on sequence similarities. Also critically, the rapid and continuing drop in sequencing cost allowed for more research groups to work on diverse phytoplasmas collected from different hosts and geographical regions (Table 5.1).

In 2013, in Taiwan it was published a draft genome assembly of '*Ca*. P. aurantifolia' NTU2011 associated with peanut witches' broom (Chung et al., 2013), which was the first Asian phytoplasma genome project based on an NGS method. Soon after, NGS methods were adopted by other Asian phytoplasma genome projects, which greatly improved the sampling of more diverse strains. These include '*Ca*. P. asteris' witches' broom disease (WBD) from wheat in China (Chen et al., 2014) ('*Ca*. P. tritici' Zhao et al., 2021), '*Ca*. P. asteris' OY-V from onion in Japan (Kakizawa et al., 2014), '*Ca*. P. phoenicium' SA213 from almond in South Lebanon (Quaglino et al., 2015), '*Ca*. P. aurantifolia' NCHU2014 from purple coneflower in Taiwan (Chang et al., 2015), '*Ca*. P. aurantifolia' WBDL from lime in Omen (GenBank accession GCF_002009625.1; released in 2017), '*Ca*. P. asteris' LD1 from rice in China (Zhu et al., 2017), '*Ca*. P. asteris' DY2014 from periwinkle and '*Ca*. P. asteris' SS2016 from green onion in Taiwan (Cho et al., 2019b), and '*Ca*. P. asteris' HP from hydrangea in Japan (Nijo et al., 2021).

Table 5.1 Characteristics of the Asian phytoplasma strains with genome sequences available.

Strain	16Sr group	Geographic origin	Host	Accession	Assembly status	Genome size (kb)	Cds (intact)	Cds (pseudo)	Year	Sequencing platform	References
'Ca. P. asteris' OY-W	16SrI-B	Japan	Onion (Allium cepa)	Not available	20 contigs	752	602	?	2002	Sanger	Oshima et al. (2002)
'Ca. P. asteris' OY-M	16SrI-B	Japan	Onion (Allium cepa)	GCA_000009845.1	Complete	853	752	46	2004	Sanger	Oshima et al. (2004)
'Ca. P. aurantifolia' NTU2011	16SrII	Taiwan	Peanut (Arachis hypogaea)	GCA_000364425.1	14 contigs	567	425	29	2013	Illumina	Chung et al. (2013)
'Ca. P. tritici' WBD	16SrI-C	China	Wheat (Triticum aestivum)	GCF_000495255.1	6 contigs	611	471	76	2014	Illumina	Chen et al. (2014)
'Ca. P. asteris' OY-V	16SrI-B	Japan	Onion (Allium cepa)	GCF_000744065.1	170 contigs	740	705	160	2014	Illumina and 454	Kakizawa et al. (2014)
'Ca. P. phoenicium' SA213	16SrIX-B	Lebanon	Almond (Prunus dulcis)	GCF_001189415.1	78 contigs	346	311	29	2015	Illumina	Quaglino et al. (2015)
'Ca. P. aurantifolia' NCHU2014	16SrII	Taiwan	Purple coneflower (Echinacea purpurea)	GCA_001307505.2	Complete	640	475	35	2015 (draft); 2020 (complete)	Illumina and ONT	Chang et al. (2015); Tan et al. (2021)
'Ca. P. aurantifolia' WBDL	16SrII-B	Oman	Lime (Citrus aurantifolia)	GCF_002009625.1	98 contigs	475	386	44	2017	Illumina	Unpublished
'Ca. P. asteris' LD1	16SrI-B	China	Rice (Oryza sativa)	GCF_001866375.1	8 contigs	599	522	32	2017	Illumina	Zhu et al. (2017)
'Ca. P. ziziphi' Jwb-nky	16SrV-B	China	Jujube (Ziziphus jujuba)	GCF_003640545.1	Complete	751	652	26	2018	Illumina and PacBio	Wang et al. (2018)

'Ca. P. asteris' DY2014	16SrI-B	Taiwan	Periwinkle (Catharanthus roseus)	GCA_005093185.1	8 contigs	825	775	63	2019	Illumina	Cho et al. (2019b)
'Ca. P. cynodontis' LW01	16SrXIV-A	India	Bermuda grass (Cynodon dactylon)	GCF_009268075.1	23 contigs	484	433	9	2020	Illumina and ONT	Kirdat et al. (2020)
'Ca. P. sacchari' SCGS	16SrXI-A	India	Sugarcane (Saccharum officinarum)	GCF_009268105.1	29 contigs	505	411	10	2020	Illumina and ONT	Kirdat et al. (2020)
'Ca. P. aurantifolia' PR08	16SrII	India	Carrot grass (Parthenium hysterophorus)	GCA_015239935.1	Complete	593	361	255	2020	ONT	Unpublished
'Ca. P. aurantifolia' PR34	16SrII	India	Carrot grass (Parthenium hysterophorus)	GCF_015100165.1	134 contigs	740	580	88	2020	Illumina and ONT	Unpublished
'Ca. P. asteris' HP	16SrI-B	Japan	Hydrangeas (Hydrangea sp.)	GCF_018327665.1	36 contigs	598	510	23	2021	Illumina	Nijo et al. (2021)
'Ca. P. asteris' Zhengzhou	16SrI-B	China	Paulownia (Paulownia fortunei)	GCF_019396865.1	Complete	592	749	146	2021	PacBio	Cao et al. (2021)
'Ca. P luffae' NCHU2019	16SrVIII-A	Taiwan	Loofah (Luffa aegyptiaca)	GCA_018024475.1	Complete	769	725	13	2021	Illumina and ONT	Huang et al. (2022)
'Ca. P. asteris' SW86	16SrI-B	India	Indian sandalwood (Santalum album)	GCF_018283495.1	20 contigs	554	444	27	2021	Illumina and ONT	Unpublished
'Ca. P. aurantifolia' SS02	16SrII	India	Sesame (Sesamum indicum)	GCF_018390775.1	47 contigs	536	432	25	2021	Illumina and ONT	Unpublished
'Ca. P. asteris' TBZ1	16SrI	Iran	Russian olive tree (Elaeagnus angustifolia)	GCA_018598675.1	1404 contigs	833	477	21	2021	Illumina	Unpublished

However, although NGS methods improved the availability of phytoplasma genome sequences, the short sequencing reads often cannot resolve the highly repetitive phytoplasma genomes to produce complete assemblies. For example, despite having a 3648-fold sequencing depth, the genome assembly of '*Ca.* P. asteris' OY-V is still highly fragmented with 170 contigs (Kakizawa et al., 2014). Furthermore, even when PCR and Sanger sequencing were used to improve draft assemblies produced by NGS methods, such effort may be hampered by the extreme nucleotide composition bias and abundance of homopolymers in phytoplasma genomes (Cho et al., 2019b). Fortunately, the third-generation sequencing technologies, such as those provided by Pacific Biosciences (PacBio) and Oxford Nanopore Technologies (ONT), could complement NGS methods to produce high-quality genome assemblies. With single-molecule long reads, the third-generation sequencing technologies can resolve repetitive and complex regions of genomes, leading to accurate scaffolding of contigs (Koren et al., 2013; Koren and Phillippy, 2015; Miyamoto et al., 2014), while the high accuracy of NGS short reads could help to reduce the error rates of the finished assemblies (Kube et al., 2019).

In 2018, the genome of '*Ca.* P. ziziphi' Jwb-nky, which was from Chinese jujube in China, became the first complete phytoplasma genome assembly generated by combining Illumina and PacBio reads (Wang et al., 2018). Subsequently, several more complete assemblies were generated by utilizing the third-generation sequencing technologies, including '*Ca.* P. aurantifolia' NCHU2014 from purple coneflower in Taiwan (GenBank accession GCA_001307505.2; released in 2020) (Tan et al., 2021), '*Ca.* P. aurantifolia' PR08 from carrot grass in India (GenBank accession GCA_015239935.1; released in 2020), '*Ca.* P. asteris' Zhengzhou from paulownia in China (Cao et al., 2021), and '*Ca.* P. luffae' NCHU2019 from loofah in Taiwan (GenBank accession GCA_018024475.2; released in 2021) (Huang et al., 2022). However, the availability of long reads does not necessarily guarantee that a complete genome assembly could be produced. For example, five phytoplasma genome assemblies from India that utilized Illumina and ONT reads remained incomplete drafts, including '*Ca.* P. aurantifolia' PR34 from carrot grass (GenBank accession GCF_015100165.1; released in 2020), '*Ca.* P. cynodontis' LW01 from Bermuda grass (Kirdat et al., 2020), '*Ca.* P. sacchari' SCGS from sugarcane (Kirdat et al., 2021), '*Ca.* P. asteris' SW86 from Indian sandalwood (GenBank accession GCF_018283495.1; released in 2021), and '*Ca.* P. aurantifolia' SS02 from sesame (GenBank accession GCF_018390775.1; released in 2021). These results further demonstrated the challenges in phytoplasma genomic studies.

In summary, although phytoplasma genomics research had an early start in Asia, the initial progress was slow due to many technical challenges. Nevertheless, the pace accelerated considerably as sequencing technologies and bioinformatic tools improved. By early 2021, 16 Asian phytoplasma genome assemblies are available (Table 5.1). These resources facilitated the investigation of phytoplasma biology in multiple aspects. However, despite the significant progress that has been made, the sampling remained limited. Among the 49 '*Ca.* Phytoplasma' species that have been described or proposed (Bertaccini and Lee, 2018; Bertaccini et al., 2022; IRPCM, 2004), only 15 have been subjected to whole-genome sequencing (Fig. 5.1). Out of these 15, only seven were from Asia, representing a small fraction of known Asian phytoplasma diversity (Kumari et al., 2019; Rao et al., 2011). Moreover, the available sampling is highly biased toward phytoplasmas that infect herbaceous plants (Table 5.1). Thus, future effort in expanding the sequencing of phytoplasma genomes is critical.

3. Biological insights
3.1 Gene content and metabolism

In bacteria, genome size is strongly correlated with the number of coding sequences (CDSs) (Kuo et al., 2009; Lo et al., 2016; McCutcheon and Moran, 2012; Ochman and Davalos, 2006). Compared to free-living bacteria that often have genomes larger than 4,000 kb and containing >4000 CDSs, phytoplasma genomes are much smaller in the range of ~530–1350 kb (Marcone et al., 1999). As expected from the small genome sizes, the Asian phytoplasma genomes characterized to date all have <1,000 CDSs (Table 5.1). With the small genome sizes and reduced gene sets, phytoplasmas lack many metabolic pathways found in other bacteria. When the first draft genome of 'Ca. P. asteris' became available, it was found to lack the genes for amino acid biosynthesis, fatty acid biosynthesis, oxidative phosphorylation, and the tricarboxylic acid cycle (Oshima et al., 2002). Furthermore, compared to the closely related mycoplasmas, which were thought to have the minimal gene set required for a cellular organism (Fraser et al., 1995; Koonin, 2000; Maniloff, 1996), phytoplasmas also lack the pentose phosphate cycle genes. When the first complete phytoplasma genome sequence became available (Oshima et al., 2004), the absence of these genes was confirmed. More surprisingly, it was found to lack genes for all adenosine triphosphate–synthase subunits, which were considered essential for cellular life. These observations explained the difficulty of cultivating phytoplasmas outside their hosts.

The obligate parasitic lifestyle and highly reduced gene content for metabolic pathways suggested that phytoplasmas are dependent on their hosts for various metabolites. Consistent with this expectation, phytoplasma genomes usually contain relatively large transporter genes (Chung et al., 2013; Kube et al., 2012). For example, there are 44 and 46 transporter genes in the genomes of 'Ca. P. asteris' DY2014 and 'Ca. P. tritici', accounting for 6% and 9% of their CDSs, respectively (Chen et al., 2014; Cho et al., 2019b). These genes are involved in the import of various inorganic and organic compounds.

Some interesting patterns emerged when the gene content and metabolic pathways were examined based on a phylogenetic framework. Based on the molecular phylogenies inferred using either the 16S rRNA genes or conserved protein-coding genes (Cho et al., 2019b; Chung et al., 2013; Hogenhout et al., 2008; Hogenhout and Seruga Music, 2009; Seruga Music et al., 2019), phytoplasmas can be classified into three major phylogenetic clusters (Fig. 5.1). Early genomic characterizations that focused on 'Ca. P. asteris' and 'Ca. P. australiense' of the cluster I suggested that the glycolysis pathway may be the major conserved mechanism for phytoplasma energy production (Bai et al., 2006; Oshima et al., 2002, 2004; Tran-Nguyen et al., 2008). Moreover, duplication of the glycolysis pathway genes was found for some but not all strains of 'Ca. P. asteris' (Cho et al., 2019b; Oshima et al., 2007). Comparison between two onion yellows strains indicated that the duplication of these genes is associated with higher phytoplasma cell densities and more severe disease symptoms in the infected plants, suggesting that this gene copy number variation may influence carbohydrate consumption and contribute to pathogenesis (Oshima et al., 2007). However, as more diverse lineages were characterized, it was found that the energy-yielding part of the glycolysis pathway is absent in 'Ca. P. mali' of cluster II (Kube et al., 2008, 2012), and the entire glycolysis pathway is absent in 'Ca. P. aurantifolia' (Chung et al., 2013) and 'Ca. P. oryzae' (Fischer et al., 2016) of the cluster III. Intriguingly, those cluster III lineages harbor the genes for citrate fermentation (i.e., *citC*, *citE*, *citF*, and *citG*), absent in the clusters I and II lineages. Because citrate is one of the predominant organic acids in phloem sap, this difference in gene content suggests that even though all phytoplasmas are phloem-limited pathogens, different strategies have evolved to exploit similar ecological niches.

3.2 Effectors and putative secreted proteins

As plants possess several defense systems against pathogens (Cui et al., 2015; Serrano et al., 2014; Zipfel, 2014), pathogens require unique strategies to invade plant hosts. Various plant pathogens are known to secrete small proteins known as effectors to manipulate plant immunity and facilitate infection (Hogenhout et al., 2009; Toruño et al., 2016; Win et al., 2012). Although the mechanisms of various effectors have been reported for other plant pathogens (Macho and Zipfel, 2015; Toruño et al., 2016), the study of phytoplasma effectors remains a vital topic in the field of molecular plant—microbe interactions because phytoplasmas are associated with distinct symptoms (Hogenhout et al., 2008; Sugio et al., 2011b). Furthermore, the life cycle of phytoplasmas involves organisms of two different kingdoms, namely host plants (*Plantae*) and insect vectors (*Animalia*). Thus, studies of phytoplasma effectors provide tools to dissect plant developmental processes and improve the understanding of how symbiotic bacteria adapt to two different eukaryotic hosts (Bray Speth et al., 2007; Sugio et al., 2011b; Win et al., 2012).

For the study of phytoplasma effectors, availability of genome sequences provided complete catalogs for data mining of candidate genes. Conceptually, the procedure is based on bioinformatic prediction of genes that encode putative secreted proteins and then use these genes as candidate effectors. For validation and investigation of effector functions, these candidate genes may be cloned and expressed in plants. In 2008, this approach was first developed and applied to the '*Ca.* P. asteris' AY-WB genome (Bai et al., 2009). Four genes that encode secreted proteins have been validated as phytoplasma effectors and experimentally investigated for their functions through the multiple studies. These include SAP05 (Gamboa et al., 2019; Huang et al., 2021), SAP11/SWP1 (Bai et al., 2009; Chang et al., 2018; Sugio et al., 2011a; Tan et al., 2016; N. Wang et al., 2018), and SAP54/PHYL1 (MacLean et al., 2011, 2014; Maejima et al., 2014; Orlovskis and Hogenhout, 2016; Yang et al., 2015) that interfere with plant transcription factors and cause developmental changes in the transgenic hosts. Another effector, TENGU, functions by affecting the signaling pathways of auxin and jasmonic acid in plants (Hoshi et al., 2009; Minato et al., 2014; Sugawara et al., 2013). These studies provide novel insights into the molecular mechanisms that phytoplasmas utilize to manipulate their hosts. With further improvements in the bioinformatic methods for the prediction of secreted proteins (Garcion et al., 2021), additional novel phytoplasma effector genes are expected to be discovered through future phytoplasma genome analysis. However, experimental validation and characterization of these candidate genes remain challenging.

For a global view of phytoplasma genes that encode putative secreted proteins, an early cross-species comparison among representatives from all three phylogenetic clusters revealed no core genes shared by all phytoplasma lineages (Chung et al., 2013). Instead, most of these genes are species- or strain-specific (Chung et al., 2013). More recently, a comparison of multiple '*Ca.* P. asteris' strains revealed that at the within-species level, the number of genes that encode putative secreted proteins ranges from 10 to 45 (Cho et al., 2020). These findings demonstrated the high levels of genetic diversity concerning the distribution of putative effector genes among different lineages. Moreover, such diversity may be attributed to horizontal gene transfer (HGT) between lineages and lineage-specific gene expansion or reduction (Cho et al., 2020). The variation of effector repertoires was hypothesized to be associated with symptoms from different phytoplasmas (Sugio and Hogenhout, 2012). However, more investigation is necessary to understand the genotype-phenotype correspondence.

3.3 Mobile genetic elements

One unique feature of phytoplasma genomes is large mobile genetic elements (MGEs) that are up to ~25 kb in size (Chen et al., 2012; Cho et al., 2019b; Hogenhout and Seruga Music, 2009; Sugio and Hogenhout, 2012). These MGEs were described as either potential mobile units (PMUs) that originated from transposons (Bai et al., 2006) or sequence-variable mosaics (SVMs) that originated from phages (Jomantiene et al., 2007; Jomantiene and Davis, 2006; Wei et al., 2008). The exact boundaries for defining these MGEs differ between these two classification schemes. For example, the transposase gene *tra5* often found in these regions is considered part of PMU but not SVM. Nevertheless, the signature genes associated with these MGEs include those encode DNA helicase (*dnaB*), DNA primase (*dnaG*), protease (*hflB/pmp*), DNA-binding protein (*himA*), single-stranded DNA-binding protein (*ssb*), thymidylate kinase (*tmk*), and sigma factor (*sigF/rpoD/fliA*).

Regardless of the exact origin, these MGEs play essential roles in phytoplasma genome modification. First of all, the abundance of these MGEs is correlated with genome size at between- and within-species levels (Huang et al., 2022). For example, only 23 PMU-associated genes were found in the relatively small chromosome (i.e., 636 kb) of '*Ca.* P. aurantifolia' NCHU 2014 (Chang et al., 2015), while '*Ca.* P. solani' SA-1 has a relatively large chromosome (i.e., >821 kb) and contains at least 66 PMU-associated genes (Seruga Music et al., 2019). Within '*Ca.* P. asteris', strain M3 has a chromosome size of 576 kb and contains 27 PMU-associated genes (Orlovskis et al., 2017), while strain OY-M has a chromosome size of 853 kb and contains 137 PMU-associated genes (Oshima et al., 2004).

In addition to the contribution to genome size variation, PMUs are often found at the synteny breakpoints in the genome alignments between different phytoplasmas (Cho et al., 2019b; Orlovskis et al., 2017; Sugio and Hogenhout, 2012). These findings suggest that PMUs may facilitate genome rearrangements and contribute to genome instability (Bai et al., 2006; Seruga Music et al., 2019). Consistent with this hypothesis, the PMU1 in the '*Ca.* P. asteris' AY-WB genome was found to exist in a linear form as a part of the chromosomal (i.e., L-PMU1) and an extrachromosomal circular molecule (C-PMU1) (Toruño et al., 2010), indicating that this PMU is a functional MGE. Intriguingly, the C-PMU1 copy number and PMU1 gene expression are consistently higher in insects compared with plants, suggesting that C-PMU1 synthesis and expression are regulated (Toruño et al., 2010).

Another critical aspect of harboring active MGEs is that the movements of these elements between different phytoplasmas can promote HGT. Consistent with this hypothesis, PMU-mediated HGT was first reported in Asian phytoplasmas between distantly related '*Ca.* P. aurantifolia' and '*Ca.* P. asteris' (Chung et al., 2013; Ku et al., 2013). Later on, additional HGT events were inferred for other phytoplasmas distantly (Seruga Music et al., 2019; Wang et al., 2018) or closely related (Cho et al., 2019b). These PMU-mediated HGT events are thought to be important in the phytoplasma virulence because several genes that encode confirmed or putative effectors are associated with PMUs (Bai et al., 2009; Cho et al., 2019b; Chung et al., 2013; Orlovskis et al., 2017; Seruga Music et al., 2019; Wang et al., 2018). The horizontal transfer of PMUs can result in novel combinations of effector gene content, which may, in turn, facilitate adaptation to different hosts (Chung et al., 2013; Ku et al., 2013).

4. Linking genomics to taxonomy

Taxonomy of uncultured bacteria is a critical topic that affects the entire field of microbiology (Konstantinidis et al., 2017; Murray et al., 2020; Roselló-Móra and Whitman, 2019). For uncultured pathogens accurate classification and identification, as well as unambiguous nomenclature, are not only crucial for research efforts but also vital for effective disease management.

Since the 1990s, a system based on restriction fragment length polymorphism (RFLP) analysis of 16S rRNA genes (Gundersen and Lee, 1996; Lee et al., 1993, 1998) has been widely used for phytoplasma differentiation. The development of a web-tool *i*PhyClassifier (Zhao et al., 2009) further promoted the usage of this system, particularly in disease reports. To date, at least 36 16S rRNA gene RFLP (16Sr) groups and >150 subgroups have been reported (Naderali et al., 2017; Zhao and Davis, 2016). However, despite its popularity, there are several shortcomings for this system (Cho et al., 2020). For example, the RFLP method assesses only the restriction sites and has lower resolution power compared to whole sequence analysis. Additionally, intragenomic sequence variations between the two copies of 16S rRNA genes may cause complications. Most critically, the reliance of a single locus for classification and identification is inherently unreliable (Glaeser and Kämpfer, 2015).

'*Ca*. Phytoplasma' species are used for provisionally classification based on the sequence of the entire 16S ribosomal gene (Bertaccini and Lee, 2018; Bertaccini et al., 2022; IRPCM, 2004) and 49 '*Ca*. Phytoplasma', species have been described (Fig. 5.1).

For cultured bacteria, integrating genome data in taxonomy is now widely accepted (Chun et al., 2018). The genome-based taxonomy, particularly the use of genome-wide average nucleotide identity (ANI), was demonstrated as universally applicable to all bacteria, regardless of the status of cultivation (Jain et al., 2018; Kim et al., 2014). In several recent works, the 95% ANI cutoff suggested to delineate bacterial species was demonstrated as applicable to phytoplasmas (Cho et al., 2019a, 2020) and utilized to support the description of, '*Ca*. P. sacchari' (Kirdat et al., 2021) and '*Ca*. P. tritici' (Zhao et al., 2021).

While the integration of genomic analysis can improve the resolution and confidence of phytoplasma taxonomy, such transition involves multiple challenges that may not be resolvable in the near future. In the interim, multilocus sequence analysis (MLSA) (Glaeser and Kämpfer, 2015) can supplement the commonly used 16S rRNA gene analysis (Bertaccini et al., 2022; Zhao et al., 2009). Extensive efforts have been made in developing molecular markers to refine phytoplasma differentiation (Cimerman et al., 2009; Dumonceaux et al., 2014; Hodgetts et al., 2008; Lee et al., 2010; Marcone et al., 2000; Martini et al., 2007; Mitrović et al., 2011; Muirhead et al., 2019; Schneider and Gibb, 1997). However, systematic evaluation of these existing markers is still lacking. In this aspect, the genomic analysis provides a powerful tool to compare the relative performance of individual markers in terms of resolving power and cost-effectiveness (Cho et al., 2020; Kuo and Ochman, 2009), as well as to develop new PCR primers efficiently when needed (Cho et al., 2020; Kakizawa and Kamagata, 2014).

5. Conclusions and perspectives

Since the first phytoplasma genome was published, genomic studies have made substantial contributions to the understanding of these bacteria. Even though the phytoplasma cultivation in artificial media remains a challenge, genomics works revealed distinct metabolism, resolved phylogenetic relationships, provided insights into evolution, contributed to provisional taxonomy designation, and facilitated investigations of virulence mechanisms of these bacteria. For practical purposes, genomics studies also contributed to developing molecular markers that improve detection and identification, which are critical for disease management. However, many phytoplasma groups still lack genomic studies, including several ones that are associated with important plant diseases in Asia. Moreover, producing high-quality genome assemblies of phytoplasmas remains challenging. These two issues are some of the most critical aspects for future phytoplasma genomics studies.

References

Andersen, M.T., Liefting, L.W., Havukkala, I., Beever, R.E., 2013. Comparison of the complete genome sequence of two closely related isolates of '*Candidatus* Phytoplasma australiense' reveals genome plasticity. BMC Genom. 14, 529.

Anuradha, S., Kashyap, L., Kumar, R., Singh, P., 2019. Sugarcane grassy shoot (SCGS) disease - an overview. Int. J. Pure App. Biosci. 7, 371–378.

Bai, X., Correa, V.R., Toruño, T.Y., Ammar, E-D., Kamoun, S., Hogenhout, S.A., 2009. AY-WB phytoplasma secretes a protein that targets plant cell nuclei. Mol. Plant Microbe Interact. 22, 18–30.

Bai, X., Zhang, J., Ewing, A., Miller, S.A., Jancso Radek, A., Shevchenko, D.V., Tsukerman, K., Walunas, T., Lapidus, A., Campbell, J.W., Hogenhout, S.A., 2006. Living with genome instability: the adaptation of phytoplasmas to diverse environments of their insect and plant hosts. J. Bacteriol. 188, 3682–3696.

Bertaccini, A., Arocha-Rosete, Y., Contaldo, N., Duduk, B., Fiore, N., Montano, H.G., Kube, M., Kuo, C-H., Martini, M., Oshima, K., Quaglino, F., Schneider, B., Wei, W., Zamorano, A., 2022. Revision of the '*Candidatus* Phytoplasma' species description guidelines. Int. J. Syst. Evol. Microbiol. 72 (4), 005353.

Bertaccini, A., Lee, I-M., 2018. Phytoplasmas: an update. In: Rao, G.P., Bertaccini, A., Fiore, N., Liefting, L.W. (Eds.), Phytoplasmas: Plant Pathogenic Bacteria-I: Characterisation and Epidemiology of Phytoplasma - Associated Diseases. Springer Singapore, Singapore, pp. 1–29.

Bray Speth, E., Lee, Y.N., He, S.Y., 2007. Pathogen virulence factors as molecular probes of basic plant cellular functions. Curr. Opin. Plant Biol., Cell Biol. 10, 580–586.

Cao, Y., Sun, G., Zhai, X., Xu, P., Ma, L., Deng, M., Zhao, Z., Yang, H., Dong, Y., Shang, Z., Lv, Y., Yan, L., Liu, H., Cao, X., Li, B., Wang, Z., Zhao, X., Yu, H., Wang, F., Ma, W., Huang, J., Fan, G., 2021. Genomic insights into the fast growth of paulownias and the formation of *Paulownia* witches' broom. Mol. Plant.

Chang, S-H., Cho, S-T., Chen, C-L., Yang, J-Y., Kuo, C-H., 2015. Draft genome sequence of a 16SrII-A subgroup phytoplasma associated with purple coneflower (*Echinacea purpurea*) witches' broom disease in Taiwan. Genome Announc. 3, 013988-e1415.

Chang, S.H., Tan, C.M., Wu, C-T., Lin, T-H., Jiang, S-Y., Liu, R-C., Tsai, M-C., Su, L-W., Yang, J-Y., 2018. Alterations of plant architecture and phase transition by the phytoplasma virulence factor SAP11. J. Exp. Bot. 69, 5389–5401.

Chen, L-L., Chung, W-C., Lin, C-P., Kuo, C-H., 2012. Comparative analysis of gene content evolution in phytoplasmas and mycoplasmas. PLoS One 7, e34407.

Chen, W., Li, Y., Wang, Q., Wang, N., Wu, Y., 2014. Comparative genome analysis of wheat blue dwarf phytoplasma, an obligate pathogen that causes wheat blue dwarf disease in China. PLoS One 9, e96436.

Cho, S-T., Kung, H-J., Huang, W., Hogenhout, S.A., Kuo, C-H., 2020. Species boundaries and molecular markers for the classification of 16SrI phytoplasmas inferred by genome analysis. Front. Microbiol. 11, 1531.

Cho, S-T., Kung, H-J., Kuo, C-H., 2019a. Genomic similarities among 16SrI phytoplasmas and implications on species boundaries. Phytopathog. Mollic. 9, 73.

Cho, S-T., Lin, C-P., Kuo, C-H., 2019b. Genomic characterization of the periwinkle leaf yellowing (PLY) phytoplasmas in Taiwan. Front. Microbiol. 10, 2194.

Chun, J., Oren, A., Ventosa, A., Christensen, H., Arahal, D.R., da Costa, M.S., Rooney, A.P., Yi, H., Xu, X-W., De Meyer, S., Trujillo, M.E.Y., 2018. Proposed minimal standards for the use of genome data for the taxonomy of prokaryotes. Int. J. Syst. Evol. Microbiol. 68, 461–466.

Chung, W-C., Chen, L-L., Lo, W-S., Lin, C-P., Kuo, C-H., 2013. Comparative analysis of the peanut witches' broom phytoplasma genome reveals horizontal transfer of potential mobile units and effectors. PLoS One 8, e62770.

Cimerman, A., Pacifico, D., Salar, P., Marzachì, C., Foissac, X., 2009. Striking diversity of *vmp1*, a variable gene encoding a putative membrane protein of the stolbur phytoplasma. Appl. Environ. Microbiol. 75, 2951–2957.

Cui, H., Tsuda, K., Parker, J.E., 2015. Effector-triggered immunity: from pathogen perception to robust defense. Annu. Rev. Plant Biol. 66, 487–511.

Dumonceaux, T.J., Green, M., Hammond, C., Perez, E., Olivier, C., 2014. Molecular diagnostic tools for detection and differentiation of phytoplasmas based on chaperonin-60 reveal differences in host plant infection patterns. PLoS One 9, e116039.

Fischer, A., Santana-Cruz, I., Wambua, L., Olds, C., Midega, C., Dickinson, M., Kawicha, P., Khan, Z., Masiga, D., Jores, J., Schneider, B., 2016. Draft genome sequence of '*Candidatus* Phytoplasma oryzae' strain Mbita1, the causative agent of Napier grass stunt disease in Kenya. Genome Announc. 4, 002977-e316.

Fraser, C.M., Gocayne, J.D., White, O., Adams, M.D., Clayton, R.A., Fleischmann, R.D., Bult, C.J., Kerlavage, A.R., Sutton, G., Kelley, J.M., Fritchman, J.L., Weidman, J.F., Small, K.V., Sandusky, M., Fuhrmann, J., Nguyen, D., Utterback, T.R., Saudek, D.M., Phillips, C.A., Merrick, J.M., Tomb, J-F., Dougherty, B.A., Bott, K.F., Hu, P-C., Lucier, T.S., Peterson, S.N., Smith, H.O., Hutchison, C.A., Venter, J.C., 1995. The minimal gene complement of *Mycoplasma genitalium*. Science 270, 397–404.

Gamboa, C., Cui, W., Quiroga, N., Fernández, C., Fiore, N., Zamorano, A., 2019. Identification of 16SrIII-J phytoplasma effectors using a viral vector. Phytopath. Moll. 9, 229.

Garcion, C., Béven, L., Foissac, X., 2021. Comparison of current methods for signal peptide prediction in phytoplasmas. Front. Microbiol. 12, 661524.

Glaeser, S.P., Kämpfer, P., 2015. Multilocus sequence analysis (MLSA) in prokaryotic taxonomy. Syst Appl Microbiol, Taxonomy in the age of genomics 38, 237–245.

Gundersen, D.E., Lee, I-M., 1996. Ultrasensitive detection of phytoplasmas by nested-PCR assays using two universal primer pairs. Phytopathol. Mediterr. 35, 144–151.

Hoat, T.X., Quan, M.V., Anh, D.T.L., Cuong, N.N., Vuong, P.T., Alvarez, E., Nguyen, T.T.D., Wyckhuys, K., Paltrinieri, S., Pardo, J.M., Mejia, J.F., Thanh, N.D., Dickinson, M., Duong, C.A., Kumasaringhe, N.C., Bertaccini, A., 2015. Phytoplasma diseases on major crops in Vietnam. Phytopath. Moll. 5, S69.

Hodgetts, J., Boonham, N., Mumford, R., Harrison, N., Dickinson, M., 2008. Phytoplasma phylogenetics based on analysis of *secA* and 23S rRNA gene sequences for improved resolution of candidate species of '*Candidatus* Phytoplasma'. Int. J. Syst. Evol. Microbiol. 58, 1826–1837.

Hogenhout, S.A., Oshima, K., Ammar, E-D., Kakizawa, S., Kingdom, H.N., Namba, S., 2008. Phytoplasmas: bacteria that manipulate plants and insects. Mol. Plant Pathol. 9, 403–423.

Hogenhout, S.A., Seruga Music, M., 2009. Phytoplasma genomics, from sequencing to comparative and functional genomics - what have we learnt? In: Phytoplasmas: Genomes, Plant Hosts and Vectors. CABI, pp. 19–36.

Hogenhout, S.A., Van der Hoorn, R.A.L., Terauchi, R., Kamoun, S., 2009. Emerging concepts in effector biology of plant-associated organisms. Mol. Plant Microbe. Interact. 22, 115–122.

Hoshi, A., Oshima, K., Kakizawa, S., Ishii, Y., Ozeki, J., Hashimoto, M., Komatsu, K., Kagiwada, S., Yamaji, Y., Namba, S., 2009. A unique virulence factor for proliferation and dwarfism in plants identified from a phytopathogenic bacterium. Proc. Natl. Acad. Sci. U.S.A. 106, 6416–6421.

Huang, C-T., Cho, S-T., Lin, Y-C., Tan, C-M., Chiu, Y-C., Yang, J-Y., Kuo, C-H., 2022. Comparative genome analysis of 'Candidatus Phytoplasma luffae' reveals the influential roles of potential mobile units in phytoplasma evolution. Front Microbiol. 13, 773608.

Huang, W., MacLean, A.M., Sugio, A., Maqbool, A., Busscher, M., Cho, S-T., Kamoun, S., Kuo, C-H., Immink, R.G.H., Hogenhout, S.A., 2021. Parasite co-opts a ubiquitin receptor to induce a plethora of developmental changes. bioRxiv, 02.15.430920.

IRPCM, 2004. 'Candidatus Phytoplasma', a taxon for the wall-less, non-helical prokaryotes that colonize plant phloem and insects. Int. J. Syst. Evol. Microbiol. 54, 1243–1255.

Jain, C., Rodriguez-R, L.M., Phillippy, A.M., Konstantinidis, K.T., Aluru, S., 2018. High throughput ANI analysis of 90K prokaryotic genomes reveals clear species boundaries. Nat. Commun. 9, 5114.

Jomantiene, R., Davis, R.E., 2006. Clusters of diverse genes existing as multiple, sequence-variable mosaics in a phytoplasma genome. FEMS Microbiol. Lett. 255, 59–65.

Jomantiene, R., Zhao, Y., Davis, R.E., 2007. Sequence-variable mosaics: composites of recurrent transposition characterizing the genomes of phylogenetically diverse phytoplasmas. DNA Cell Biol. 26, 557–564.

Kakizawa, S., Kamagata, Y., 2014. A multiplex-PCR method for strain identification and detailed phylogenetic analysis of AY-group phytoplasmas. Plant Dis. 98, 299–305.

Kakizawa, S., Makino, A., Ishii, Y., Tamaki, H., Kamagata, Y., 2014. Draft genome sequence of 'Candidatus Phytoplasma asteris' strain OY-V, an unculturable plant-pathogenic bacterium. Genome Announc. 2, 009444-e1014.

Kim, M., Oh, H-S., Park, S-C., Chun, J., 2014. Towards a taxonomic coherence between average nucleotide identity and 16S rRNA gene sequence similarity for species demarcation of prokaryotes. Int. J. Syst. Evol. Microbiol. 64, 346–351.

Kirdat, K., Tiwarekar, B., Thorat, V., Narawade, N., Dhotre, D., Sathe, S., Shouche, Y., Yadav, A., 2020. Draft genome sequences of two phytoplasma strains associated with sugarcane grassy shoot (SCGS) and Bermuda grass white leaf (BGWL) diseases. Mol. Plant Microbe Interact. 33, 715–717.

Kirdat, K., Tiwarekar, B., Thorat, V., Sathe, S., Shouche, Y., Yadav, A., 2021. 'Candidatus Phytoplasma sacchari', a novel taxon - associated with Sugarcane Grassy Shoot (SCGS) disease. Int. J. Syst. Evol. Microbiol. 71, 004591.

Konstantinidis, K.T., Rosselló-Móra, R., Amann, R., 2017. Uncultivated microbes in need of their own taxonomy. ISME J. 11, 2399–2406.

Koonin, E.V., 2000. How many genes can make a cell: the minimal-gene-set concept. Annu. Rev. Genom. Hum. Genet. 1, 99–116.

Koren, S., Harhay, G.P., Smith, T.P., Bono, J.L., Harhay, D.M., Mcvey, S.D., Radune, D., Bergman, N.H., Phillippy, A.M., 2013. Reducing assembly complexity of microbial genomes with single-molecule sequencing. Genome Biol. 14, R101.

Koren, S., Phillippy, A.M., 2015. One chromosome, one contig: complete microbial genomes from long-read sequencing and assembly. Curr. Opin. Microbiol., Host–Microbe Interact.: Bact. Genom. 23, 110–120.

Ku, C., Lo, W-S., Kuo, C-H., 2013. Horizontal transfer of potential mobile units in phytoplasmas. Mobile Genet. Elem. 3, e26145.

Kube, M., Duduk, B., Oshima, K., 2019. Genome sequencing. In: Bertaccini, A., Oshima, K., Kube, M., Rao, G.P. (Eds.), Phytoplasmas: Plant Pathogenic Bacteria - III: Genomics, Host Pathogen Interactions and Diagnosis. Springer, Singapore, pp. 1–16.

Kube, M., Mitrovic, J., Duduk, B., Rabus, R., Seemüller, E., 2012. Current view on phytoplasma genomes and encoded metabolism. Sci. World J. 1–25.

Kube, M., Schneider, B., Kuhl, H., Dandekar, T., Heitmann, K., Migdoll, A., Reinhardt, R., Seemuller, E., 2008. The linear chromosome of the plant-pathogenic mycoplasma 'Candidatus Phytoplasma mali'. BMC Genom. 9, 306.

Kumari, S., Nagendran, K., Rai, A.B., Singh, B., Rao, G.P., Bertaccini, A., 2019. Global status of phytoplasma diseases in vegetable crops. Front. Microbiol. 10, 1349.

Kuo, C-H., Moran, N.A., Ochman, H., 2009. The consequences of genetic drift for bacterial genome complexity. Genome Res. 19, 1450–1454.

Kuo, C-H., Ochman, H., 2009. Inferring clocks when lacking rocks: the variable rates of molecular evolution in bacteria. Biol. Direct. 4, 35.

Lee, I-M., Bottner-Parker, K.D., Zhao, Y., Davis, R.E., Harrison, N.A., 2010. Phylogenetic analysis and delineation of phytoplasmas based on *secY* gene sequences. Int. J. Syst. Evol. Microbiol. 60, 2887–2897.

Lee, I-M., Davis, R.E., Gundersen-Rindal, D.E., 2000. Phytoplasma: phytopathogenic mollicutes. Annu. Rev. Microbiol. 54, 221–255.

Lee, I-M., Gundersen-Rindal, D.E., Davis, R.E., Bartoszyk, I.M., 1998. Revised classification scheme of phytoplasmas based on RFLP analyses of 16S rRNA and ribosomal protein gene sequences. Int. J. Syst. Evol. Microbiol. 48, 1153–1169.

Lee, I-M., Hammond, R., Davis, R., Gundersen, D., 1993. Universal amplification and analysis of pathogen 16S rDNA for classification and identification of mycoplasma organisms. Phytopathology 83, 834–842.

Lo, W-S., Huang, Y-Y., Kuo, C-H., 2016. Winding paths to simplicity: genome evolution in facultative insect symbionts. FEMS Microbiol. Rev. 40, 855–874.

Macho, A.P., Zipfel, C., 2015. Targeting of plant pattern recognition receptor-triggered immunity by bacterial type-III secretion system effectors. Curr. Opin. Microbiol., Host–Microbe Interact.: Bacteria Genom. 23, 14–22.

MacLean, A.M., Orlovskis, Z., Kowitwanich, K., Zdziarska, A.M., Angenent, G.C., Immink, R.G.H., Hogenhout, S.A., 2014. Phytoplasma effector SAP54 hijacks plant reproduction by degrading MADS-box proteins and promotes insect colonization in a RAD23-dependent manner. PLoS Biol. 12, e1001835.

MacLean, A.M., Sugio, A., Makarova, O.V., Findlay, K.C., Grieve, V.M., Tóth, R., Nicolaisen, M., Hogenhout, S.A., 2011. Phytoplasma effector SAP54 induces indeterminate leaf-like flower development in *Arabidopsis* plants. Plant Physiol. 157, 831–841.

Maejima, K., Iwai, R., Himeno, M., Komatsu, K., Kitazawa, Y., Fujita, N., Ishikawa, K., Fukuoka, M., Minato, N., Yamaji, Y., Oshima, K., Namba, S., 2014. Recognition of floral homeotic MADS domain transcription factors by a phytoplasmal effector, phyllogen, induces phyllody. Plant J. 78, 541–554.

Maniloff, J., 1996. The minimal cell genome: "on being the right size". Proc. Natl. Acad. Sci. U.S.A. 93, 10004–10006.

Marcone, C., Lee, I.M., Davis, R.E., Ragozzino, A., Seemüller, E., 2000. Classification of aster yellows-group phytoplasmas based on combined analyses of rRNA and *tuf* gene sequences. Int. J. Syst. Evol. Microbiol. 50, 1703–1713.

Marcone, C., Neimark, H., Ragozzino, A., Lauer, U., Seemüller, E., 1999. Chromosome sizes of phytoplasmas composing major phylogenetic groups and subgroups. Phytopathology 89, 805–810.

Martini, M., Lee, I-M., Bottner, K.D., Zhao, Y., Botti, S., Bertaccini, A., Harrison, N.A., Carraro, L., Marcone, C., Khan, A.J., Osler, R., 2007. Ribosomal protein gene-based phylogeny for finer differentiation and classification of phytoplasmas. Int. J. Syst. Evol. Microbiol. 57, 2037–2051.

McCutcheon, J.P., Moran, N.A., 2012. Extreme genome reduction in symbiotic bacteria. Nat. Rev. Microbiol. 10, 13–26.

Minato, N., Himeno, M., Hoshi, A., Maejima, K., Komatsu, K., Takebayashi, Y., Kasahara, H., Yusa, A., Yamaji, Y., Oshima, K., Kamiya, Y., Namba, S., 2014. The phytoplasmal virulence factor TENGU causes plant sterility by downregulating of the jasmonic acid and auxin pathways. Sci. Rep. 4, 7399.

Mitrović, J., Kakizawa, S., Duduk, B., Oshima, K., Namba, S., Bertaccini, A., 2011. The *groEL* gene as an additional marker for finer differentiation of '*Candidatus* Phytoplasma asteris'-related strains. Ann. Appl. Biol. 159, 41–48.

Miyamoto, M., Motooka, D., Gotoh, K., Imai, T., Yoshitake, K., Goto, N., Iida, T., Yasunaga, T., Horii, T., Arakawa, K., Kasahara, M., Nakamura, S., 2014. Performance comparison of second- and third-generation sequencers using a bacterial genome with two chromosomes. BMC Genom. 15, 699.

Muirhead, K., Pérez-López, E., Bahder, B.W., Hill, J.E., Dumonceaux, T., 2019. The CpnClassiPhyR Is a resource for cpn60 universal target-based classification of phytoplasmas. Plant Disease. 103, 2494–2497.

Murray, A.E., Freudenstein, J., Gribaldo, S., Hatzenpichler, R., Hugenholtz, P., Kämpfer, P., Konstantinidis, K.T., Lane, C.E., Papke, R.T., Parks, D.H., Rossello-Mora, R., Stott, M.B., Sutcliffe, I.C., Thrash, J.C., Venter, S.N., Whitman, W.B., Acinas, S.G., Amann, R.I., Anantharaman, K., Armengaud, J., Baker, B.J., Barco, R.A., Bode, H.B., Boyd, E.S., Brady, C.L., Carini, P., Chain, P.S.G., Colman, D.R., DeAngelis, K.M., de los Rios, M.A., Estrada-de los Santos, P., Dunlap, C.A., Eisen, J.A., Emerson, D., Ettema, T.J.G., Eveillard, D., Girguis, P.R., Hentschel, U., Hollibaugh, J.T., Hug, L.A., Inskeep, W.P., Ivanova, E.P., Klenk, H-P., Li, W-J., Lloyd, K.G., Löffler, F.E., Makhalanyane, T.P., Moser, D.P., Nunoura, T., Palmer, M., Parro, V., Pedrós-Alió, C., Probst, A.J., Smits, T.H.M., Steen, A.D., Steenkamp, E.T., Spang, A., Stewart, F.J., Tiedje, J.M., Vandamme, P., Wagner, M., Wang, F-P., Yarza, P., Hedlund, B.P., Reysenbach, A-L., 2020. Roadmap for naming uncultivated archaea and bacteria. Nat. Microbiol. 5, 987–994.

Naderali, N., Nejat, N., Vadamalai, G., Davis, R.E., Wei, W., Harrison, N.A., Kong, L., Kadir, J., Tan, Y-H., Zhao, Y., 2017. '*Candidatus* Phytoplasma wodyetiae', a new taxon associated with yellow decline disease of foxtail palm (*Wodyetia bifurcata*) in Malaysia. Int. J. Syst. Evol. Microbiol. 67, 3765–3772.

Namba, S., 2019. Molecular and biological properties of phytoplasmas. Proc. Jpn. Acad. Ser. B 95, 401–418.

Nijo, T., Iwabuchi, N., Tokuda, R., Suzuki, T., Matsumoto, O., Miyazaki, A., Maejima, K., Oshima, K., Namba, S., Yamaji, Y., 2021. Enrichment of phytoplasma genome DNA through a methyl-CpG binding domain-mediated method for efficient genome sequencing. J. Gen. Plant Pathol. 87, 154–163.

Ochman, H., Davalos, L.M., 2006. The nature and dynamics of bacterial genomes. Science 311, 1730–1733.

Orlovskis, Z., Canale, M.C., Haryono, M., Lopes, J.R.S., Kuo, C-H., Hogenhout, S.A., 2017. A few sequence polymorphisms among isolates of maize bushy stunt phytoplasma associate with organ prolifeation symptoms of infected maize plants. Ann. Bot. 119, 869–884.

Orlovskis, Z., Hogenhout, S.A., 2016. A bacterial parasite effector mediates insect vector attraction in host plants independently of developmental changes. Front. Plant Sci. 7, 885.

Oshima, K., Kakizawa, S., Arashida, R., Ishii, Y., Hoshi, A., Hayashi, Y., Kagiwada, S., Namba, S., 2007. Presence of two glycolytic gene clusters in a severe pathogenic line of '*Candidatus* Phytoplasma asteris'. Mol. Plant Pathol. 8, 481–489.

Oshima, K., Kakizawa, S., Nishigawa, H., Jung, H-Y., Wei, W., Suzuki, S., Arashida, R., Nakata, D., Miyata, S., Ugaki, M., Namba, S., 2004. Reductive evolution suggested from the complete genome sequence of a plant-pathogenic phytoplasma. Nat. Genet. 36, 27–29.

Oshima, K., Maejima, K., Namba, S., 2013. Genomic and evolutionary aspects of phytoplasmas. Front. Microbiol. 4, 230.

Oshima, K., Miyata, S., Sawayanagi, T., Kakizawa, S., Nishigawa, H., Jung, H-Y., Furuki, K., Yanazaki, M., Suzuki, S., Wei, W., Kuboyama, T., Ugaki, M., Namba, S., 2002. Minimal set of metabolic pathways suggested from the genome of onion yellows phytoplasma. J. Gen. Plant Pathol. 68, 225–236.

Quaglino, F., Kube, M., Jawhari, M., Abou-Jawdah, Y., Siewert, C., Choueiri, E., Sobh, H., Casati, P., Tedeschi, R., Lova, M.M., Alma, A., Bianco, P.A., 2015. 'Candidatus Phytoplasma phoenicium' associated with almond witches' broom disease: from draft genome to genetic diversity among strain populations. BMC Microbiol. 15, 148.

Rao, G., Mall, S., Raj, S., Snehi, S., 2011. Review article: phytoplasma diseases affecting various plant species in India. Acta Phytopathol. Entomol. Hung. 46, 59—99.

Rosselló-Móra, R., Whitman, W.B., 2019. Dialogue on the nomenclature and classification of prokaryotes. Syst. Appl. Microbiol., Tax. Uncultivat. Bacteria Archaea 42, 5—14.

Schneider, B., Gibb, K.S., 1997. Sequence and RFLP analysis of the elongation factor Tu gene used in differentiation and classification of phytoplasmas. Microbiology 143, 3381—3389.

Serrano, M., Coluccia, F., Torres, M., L'Haridon, F., Métraux, J-P., 2014. The cuticle and plant defense to pathogens. Front. Plant Sci. 5, 274.

Seruga Music, M., Samarzija, I., Hogenhout, S.A., Haryono, M., Cho, S-T., Kuo, C-H., 2019. The genome of 'Candidatus Phytoplasma solani' strain SA-1 is highly dynamic and prone to adopting foreign sequences. Syst. Appl. Microbiol. 42, 117—127.

Sugawara, K., Honma, Y., Komatsu, K., Himeno, M., Oshima, K., Namba, S., 2013. The alteration of plant morphology by small peptides released from the proteolytic processing of the bacterial peptide TENGU. Plant Physiol. 162, 2005—2014.

Sugio, A., Hogenhout, S.A., 2012. The genome biology of phytoplasma: modulators of plants and insects. Curr. Opin. Microbiol. 15, 247—254.

Sugio, A., Kingdom, H.N., MacLean, A.M., Grieve, V.M., Hogenhout, S.A., 2011a. Phytoplasma protein effector SAP11 enhances insect vector reproduction by manipulating plant development and defense hormone biosynthesis. Proc. Natl. Acad. Sci. U. S. A. 108, E1254—E1263.

Sugio, A., MacLean, A.M., Kingdom, H.N., Grieve, V.M., Manimekalai, R., Hogenhout, S.A., 2011b. Diverse targets of phytoplasma effectors: from plant development to defense against insects. Annu. Rev. Phytopathol. 49, 175—195.

Tan, C.M., Li, C-H., Tsao, N-W., Su, L-W., Lu, Y-T., Chang, S.H., Lin, Y.Y., Liou, J-C., Hsieh, L-C., Yu, J-Z., Sheue, C-R., Wang, S-Y., Lee, C-F., Yang, J-Y., 2016. Phytoplasma SAP11 alters 3-isobutyl-2-methoxypyrazine biosynthesis in *Nicotiana benthamiana* by suppressing *NbOMT1*. J. Exp. Bot. 67, 4415—4425.

Tan, C.M., Lin, Y-C., Li, J-R., Chien, Y-Y., Wang, C-J., Chou, L., Wang, C-W., Chiu, Y-C., Kuo, C-H., Yang, J-Y., 2021. Accelerating complete phytoplasma genome assembly by immunoprecipitation-based enrichment and MinION-based DNA sequencing for comparative analyses. Front Microbiol 12, 766221.

Toruño, T.Y., Musić, M.S., Simi, S., Nicolaisen, M., Hogenhout, S.A., 2010. Phytoplasma PMU1 exists as linear chromosomal and circular extrachromosomal elements and has enhanced expression in insect vectors compared with plant hosts. Mol. Microbiol. 77, 1406—1415.

Toruño, T.Y., Stergiopoulos, I., Coaker, G., 2016. Plant-pathogen effectors: cellular probes interfering with plant defenses in spatial and temporal manners. Annu. Rev. Phytopathol. 54, 419—441.

Tran-Nguyen, L.T.T., Kube, M., Schneider, B., Reinhardt, R., Gibb, K.S., 2008. Comparative genome analysis of 'Candidatus Phytoplasma australiense' (subgroup *tuf*-Australia I; *rp*-A) and 'Ca. Phytoplasma asteris' strains OY-M and AY-WB. J. Bacteriol. 190, 3979—3991.

Wang, J., Song, L., Jiao, Q., Yang, S., Gao, R., Lu, X., Zhou, G., 2018a. Comparative genome analysis of jujube witches' broom phytoplasma, an obligate pathogen that causes jujube witches' broom disease. BMC Genom. 19, 689.

Wang, N., Yang, H., Yin, Z., Liu, W., Sun, L., Wu, Y., 2018b. Phytoplasma effector SWP1 induces witches' broom symptom by destabilizing the TCP transcription factor BRANCHED1. Mol. Plant Pathol. 19, 2623—2634.

Wei, W., Davis, R.E., Jomantiene, R., Zhao, Y., 2008. Ancient, recurrent phage attacks and recombination shaped dynamic sequence-variable mosaics at the root of phytoplasma genome evolution. Proc. Natl. Acad. Sci. U. S. A. 105, 11827–11832.

Win, J., Chaparro-Garcia, A., Belhaj, K., Saunders, D.G.O., Yoshida, K., Dong, S., Schornack, S., Zipfel, C., Robatzek, S., Hogenhout, S.A., Kamoun, S., 2012. Effector biology of plant-associated organisms: concepts and perspectives. Cold Spring Harbor Symp. Quant. Biol. 77, 235–247.

Yang, C-Y., Huang, Y-H., Lin, C-P., Lin, Y-Y., Hsu, H-C., Wang, C-N., Liu, L-Y.D., Shen, B-N., Lin, S-S., 2015. MicroRNA396-targeted SHORT VEGETATIVE PHASE is required to repress flowering and is related to the development of abnormal flower symptoms by the phyllody symptoms1 effector. Plant Physiol. 168, 1702–1716.

Zhao, J., Liu, Z., Liu, M., 2019. The resistance of jujube trees to jujube witches' broom disease in China. In: Olivier, C.Y., Dumonceaux, T.J., Pérez-López, E. (Eds.), Sustainable Management of Phytoplasma Diseases in Crops Grown in the Tropical Belt: Biology and Detection, Sustainability in Plant and Crop Protection. Springer International Publishing, Cham, pp. 219–232.

Zhao, Y., Davis, R.E., 2016. Criteria for phytoplasma 16Sr group/subgroup delineation and the need of a platform for proper registration of new groups and subgroups. Int. J. Syst. Evol. Microbiol. 66, 2121–2123.

Zhao, Y., Wei, W., Davis, R.E., Lee, I-M., Bottner-Parker, K.D., 2021. The agent associated with blue dwarf disease in wheat represents a new phytoplasma taxon, 'Candidatus Phytoplasma tritici'. Int. J. Syst. Evol. Microbiol. 71, 1.

Zhao, Y., Wei, W., Lee, I-M., Shao, J., Suo, X., Davis, R.E., 2009. Construction of an interactive online phytoplasma classification tool, iPhyClassifier, and its application in analysis of the peach X-disease phytoplasma group (16SrIII). Int. J. Syst. Evol. Microbiol. 59, 2582–2593.

Zhu, Y., He, Y., Zheng, Z., Chen, J., Wang, Z., Zhou, G., 2017. Draft genome sequence of rice orange leaf phytoplasma from Guangdong, China. Genome Announc. 5, 004300-e517.

Zipfel, C., 2014. Plant pattern-recognition receptors. Trends Immunol. 35, 345–351.

CHAPTER 6

Cross-boundary movement of phytoplasmas in Asia and status of plant quarantine

V. Celia Chalam, Pooja Kumari, D.D. Deepika, Priya Yadav, K. Kalaiponmani and A.K. Maurya

Division of Plant Quarantine, ICAR-National Bureau of Plant Genetic Resources, New Delhi, Delhi, India

1. Introduction

Plant diseases substantially reduce crop production every year, resulting in massive economic losses throughout the world. Trade and exchange of germplasm at international level play a key role in the long-distance dissemination of a destructive pathogen or its virulent pathotype/race/strain along with agri-horticultural produce. Biosecurity encompasses all policy and regulatory frameworks (including instruments and activities) to manage risks associated with food and agriculture (including relevant environmental risks), including fisheries and forestry. It is composed of three sectors, namely food safety, plant life and health, and animal life and health. Due to liberalization under World Trade Organization (WTO), the recent years have seen a significant growth in trade and exchange of agri-horticultural crops. The global movement of seed material has the potential of introducing new pathogens which may pose potential risk to the agriculture of the importing country. The present-day definition of a pest is any species, strain, or biotype of plant, animal, or pathogenic agent injurious to plants or plant products. A quarantine pest is the pest of potential economic importance to the area endangered thereby and not yet present there, or present but not widely distributed, and being officially controlled.

The devastating effects resulting from pathogens introduced along with international movement of seed and other planting material are well documented. The Irish famine of 1845, which forced the people to migrate en masse from Europe, was the result of almost total failure of potato crop due to infections of late blight pathogen *(Phytophthora infestans)* introduced from Central America. Coffee rust (*Hemileia vastatrix*) appeared in Sri Lanka in 1875 and reduced the coffee production by >90% in 1889. The disease entered India in 1876 from Sri Lanka, and within a decade, the coffee industry of South India was badly affected. Bulk import of seeds and other planting material without proper phytosanitary measures, indiscriminate exchange of germplasm, and the distribution of seed and other planting material by international agencies have increased the possibility of dissemination of pathogens in areas previously considered pathogen-free (Khetarpal et al., 2006; Dubey et al., 2021). Further, the threat may become severe, if more virulent strains or races of the pathogen are introduced into previously disease-free areas.

Like in other countries, a number of exotic plant pathogens got introduced into India along with imported planting material causing serious crop losses from time to time. These included potato late blight (*P. infestans* in 1883), coffee rust (*H. vastatrix* in 1879), *banana bunchy top virus (BBTV) in* 1940, and flag smut of wheat (*Urocystis tritici*) in 1906. These introductions highlighted the fact that increased pace of international travel and trade had exposed countries to the danger of infiltration of exotic pathogens harmful to the agriculture.

The most fundamental approach to the management of a disease is to ensure that it is not present through exclusion (quarantine) or eradication. National Plant Protection Organizations (NPPOs) assume responsibility for protecting their countries from the unwanted entry of new pests and for coordinating programmers to eradicate those that have recently arrived and are still sufficiently confined for their elimination to be realistic. The strategies for plant health management include: certified disease-free seed and other planting material, chemical control, biological control, cultural control, and use of resistant varieties. In the context of quality control, bulk samples of seed lots need to be tested by drawing workable samples as per norms. The detection of pathogens is then carried out by the approved or available techniques. Over the years, a great variety of methods have been developed that permit the detection and identification of pathogens. The successful detection and control of pathogens in seed and other planting material depend upon the availability of rapid, reliable, robust, specific, and sensitive methods for the detection and identification of pathogens.

The detection and diagnosis of pathogens are crucial for trade and for exchange of germplasm. Early, sensitive and accurate diagnosis is indispensable for certification of seed and other planting material under exchange. The selection of a diagnostic method for evaluating plant health depends on the host to be tested and the type of pathogens. The technique should be reliable for quarantine requirements, reproducible within statistical limits, economical with regard to time, labor, and equipment, and should be rapid to provide results of large samples in the shortest time.

The bulk samples need to be tested by drawing workable samples as per norms. The detection of pathogens is then carried out by the approved or available techniques. Over the years, a great variety of methods have been developed that permit the detection and identification of pathogens including phytoplasma.

2. Exclusion of phytoplasma through quarantine
2.1 International scenario: imports and exports

The recent trade related developments in international activities and the thrust of the WTO Agreements imply that countries need to update their quarantine or plant health services to facilitate pest-free import/export.

The establishment of the WTO in 1995 has provided unlimited opportunities for international trade of agricultural products. History has witnessed the devastating effects resulting from diseases and insect pests introduced along with the international movement of planting material, agricultural produce and products. It is only recently, however, that legal standards have come up in the form of Sanitary and Phytosanitary (SPS) Measures for regulating the international trade. The WTO Agreement on the Application of SPS measures concerns the application of food safety and animal and plant health regulations. It recognizes government's rights to take SPS measures but stipulates that they must be based on science, should be applied to the extent necessary to protect human, animal, or plant life or health, and should not unjustifiably discriminate between members where identical or similar conditions prevail.

The SPS Agreement aims to overcome health-related impediments of plants and animals to market access by encouraging the "establishment, recognition, and application of common SPS measures by different members." The primary incentive for the use of common international norms is that these provide the necessary health protection based on scientific evidence and improve trade flow at the same time.

SPS measures are defined as any measure applied within the territory of the Member State:

- to protect animal or plant life or health from risks arising from the entry, establishment, or spread of pests, diseases, disease-carrying/causing organisms;
- to protect human or animal life or health from risks arising from additives, contaminants, toxins, or disease-causing organisms in food, beverages, or foodstuffs;
- to protect human life or health from risks arising from diseases carried by animals, plants, or their products, or from the entry, establishment/spread of pests; or
- to prevent or limit other damage from the entry, establishment, or spread of pests.

The SPS Agreement explicitly refers to three standard-setting international organizations commonly called as the "three sisters" whose activities are considered to be particularly relevant to its objectives: International Plant Protection Convention (IPPC) of Food and Agriculture Organization (FAO) of the United Nations, World Organization for Animal Health (OIE), and Codex Alimentarius Commission of Joint FAO/WHO. The IPPC develops the International Standards for Phytosanitary Measures (ISPMs) which provide guidelines on pest prevention, detection, and eradication. To date, 47 ISPMs have been developed as mentioned below:

1. ISPM 1: Phytosanitary principles for the protection of plants and the application of phytosanitary measures in international trade
2. ISPM 2: Framework for pest risk analysis (PRA)
3. ISPM 3: Guidelines for the export, shipment, import, and release of biological control agents and other beneficial organisms
4. ISPM 4: Requirements for the establishment of pest-free areas
5. ISPM 5: Glossary of phytosanitary terms
6. ISPM 6: Guidelines for surveillance
7. ISPM 7: Phytosanitary certification system
8. ISPM 8: Determination of pest status in an area
9. ISPM 9: Guidelines for pest eradication programs
10. ISPM 10: Requirements for the establishment of pest-free places of production and pest-free production sites
11. ISPM 11: PRA for quarantine pests
12. ISPM 12: Phytosanitary certificates

13. ISPM 13: Guidelines for the notification of noncompliance and emergency action
14. ISPM 14: The use of integrated measure in a systems approach for pest risk management
15. ISPM 15: Regulation of wood packaging material in international trade
16. ISPM 16: Regulated nonquarantine pests: concept and application
17. ISPM 17: Pest reporting
18. ISPM 18: Guidelines for the use of irradiation as a phytosanitary measure
19. ISPM 19: Guidelines on lists of regulated pests
20. ISPM 20: Guidelines for phytosanitary import regulatory system
21. ISPM 21: PRA for regulated nonquarantine pests
22. ISPM 22: Requirements for the establishment of areas of low pest prevalence
23. ISPM 23: Guidelines for inspection
24. ISPM 24: Guidelines for the determination and recognition of equivalence of phytosanitary measures
25. ISPM 25: Consignments in transit
26. ISPM 26: Establishment of pest-free areas for fruit flies (Tephritidae)
27. ISPM 27: Diagnostic protocols for regulated pests
28. ISPM 28: Phytosanitary treatments for regulated pests
29. ISPM 29: Recognition of pest-free areas and areas of low pest prevalence
30. ISPM 30: Establishment of areas of low pest prevalence for fruit flies (Tephritidae)
31. ISPM 31: Methodologies for sampling of consignments
32. ISPM 32: Categorization of commodities according to their pest risk
33. ISPM 33: Pest-free potato (*Solanum* spp.) micropropagative material and minitubers for international trade
34. ISPM 34: Design and operation of postentry quarantine stations for plants
35. ISPM 35: Systems approach for pest risk management of fruit flies (Tephritidae)
36. ISPM 36: Integrated measures for plants for planting
37. ISPM 37: Determination of host status of fruit-to-fruit flies (Tephritidae)
38. ISPM 38. International movement of seeds
39. ISPM 39: International movement of wood
40. ISPM 40: International movement of growing media in association with plants for planting
41. ISPM 41: International movement of used vehicles, machinery, and equipment
42. ISPM 42: Requirements for the use of temperature treatments as phytosanitary measures
43. ISPM 43: Requirements for the use of fumigation as a phytosanitary measure
44. ISPM-44: Requirements for the use of modified atmosphere treatments as phytosanitary measures
45. ISPM 45: Requirements for NPPOs if authorizing entities to perform phytosanitary actions
46. ISPM 46: Commodity-specific standards for phytosanitary measures
47. ISPM 47: Audit in the phytosanitary context

Prior to the establishment of WTO, governments on a voluntary basis could adopt international standards, guidelines, recommendations, and other advisory texts. Although these norms shall remain voluntary, a new status has been conferred upon them by the SPS Agreement. A WTO Member adopting such norms is presumed to be in full compliance with the SPS Agreement.

2.2 Status of phytoplasmas as regulated pest in Asia

The important crops of Asia are host to a large number devastating pests including phytoplasma inflicting severe losses to the tune of billions of dollars annually, though the exact figure is not available. Country-wise list of regulated phytoplasmas in different countries of Asia is given in Table 6.1. Majority of the countries in Asia are members of the WTO and are parties to the Agreement on Application of SPS Measures (known as SPS Agreement). This contributed for majority to amend their legislative measures to include regulated pests list including phytoplasmas and made available on website of IPPC (http://www.ippc.org). However, the implementation of legislative provisions, leaving aside some developed countries appear to be far from satisfactory in the region due to economic and other reasons.

Phytoplasmas are mentioned as regulated pest for 30 countries in Asia viz., Armenia, Azerbaijan, Bahrain, Belarus, Cambodia, China, Cyprus, Georgia, India, Indonesia, Iran, Iraq, Japan, Jordan, Korea, Lao, Malaysia, Myanmar, Nepal, Oman, Pakistan, Philippines, Qatar, Singapore, Sri Lanka, Syria, Thailand, Turkey, Ukraine, and Vietnam. Regulated pest list including phytoplasmas is not available for 20 countries viz., Afghanistan, Bangladesh, Bhutan, Brunei, Israel, Kazakhstan, Kuwait, Kyrgyzstan, Lebanon, Maldives, Mongolia, North Korea, Russia, Saudi Arabia, Serbia, Tajikistan, Timor-Leste, Turkmenistan, Yemen, Uzbekistan, and the United Arab Emirates. Four countries viz., Hongkong, Macau, Palestine, and Taiwan are not listed in IPPC website.

Malaysia is mentioning phytoplasmas as mycoplasma-like organisms. Most of the countries are not following nomenclature as per International Committee on Systematic Bacteriology (ICSB) Sub-committee on the Taxonomy of Mollicutes, and all countries need to follow ICSB norms (IRPCM, 2004; Bertaccini et al., 2022).

2.3 National scenario: imports

Plant quarantine is defined as all activities designed to prevent the introduction and/or spread of quarantine pests or to ensure their official control. Quarantine pest is a pest of potential economic importance to the area endangered thereby and not yet present there, or present but not widely distributed and being officially controlled.

As early as in 1914, the Government of India passed a comprehensive Act, known as Destructive Insects and Pests (DIP) Act, to regulate or prohibit the import of any article into India likely to carry any pest that may be destructive to any crop or from one state to another. The DIP Act has since undergone several amendments. In October 1988, New Policy on Seed Development was announced, liberalizing the import of seeds and other planting material. In view of this, Plants, Fruits, and Seeds (Regulation of import into India) Order (PFS Order) first promulgated in 1984 was revised in 1989. The PFS Order was further revised in the light of WTO Agreements and the Plant Quarantine (Regulation of Import into India). Order 2003 (hereafter referred to as PQ Order) came into force on January 1, 2004 to comply with the SPS Agreement (Khetarpal et al., 2006). A number of amendments of the PQ Order were notified, revising definitions, clarifying specific queries raised by quarantine authorities of various countries, with revised lists of crops under the Schedules VI and VII and quarantine weed species under Schedule VIII. The revised list under Schedules VI and VII now include 700 and 519 crops/commodities, respectively, and Schedule VIII now include 57 quarantine weed species. The PQ Order ensures the incorporation of

Table 6.1 Phytoplasmas as regulated pest in Asia region.

S. No.	Phytoplasma	Ar	Az	Ba	Be	Ca	Ch	Cy	G	In	Is	Ir	Iq
1.	'Candidatus Phytoplasma asteris'	–	–	–	–	–	–	–	–	–	✔	–	–
2.	'Candidatus Phytoplasma aurantifolia'	–	–	–	–	–	–	–	–	–	–	–	–
3.	'Candidatus Phytoplasma australiense'	–	–	–	–	–	✔	–	–	–	–	–	–
4.	'Candidatus Phytoplasma mali'	–	–	–	✔	–	–	–	✔	–	✔	✔	–
5.	'Candidatus Phytoplasma oryzae'	–	–	–	–	–	–	–	–	–	–	–	–
6.	'Candidatus Phytoplasma palmae'	–	–	–	–	–	–	–	–	–	✔	–	–
7.	'Candidatus Phytoplasma palmicola'	–	–	✔	–	–	–	–	–	✔	–	✔	–
8.	'Candidatus Phytoplasma pini'	–	–	–	–	–	–	–	–	–	–	–	–
9.	'Candidatus Phytoplasma prunorum'	–	–	✔	–	–	–	–	–	–	–	✔	–
10.	'Candidatus Phytoplasma pyri'	✔	–	✔	✔	–	✔	–	–	–	–	✔	–
11.	'Candidatus Phytoplasma rubi'	–	–	–	–	–	–	–	–	–	–	–	–
12.	'Candidatus Phytoplasma ulmi'	–	–	–	–	–	✔	–	✔	✔	–	✔	–
13.	'Flavescence dorée'	–	–	–	✔	–	–	–	–	–	–	✔	–
14.	'Candidatus Phytoplasma aurantifolia'	–	–	–	–	–	✔	–	–	–	–	–	–
15.	'Candidatus Phytoplasma australasia'	–	–	–	–	–	–	–	–	–	–	–	–
16.	'Candidatus Phytoplasma pruni'	–	–	✔	–	–	✔	–	–	–	–	✔	–
17.	'Candidatus Phytoplasma solani'	–	–	✔	–	–	–	–	–	✔	–	✔	–
18.	'Candidatus Phytoplasma rubi'	–	–	–	–	–	–	–	–	–	–	–	–
19.	'Candidatus Phytoplasma sacchari'	–	–	–	–	–	–	–	–	✔	–	–	–

Ar, Armenia; Az, Azerbaijan; Ba, Bahrain; Be, Belarus; Ca, Cambodia; Ch, China; Cy, Cyprus; G, Georgia; In, India; Is, Indonesia; Ir, Iran; Iq, Iraq; Ja, Japan; Jo, Jordan; K, Korea; L, Lao; M, Malaysia; My, Myanmar; N, Nepal; O, Oman; P, Pakistan; Ph, Philippines; Q, Qatar; S, Singapore; Sr, Sri Lanka; Sy, Syria; T, Thailand; Tu, Turkey; U, Ukraine; V, Vietnam.

Table 6.1 Phytoplasmas as regulated pest in Asia region.

J	Jo	K	L	M	My	N	O	P	Ph	Q	S	Sr	Sy	T	Tu	U	V
–	–	–	–	–	–	–	–	–	✓	–	–	–	–	–	–	–	–
✓	–	✓	–	–	–	–	–	–	✓	–	–	–	–	–	–	–	–
✓	–	✓	–	–	–	–	–	–	–	–	–	–	–	–	–	–	–
✓	–	–	–	–	–	–	–	–	–	–	–	–	–	–	–	–	–
–	–	✓	–	–	–	–	–	–	–	–	–	–	–	–	–	–	–
–	–	–	–	–	–	–	–	–	–	–	–	–	–	–	–	–	–
–	–	✓	–	✓	–	–	✓	–	–	✓	–	✓	✓	–	✓	–	–
–	–	✓	–	–	–	–	–	–	–	–	–	–	–	–	–	–	–
✓	–	✓	–	–	–	–	–	–	–	–	–	–	–	–	–	–	–
✓	–	✓	–	–	–	–	✓	–	–	✓	–	–	✓	–	–	–	–
–	–	✓	–	–	–	–	–	–	–	–	–	–	–	–	–	–	–
–	–	✓	–	–	–	–	–	–	–	–	–	–	–	–	–	–	–
–	–	–	–	–	–	–	–	–	–	–	–	–	–	–	–	–	–
–	–	–	–	–	–	–	–	–	–	✓	–	✓	✓	✓	–	–	–
–	–	–	–	–	–	–	–	–	–	–	–	–	–	–	–	–	–
✓	✓	–	–	–	✓	–	✓	–	–	✓	–	–	✓	–	–	–	–
✓	✓	–	–	–	–	✓	✓	–	–	✓	–	✓	✓	–	–	–	–
✓	–	✓	–	–	–	–	–	–	–	–	–	–	–	–	–	–	–
✓	–	–	–	✓	–	–	–	–	–	–	–	✓	–	–	–	–	–

"Additional/Special Declarations" for import commodities free from quarantine pests, on the basis of PRA following international norms, particularly for seed/planting material.

The Directorate of Plant Protection, Quarantine and Storage (DPPQS) under the Ministry of Agriculture and Farmers Welfare is responsible for enforcing quarantine regulations and for quarantine inspection and disinfestation of agricultural commodities. The quarantine processing of bulk consignments of grain/pulses for consumption and seed/planting material for sowing are undertaken by the 73 Plant Quarantine Stations located in different parts of the country, and many pests were intercepted in imported consignments (Sushil, 2016). Import of bulk material for sowing/planting purposes is authorized only through six Regional Plant Quarantine Stations. There are 45 Inspection Authorities (IAs) including 42 IAs who inspect the consignment being grown in isolation in different parts of the country and three IAs who inspect tissue culture raised plants. Besides, DPPQS has developed 24 National standards on various phytosanitary issues such as on PRA, pest-free areas for fruit flies and stone weevils, certification of facilities for the treatment of wood packaging material, methyl bromide fumigation, etc. Also, two Standard Operating Procedures have been notified on export inspection and phytosanitary certification of plants/plant products and other regulated articles and postentry quarantine inspection.

Following are phytoplasma associated diseases of quarantine significance for India which are not reported from India/included in the PQ Order 2003 as regulated pests (Bhalla et al., 2018; Chalam et al., 2013; Chalam et al., 2018):

1. Aster yellows phytoplasmas
2. Blueberry stunt phytoplasma
3. Cameroon marbling (phytoplasma)
4. Cassava witches' broom (phytoplasma)
5. Cranberry false blossom (phytoplasma)
6. Elm phloem necrosis (phytoplasma)
7. Palm lethal yellowing phytoplasma and related strains
8. Peach rosette phytoplasma
9. Peach yellows phytoplasma
10. Phytoplasmas affecting *Pomoideae*
11. Potato purple-top wilt and "stolbur" phytoplasmas
12. Strawberry green petal, phyllody, and yellows (phytoplasmas)
13. Sweet potato witches' broom (phytoplasma)
14. Witches' broom of blueberry (phytoplasma)

The ICAR-National Bureau of Plant Genetic Resources (ICAR-NBPGR), the nodal institution for exchange of plant genetic resources (PGRs) has been empowered under the PQ Order to handle quarantine processing of germplasm including transgenic planting material imported for research purposes into the country by both public and private sectors. ICAR-NBPGR has developed well-equipped laboratories and postentry quarantine green house complex. Keeping in view the biosafety requirements, National Containment Facility of level-4 (CL-4) has been established at NBPGR to ensure that no viable biological material/pollen/pathogen enters or leaves the facility during quarantine processing of transgenics. Till date, >13,000 samples of transgenic crops comprising *Arabidopsis thaliana*, *Brassica* spp., chickpea, corn, cotton, potato, rice, soybean, tobacco, tomato, and wheat with different traits imported into India for research purposes were processed for quarantine clearance, wherein they are tested for associated exotic pests, if any, and also for ensuring the absence of terminator gene technology (embryogenesis deactivator gene) which are mandatory legislative requirements. At ICAR-NBPGR, some of the important pathogens intercepted include fungi: *Fusarium nivale*, *Peronospora manshurica*, and *Uromyces betae* and bacterium: *Xanthomonas campestris* pv. *campestris*. In the last three decades by adopting a workable strategy such as PEQ growing in PEQ greenhouses/containment facility and inspection, PEQ inspection at indenter's site, electron microscopy, enzyme-linked immunosorbent assay, and reverse transcription-polymerase chain reaction (RT-PCR), 41 viruses of great economic and quarantine importance have been intercepted in exotic germplasm including transgenics. The interceptions include 17 viruses not yet reported from India. Even though some of the intercepted viruses are not known to occur in India, their potential vectors exist and so also the congenial conditions for them to multiply, disseminate, and spread the destructive exotic viruses/strains and even native strains more efficiently. The risk of introduction of 41 seed-transmitted viruses or their strains into India was thus eliminated. All the plants infected by the viruses were uprooted and incinerated.

The infected samples were salvaged by using suitable techniques, and the disease-free germplasm was only used for further distribution and conservation. If not intercepted, some of the above quarantine pests could have been introduced into our agricultural fields and caused havoc to our productions. Thus, apart from eliminating the introduction of exotic pathogens from our crop improvement programs, the harvest obtained from disease-free plants ensured conservation of pest-free exotic germplasm in the National Genebank.

2.4 National scenario: exports

The DPPQS under the Ministry of Agriculture is responsible for enforcing quarantine regulations and for quarantine inspection and disinfestation of agri-horticultural commodities. All the material meant for export should be accompanied by Phytosanitary Certificate giving the details of the material and treatment in the model certificate prescribed under the IPPC of FAO. The Ministry of Agriculture, Government of India has notified 161 officers to grant Phytosanitary Certificate for export of plants and plant materials.

The ICAR-NBPGR, the nodal institution for exchange of PGR is vested with the authority to issue Phytosanitary Certificate for seed material and plant propagules of germplasm meant for export for research purposes after getting approval from DARE. NBPGR has developed well-equipped laboratories and green house complex. ICAR-NBPGR undertakes detailed examination of germplasm meant for export for the presence of various pests using general and pest-specific detection techniques and issues Phytosanitary Certificate giving the details of the material and treatment in the model certificate prescribed under the IPPC.

2.5 National domestic quarantine

Domestic quarantine or internal quarantine is aimed to prevent the spread of introduced exotic species or an indigenous key pest to clean (pest-free) areas within the country, and this has its provisions in the DIP Act, 1914 and is enforced by the notification issued by the Central and State Governments. More than 30 pest species seem to have been introduced into India while notifications have been issued against the spread of nine introduced pests only namely fluted scale, San Jose scale, codling moth, coffee berry borer, potato wart disease, potato cyst nematode, apple, BBTV, and banana mosaic virus (Khetarpal et al., 2006), and none of phytoplasmas are listed under domestic quarantine.

According to notifications issued under the DIP Act, an introduced pest, for example, BBTV, has been declared a pest in states of Assam, Kerala, Orissa, Tamil Nadu, and West Bengal and bananas, which come out of these states have to be accompanied by a health certificate from the state Pathologist or other competent authorities that the plants are free from it. However, due to the absence of domestic quarantine, BBTV has spread to most banana-growing areas in the country. The limitations and constraints of domestic quarantine include lack of basic information on the occurrence and distribution of major key pests in the country, and in other words, pest distribution maps are lacking for most of the key pests; absence of concerted action and enforcement of internal quarantine regulations by the state governments; lack of interstate border quarantine check-posts at rail and road lines greatly added to the free movement of planting material across the states; lack of close cooperation and effective coordination between state governments and center for timely notification of introduced pests, organizing pest detection surveys for delineating the affected areas and immediate launching of eradication campaigns in affected areas; lack of public awareness; lack of rapid diagnostic tools/kits for quick detection/identification of exotic pests at the field level; and lack of rigorous seed/stock certification or nursery inspection programs to make available the pest-free seed/planting material for farmers (Bhalla et al., 2014).

There is a dire need to revisit the existing domestic quarantine scenario for strengthening interstate quarantine check-posts and eventually for monitoring movement of viruses of significance. Also, review and update the list of phytoplasmas to be regulated under domestic quarantine. India must develop organized services of plant quarantine at state level parallel to Australia and the United States of America.

3. Perspectives

The NPPOs of developing countries need to be upgraded in terms of manpower, infrastructure, and capabilities to bring it up to the international standards as the increase in imports and the stipulation of WTO under its SPS Agreement has brought about additional challenges to be faced by the plant protection personnel. Strengthening would be for not only prevention of exotic phytoplasmas but also to check the interstate spread of indigenous pests by effective implementation of domestic quarantine regulations against certain important pests which have been introduced/detected in the country in the recent years and which are likely to spread fast (Gupta and Khetarpal, 2004; Chalam et al., 2017).

Regular survey and surveillance program needs to be undertaken to get a realistic picture of status of phytoplasmas in the region and for authentic mapping of endemic phytoplasmas present in localized pockets, and this in turn will help in the identification of pest-free areas and to include these phytoplasmas under domestic quarantine (Chalam, 2020; Chalam et al., 2020, 2022).

There is also a dire need to develop an accessible platform for getting information on biosecurity for the policy makers, administrators, and the industry groups. Need to develop the web-based information portal for exchange of official information including occurrence of phytoplasmas, diagnostics, and database on diagnosticians in the Asia region to facilitate communication among countries.

Also, it is clear that under the present international scenario, the plant protection specialists have a major role to play not only in promoting and facilitating the export and import in the interest of their respective nations but also in protecting the environment from the onslaughts of invasive alien pests. Besides, the threat to national biosecurity from the use of such instruments as bioweapons to create agroterrorism is a possibility that requires preparedness.

Biosecurity will be ensured only when there is an integrated approach to deal with its various components. Models to rationalize regulatory functions among sectors in a quest for improved effectiveness and efficiency have appeared in a number of countries. For example, New Zealand has had a Biosecurity Act since 1993 and a biosecurity minister and Council since 1999. In the United States, the Department of Homeland Security was created in 2002 with many agencies including the US Department of Agriculture's Animal and Plant Health Inspection Service. Likewise, the Australian government has established Biosecurity Australia. In Belize, food safety and animal and plant quarantine and environmental issues are dealt with by a single authority, the Belize Agricultural and Health Authority. For a biosecure Asia region, PRA and early warning systems are the most important aspects that would aid in tackling emergency situations.

The detection and diagnosis of phytoplasmas are crucial for the application of mitigation strategies, trade, and for exchange of germplasm. There is a need to accredit diagnostic laboratories in each country in Asia region for quick and accurate identification of phytoplasmas. There is an urgent need to develop National Plant Pests Diagnostic Network in each country and a Asia Plant Pests Diagnostic Network linking NPPOs in Asia region, which would be the backbone for strengthening the program on biosecurity from plant pests including phytoplasmas (Chalam and Khetarpal, 2008; Chalam et al., 2017, 2022; Chalam and Maurya, 2018).

Also, Regional Working Group of Experts for Detection and Identification of Phytoplasmas in Asia thus need to be formed to explore future cooperation in terms of sharing of expertise and facilities in Asia especially where the borders are contiguous. This would help in avoiding the introduction of phytoplasma not known in the region and also the movement of phytoplasma within the region. It may be possible to predict high-risk pests, commodities, or regions of origin by closely monitoring which pests are most frequently reported from which countries, and what commodities tend to be infected/infested would boost international trade.

The Asia region need to identify A1 pest list, that is, phytoplasmas not present in the entire Asia region and A2 pest list, that is, phytoplasmas present in one country but not present in other country so as to prevent the cross-boundary movement of phytoplasmas into Asia region and within Asia region.

References

Bertaccini, A., Arocha-Rosete, Y., Contaldo, N., Duduk, B., Fiore, N., Montano, H.G., Kube, M., Martini, C.-H., Oshima, M., Quaglino, K., Schneider, F., Wei, B., Zamorano, A., 2022. Revision of the 'Candidatus Phytoplasma' species description guidelines. Int. J. Syst. Evol. Microbiol. 72 (4), 005353.

Bhalla, S., Chalam, V.C., Tyagi, V., Lal, A., Agarwal, P.C., Bisht, I.S., 2014. Teaching Manual on Germplasm Exchange and Plant Quarantine. National Bureau of Plant Genetic Resources, New Delhi, India, p. 340 (+viii).

Bhalla, S., Chalam, V.C., Singh, B., Gupta, K., Dubey, S.C., 2018. Biosecuring Plant Genetic Resources in India: Role of Plant Quarantine. ICAR-National Bureau of Plant Genetic Resources, New Delhi, ISBN 978-81-937111-1-8, p. 216. +vol. i.

Chalam, V.C., Khetarpal, R.K., 2008. A critical appraisal of challenges in exclusion of plant viruses during transboundary movement of seeds. Indian J. Virol. 19 (2), 139−149.

Chalam, V.C., Maurya, A.K., 2018. Role of quarantine in ensuring biosecurity against transboundary plant viruses. Agric. Res. J. 55 (4), 612−626.

Chalam, V.C., Parakh, D.B., Kumar, A., Maurya, A.K., 2013. Viruses, viroids and phytoplasma of quarantine significance in edible oilseeds. In: Gupta, K., Singh, B., Akhtar, J., Singh, M.C. (Eds.), Potential Quarantine Pests for India in Edible Oilseeds. National Bureau of Plant Genetic Resources, New Delhi, India, pp. 175−212, 295pp. + vii.

Chalam, V.C., Parakh, D.B., Maurya, A.K., 2017. Role of viral diagnostics in quarantine for plant genetic resources and preparedness. Indian Journal of Plant Genet Resources 30 (3), 271−285.

Chalam, V.C., Parakh, D.B., Maurya, A.K., 2018. Viruses, viroids, phytoplasma and spiroplasma of quarantine significance in tropical and sub-tropical fruit crops. In: Bhalla, S., Chalam, V.C., Khan, Z., Kandan, A., Dubey, S.C. (Eds.), Potential Quarantine Pests for India in Tropical and Sub-tropical Fruit Crops. ICAR-National Bureau of Plant Genetic Resources, New Delhi, India, p. 510. +x.

Chalam, V.C., Sharma, R., Sharma, V.D., Maurya, A.K., 2020. Role of quarantine in management of transboundary seed-borne diseases. In: Tiwari, A.K. (Ed.), Advances in Seed Production and Management. Springer, Singapore.

Chalam, V.C., Gupta, K., Singh, M.C., Khan, Z., Akhtar, J., Gawade, B.H., Kumari, P., Kumar, P., Meena, B.R., Maurya, A.K., Meena, D.S., 2022. Role of plant quarantine in preventing entry of exotic pests. Indian Journal of Plant Genetic Resources 35 (3), 141−146.

Dubey, S.C., Gupta, K., Akhtar, J., Chalam, V.C., Singh, M.C., Khan, Z., Singh, S.P., Kumar, P., Gawade, B.H., Raj, K., Boopathi, T., Kumari, P., 2021. Plant quarantine for biosecurity during transboundary movement of plant genetic resources. Indian Phytopathol. 74, 495−508.

Gupta, K., Khetrapal, R.K., 2004. Concept of regulated pests, their risk analysis and the Indian scenario. Annual Review of Plant Pathology 3, 409−441.

IRPCM, 2004. '*Candidatus* Phytoplasma', a taxon for the wall-less, non-helical prokaryotes that colonize plant phloem and insects. Int. J. Syst. Evol. Microbiol. 54 (4), 1243−1255.

Khetarpal, R.K., Lal, A., Varaprasad, K.S., Agarwal, P.C., Bhalla, S., Chalam, V.C., Gupta, K., 2006. Quarantine for safe exchange of plant genetic resources. In: Singh, A.K., Saxena, S., Srinivasan, K., Dhillon, B.S. (Eds.), 100 Years of PGR Management in India. National Bureau of Plant Genetic Resources, New Delhi, India, pp. 83−108.

Sushil, S.N., 2016. Agricultural biosecurity system in India. In: Chalam, V.C., Akhtar, J., Dubey, S.C. (Eds.), Souvenir and Abstracts of Zonal Annual Meeting of Delhi Zone, Indian Phytopathological Society (IPS) & National Symposium on "Biosecurity in Food Value Chain." ICAR-National Bureau of Plant Genetic Resources, New Delhi, India, pp. 19−25.

CHAPTER 7

Updates on phytoplasma diseases management

Nursen Ustun[1], Maryam Ghayeb Zamharir[2] and Abdullah Mohammed Al-Sadi[3]

[1]*Laboratory of Bacteriology, Plant Diseases Deparment, Plant Protection Research Institute, Bornova, Izmir, Turkey;* [2]*Plant Diseases Department, Iranian Research Institute of Plant Protection, Agricultural Research, Education and Extension Organization (AREEO), Tehran, Iran;* [3]*Department of Crop Sciences, College of Agricultural and Marine Sciences, Sultan Qaboos University, Muscat, Oman*

1. Introduction

Phytoplasmas cell wall−less prokaryotes causing economically important diseases on annual and perennial crops and natural floras worldwide. These phloem-inhabiting plant pathogenic bacteria are classified in the genus '*Candidatus* Phytoplasma' and the class *Mollicutes* (Bové and Garnier, 2002). All phytoplasmas are transmitted by phloem-feeding insects mostly leafhoppers, planthoppers and psyllids belonging to suborder Auchenorrhyncha (Weintraub, 2007). They replicate intracellularly in plants and insects, and they need both hosts for survival (Bertaccini, 2019).

Phytoplasmas have been reported associated with diseases of more than 1000 plant species and are associated with different symptoms, including witches' brooms, phloem necrosis, phyllody, virescence, yellowing, dieback, and decline (Bertaccini et al., 2014; Bellardi et al., 2018; Bogoutdinov et al., 2019; Hemmati et al., 2021; Kumari et al., 2019; Namba, 2019). Phytoplasmas can survive in the gut and salivary glands of sap-sucking insects of the order *Hemiptera* (Ammar and Hogenhout, 2006; Hill and Sinclair, 2000; Sugio et al., 2011a; Weintraub and Beanland, 2006). These pathogens can be disseminated in the field by propagative material and phloem-feeding insects, including leafhoppers, planthoppers, and psyllids (Al-Sadi et al., 2012; Hemmati and Nikooei, 2019; Jakovljević et al., 2020; Linck and Reineke, 2019; Quaglino et al., 2019). They can also be transmitted by grafting and parasitic plants. Dodder transmission of phytoplasmas from naturally infected plant to *Catharanthus roseus* (periwinkle) has been used to maintain a wide range of phytoplasmas for further research (Přibylová and Spak, 2013). Transmission of phytoplasmas via seeds has also been reported in different crops, also in Asian countries (Çağlar et al., 2019; Manimekalai et al., 2014).

Phytoplasmas survive in diverse environments of plants and insects and due to their ability to survive and multiply in plants and insect vectors, they can spread into new hosts in a short timespan (Galetto et al., 2011; Ishii et al., 2009; Suzuki et al., 2006). Phytoplasmas also modulate and regulate plant hosts genes, hormones, and secondary metabolite biosynthesis (MacLean et al., 2011; Sugio and Hogenhout, 2012; Sugio et al., 2011b). Phytoplasmas can also produce specific proteins (called effectors) that help them overcome physical and biochemical plant defenses, multiply in plant hosts, disperse by insect vectors, and modulate plant host growth (Orlovskis, 2017; Sugio et al., 2014).

The unique nature of phytoplasmas requires different management strategy based on polyphasic integrated approach combining components of cultural, physical, biological, resistance, and chemical control (Schweigkofler et al., 2018; Bertaccini, 2019; Kumari et al., 2019). In the field, phytoplasma diseases management has been stood mainly on eradication of infected plants and insects' vector since both are sources for the phytoplasma inoculation (Weintraub and Wilson, 2010; Weintraub and Beanland, 2006). All other methods including quarantine measures, certification, thermotherapy or meristem culture, host resistance, and innovative control strategies preventing spread or reducing inoculum sources of phytoplasmas contribute to their sustainable management.

In this chapter, conventional and new phytoplasma diseases management approaches are described.

2. Quarantine

Phytoplasmas can enter into new areas by infected plant materials particularly latently infected and symptomless plants and insect vectors. To prevent the spread and entry into new areas, several mandatory requirements, rules, and laws are put into force by plant health authorities worldwide. Various phytoplasmas with high phytosanitary risks and their insect vectors categorized as quarantine pest are subject to official control in several Asian countries. "Flavescence dorée" (FD), '*Candidatus* Phytoplasma prunorum' [European stone fruit yellows (ESFY)], '*Ca.* P. mali' (apple proliferation) (AP), '*Ca.* P. pyri' (pear decline) (PD), '*Ca.* P. solani' ("stolbur"), coconut lethal yellowing, '*Ca.* P. phoenicium', '*Ca.* P. australiense', '*Ca.* P. fraxini', '*Ca.* P. hispanicum', '*Ca.* P. trifolii', '*Ca.* P. ziziphi', '*Ca.* P. aurantifolia', '*Ca.* P. pruni', are harmful quarantine organisms for many countries (https://www.eppo.int/ACTIVITIES/quarantine_activities). Regulated phytoplasma lists (quarantine list) for some Asian countries are given in Table 7.1. For these phytoplasmas, strict phytosanitary regulations and restrictions on their host movements are applied in the European Union (EFSA Panel on Plant Health, 2020) and many parts of the world. In international trade, their host plants and planting materials are inspected at border points and laboratory tested for pathogen absence. Movement of plant materials in passenger luggage presenting also the phytosanitary risk of phytoplasma spread has to be regulated for domestic and international travelers (Gurr et al., 2016).

In case of a new introduction, eradication is employed to eliminate a newly established pathogen. An early eradication is important for stopping of the disease as it was demonstrated for coconut lethal yellowing in the Dominican Republic (Gurr et al., 2016) where the early implementation of an eradication program combined with natural barriers preventing vector movement and an abundance of nonhost palms was able to slow down the disease spread (Martinez et al., 2008, 2010). Surveillance for delimitation of the pathogen present in limited areas in the country is organized in order to prevent its further spread and establishment. For surveillance and detection of infected plants, conventional inspection methods (visual examination, sampling, and identification) are commonly used. In Turkey, mandatory detection surveys for grapevine yellows (GY) ["bois noir" (BN) and FD] have been carried out since last 10 years. The vineyards in main viticulture growing areas are inspected visually during late summer-early autumn annually, and samples are taken for laboratory analyses from symptomatically suspected grapevine plants. Plants found to be infected are destroyed, and movement of the plant materials from the contaminated vineyard is prohibited. As a result of strict surveillance, pest-free areas in accordance with ISPM 4 (IPPC, 2017) and ISPM 6 (IPPC,1997) were established in many provinces of Turkey (Anonymous, 2020). Recently, aerial surveillance with unmanned aerial

Table 7.1 Regulated phytoplasma lists for some Asian countries.

Phytoplasma name	Japan[a]	Iran[b]	Turkey[c]	China[d]	India[e]	Russia[f]
Apple rubbery wood	+					
Aster yellows phytoplasma group	+				+	
Alder yellows				+		
Ash yellows				+		
Blueberry stunt				+	+	
Blueberry witches' broom					+	
'Candidatus Phytoplasma aurantifolia' (lime witches' broom)	+		+	+		
'Candidatus Phytoplasma australiense' (Australian grapevine yellows; Strawberry lethal yellows)	+					
'Candidatus Phytoplasma mali' (apple proliferation)	+	+	+	+		+
'Candidatus Phytoplasma prunorum' (apricot chlorotic leafroll; European stone fruit yellows)	+	+	+	+		
'Candidatus Phytoplasma pyri' (pear decline)	+	+	+	+		+
Cranberry false blossom	+				+	
Grapevine "flavescence dorée"	+	+	+	+		+
Cameroon marbling					+	
Citrus witches' broom					+	
Cassava witches' broom					+	
Cocoa witches' broom					+	
Grapevine yellows	+					
Peach rosette	+		+		+	
Peach X-disease	+	+	+	+		
Peach yellows	+	+	+		+	
Potato purple top wilt	+				+	
'Candidatus Phytoplasma solani' (potato stolbur)	+	+	+		+	
Potato witches' broom				+	+	
Strawberry lethal decline	+					
Rubus stunt	+					
Sugarcane grassy shoot and white leaf	+					
Sugarcane yellows	+					
Vaccinium witches' broom	+					
'Candidatus Phytoplasma ulmi' (elm phloem necrosis)		+	+	+	+	
Cherry lethal yellow		+				

Continued

Table 7.1 Regulated phytoplasma lists for some Asian countries.—cont'd

Phytoplasma name	Japan[a]	Iran[b]	Turkey[c]	China[d]	India[e]	Russia[f]
'Candidatus Phytoplasma palmae' (palm lethal yellows; coconut lethal yellowing)		+	+	+	+	
Strawberry witches' broom			+			
Strawberry multiplier				+		
Strawberry green petal					+	
Strawberry phyllody					+	
Strawberry yellows					+	

[a] Quarantine Pest List (Annexed Table 7.1 of the Ordinance for Enforcement of the Plant Protection Act, https://www.maffgo.jp/pps/j/houki/shorei/E_Annexed_Table 7.1_from_20201111.html).
[b] https://assets.ippc.int/static/media/files/reportingobligation/2019/09/05/quarantine_pest2018.pdf.
[c] http://extwprlegs1.fao.org/docs/pdf/tur164691.pdf.
[d] https://assets.ippc.int/static/media/files/reportingobligation/2019/12/18/Catalogue_of_Quarantine_Pests_for_Import_Plants_to_China_update20130306.pdf.
[e] https://plantquarantineindia.nic.in/PQISPub/pdffiles/pqorder2015.pdf.
[f] https://www.mpi.govt.nz/dmsdocument/11518/direct.

vehicles, drones, fitted with cameras and multispectral imaging is considered to be useful in the support for detection and mapping of disease (Hill et al., 2009; Gurr et al., 2016).

Sometimes local legislations are applied in the infected regions of the country. For example, in France and in Italy, directives from the local authorities or State lows, respectively regulate the mandatory control not only of FD disease but also of its vector *Scaphoideus titanus* for areas where the disease is present (https://www.cabi.org/isc/datasheet/26184).

3. Healthy planting material

The use of phytoplasma-free plants is one of the major tools in the management of phytoplasma diseases (Roddee et al., 2018). Establishing of new plantations with certified pathogen-free plant materials is of primary importance for successful phytoplasma prevention and control. In grafted plants, healthy vegetative propagation materials such as budwood, scionwood, or rootstock are essential to be used. Infected dormant bud wood can transmit phytoplasmas such as '*Ca*. P. prunorum' persisting in the stem during winter (Seemüller et al., 1998a, 1998b). Dormant scion wood can be grafted and shipped without any risk of pathogen transmission and spread only if phytoplasmas are not able to persist in the upper plant parts during dormant season such as '*Ca*. P. mali' and '*Ca*. P. pyri' (Marcone et al., 2010). For guaranteed pathogen freedom, nuclear stocks for rootstocks and scions have to be inspected visually and tested periodically during vegetation period. Adequate sampling and using of highly sensitive and specific polymerase chain reaction (PCR) procedures are necessary to ensure that they are free from phytoplasmas.

Tissue culture and thermotherapy are techniques recommended for eliminating phytoplasmas from diseased plants. For the eradication of '*Ca*. P. prunorum' from diseased *Prunus* plants, *in vitro* thermotherapy and meristem tip culture have been used (Laimer, 2003; Laimer and Bertaccini, 2008). Several tissue culture techniques coupled with thermotherapy have been proposed for the production of '*Ca* P. pnoenicum' phytoplasma-free almond plantlets. Stem-cutting culture combined with thermotherapy was reported to be the most practical and effective against this phytoplasma (Chalak et al., 2005). Subsequently shoot tip and stem-cutting cultures associated with heat treatment were found to be all suitable for phytoplasma elimination from regenerated shootlets (Chalak and Choueiri, 2015). Meristem tip culture is used to produce free grape vine and stone fruit nursery stock in Turkey which is also useful for the production of phytoplasma-free plants. In India, sugarcane is also produced through tissue culture to obtain virus and phytoplasma-free material (Viswanathan and Rao 2011; Tiwari et al., 2011), while disease-free planting material has also been used for the management of sandalwood spike disease (Mondal et al., 2020). The selection of certified seeds or planting material is one of the options for the management of witches' broom disease of acid lime in Oman and Iran (Al-Sadi et al., 2012; Siampour et al., 2019).

Another method for pathogen eradication from infected tissue is cryotherapy of shoot tips. This method has been succeeded to eliminate phytoplasmas and some other graft-transmissible pathogens from different crop plants. Cryotherapy is based on cryopreservation of shoot tips by exposing the tissues to ultra-low temperature (usually that of liquid nitrogen), storing and regenerating for multiplication. Dehydration of cells before immersing the shoot types into liquid nitrogen is a critical step to avoid lethal intracellular injury caused by the crystallization of intracellular water inside cells. Sweet potato little leaf phytoplasma has been eradicated from a sweet potato line using cryotherapy of shoot tip (Wang and Valkonen, 2008; Wang et al., 2009).

4. Antibiotics

Tetracyclines including oxytetracycline act as an inhibitor of protein synthesis interfering with the ability of bacteria to produce proteins that are essential for growth and multiplication (EPA, 1993; Borgi and Palma, 2014). Injection of tetracycline (oxytetracycline) antibiotics into the trunk was used to control phytoplasma infection in individual host plants. After antibiotic treatments, the disease symptoms can disappear for a period up to two years (Casanova et al., 1980); however, the tetracycline is not able to eliminate the pathogen completely in the host plant, and its effects are associated with only a temporary remission of symptoms (Raychaudhuri et al., 1970; Schmid, 1983). Remission of disease is characterized by developing new asymptomatic growth with normal appearance. Previously distorted tissues do not recover; only new flowers, leaves, and stems are without symptoms. Phytoplasmas degenerate during remission and are not observed in new symptom-free growth. They reappear at the end of the remission period varied depending on plant species (as short as a few weeks for herbaceous and up to two years or longer for some woody plants), disease severity, and doses of antibiotics (higher dose induces longer periods of remission) (McCoy, 1982).

Tetracycline (oxytetracycline) has been registered against coconut lethal yellowing, PD, lethal decline of pritchardia palm, peach X-disease, in the United State of America (McCoy, 1982; EPA, 1993). Also oxytetracycline used for *in vitro* chemotherapy at concentration of 75 mg/l—80mg/l was found successful in freeing infected tissues of *Catharanthus roseus* plants from yellow leaf disease in India (Singh et al., 2007b; Rao et al., 2017; Tiwari et al., 2021).

The method of tetracycline treatments is not easy applicable, feasible, and cost-efficient for commercial production; however, it can be employed for high valuable ornamental plants in touristic sites (McCoy et al., 1976; Eziashi and Omamor, 2010). The use of tetracycline can also induce resistant bacterial populations (CABI, 2021b). Moreover antibiotics in agriculture are prohibited in several countries (Europe, Turkey) in the world because of health and environmental risks (Been, 1995; Musetti et al., 2013). All these factors make tetracycline antibiotics impractical for phytoplasma disease management.

Apart from tetracycline, some other molecules such as biophenicol, chloramphenicol, enteromycelin, lycercelin, paraxin, roscillin, camphicillin, were studied for phytoplasma management and found to be quite ineffective (Kumari et al., 2019).

5. Insect vector control strategies

Insect vector control using pesticides is one of the methods recommended for limiting outbreaks of phytoplasma diseases, but it is not always as efficient as expected. The insecticides directed commonly at mobile instars (nymphs and adults) are applied in a limited number a year. They do not act immediately, which allows phytoplasmas to be acquired (Weintraub and Wilson, 2010). On the other hand, conventional insecticides, even when frequently used (Wally et al., 2004), do not control the symptoms of disease because pathogen transmission occurs faster than insecticides can act, and there is often a constant influx of new vectors from surrounding habitats. At best, use of insecticides might help control vector populations, and thus reduce intracrop transmission (Weintraub, 2007; Rao et al., 2014; Tiwari et al., 2016, 2017). Moreover, insecticides are usually recommended at economic threshold level leading to continuance of transmission. Other limiting aspects of using insecticides in insect control are related to insecticide resistance (Marcone and Rao, 2019), increasing the costs of production and negative effect on environment (due to their soil persistence and water contamination) and human health (Oliveira et al., 2019).

The most reliable means of controlling vectors is by covering the crop with insect-proof screening (Weintraub, 2007). For example, Australia is controlling phytoplasma diseases in papaya using insect-proof screening. Nets protect the plants from the visit of insect vectors and the inoculation of phytoplasmas. They are more efficient than insecticides but have negative effect on pollination resulting in the reduction of fruit production. Although this method is expensive and inapplicable for large-scale agriculture crops (sugar cane, corn, rice, fruit trees, grapes), it can be suggested for cash crops, organically grown fruits, or when a premium price can be obtained for the fruits (Walsh et al., 2006).

Another effective means of insect vector control is through physical prevention by use of a mineral coating on the plant itself. Kaolin, a nonabrasive fine-grained aluminosilicate mineral, applied as a particle film, is a new version of a very old type of inorganic chemical control which may prove to be useful (Weintraub, 2007). Puterka et al. (2003) demonstrated that physical coating of plant surfaces with mineral film protected grape vine plants from feeding, oviposition, and infestation by the glassy-winged sharpshooter, *Homalodisca coagulata* (Say), which is a vector of the bacterium, *Xylella fastidiosa* which causes Pierce's disease in grape vine.

Insect vectors of phytoplasmas utilize chemical cues for the identification of their hosts. Phytoplasma infections may influence the production of plant volatiles which indirectly influenced the behavior of vector insects. Insects species-specific attractive and repellent compounds identified for many vectors can be used in traps as lures for monitoring and mass trapping purposes. In this way, semiochemicals provide an innovative approach for phytoplasma vector controls (Gross et al., 2017).

A new and potentially very powerful tool for controlling pathogen transmission is through the manipulation of symbiotic bacteria in the vectors. Once identified, cultured, and modified, these bacteria can be used to compromise transmission by reducing vector competence, by expressing a gene product that could kill the pathogen, by inducing cytoplasmic incompatibility, and causing a high offspring mortality rate, or by creating physical competition for space that the pathogenic bacteria would normally occupy (Weintraub, 2007). This symbiont-based strategy already being applied against several insect-borne human disease pathogens was investigated also against vector transmitted bacteria and phytoplasmas (Weintraub, 2007). For example, bacterium, identified as *Cardinium hertigii* and localized in the fat bodies and salivary glands of the leafhopper *S. titanus*, the vector of FD phytoplasma is considered as promising in symbiotic control efforts of FD (Bigliardi et al., 2006; Marzorati et al., 2006). One of the major challenges in this field is the delivery of the transgenic bacteria to the target vectors without adversely affecting the environment or other insect populations.

Models for the prediction of phytoplasma and vector spread may provide an insight in disease epidemiology and control. However, a few studies incorporating information on climate change and its relation to disease and vector spread are available (Oliveira et al., 2019). Lessio et al. (2015) developed an FD epidemiological model, integrating different parameters of the transmission process (acquisition of the disease, latency and expression of symptoms, recovery rate, removal and replacement of infected plants, insecticidal treatments, and the effect of hotbeds) and showing the risks of establishing new vineyards in severely infected locations.

6. Other management practices

Cultural control is commonly practiced for the management of various plant diseases, including phytoplasma diseases. Cultural control can be defined as any production or protection practices that can help reduce disease levels. It can include off-season operations, sowing time method of planting, tillage, soil amendments, sanitation, and harvest time. Sowing time was shown to have a significant effect on infection of sesame phyllody with reduction of the disease incidence when the sowing date was delayed (Hosseini et al., 2015).

Management practices have an indirect effect on insect vector populations and phytoplasma disease spread. Implementation of appropriate prophylactic measures supports the disease management. Removal of diseased plants or wild hosts with potential to be inoculum reservoirs contributes to eradicate the disease from the plantations, especially in areas where the disease has limited distribution. In Lebanon, eradication of almond trees infected with 'Ca. P. phoenicium' has been implemented to reduce the *foci* of infection (Molino-Lova et al., 2014). Some growers tend to remove the infected trees and replace them with other species or cultivars (Al-Sadi et al., 2012). In Oman, growers usually remove severely diseased acid lime trees that are affected by witches' broom disease. When the disease is in its early years, where symptoms develop on a few branches, the regular removal of symptomatic branches usually helps reducing phytoplasma titers and vector attraction to the symptomatic branches (Al-Subhi et al., 2021). In addition, this practice helps trees get rid of the symptomatic and nonproductive branches, since they produce no or small nonmarketable fruits (Hemmati et al., 2021). Destruction of infected grape vines as well as uprooting of wild *Vitis* plants in infected vineyards are prophylactic activities for management with FD (CABI, 2021a). Destroying infected trees is suggested also to many other phytoplasma diseases. In fruit orchards, rootstocks suckers are preferred by psyllid vectors of AP, PD, and ESFY phytoplasmas. Therefore, removing root suckers before migration of the psyllids is recommended to reduce the diffusion of these vectors from weeds to trees. For the control of BN phytoplasma, the preventive removal of the grape vine suckers on which *Hyalesthes obsoletus* could feed after grass mowing is strongly proposed. Weeds also can be the potential reservoir of the vectors as well as phytoplasmas, and their control is considered as an important tool in the phytoplasma disease control.

The expression of phytoplasma symptoms has been found to be affected by geographical locations. Symptoms of witches' broom disease of lime have been found to develop in certain areas but not in others (Al-Ghaithi et al., 2017; Da Silva et al., 2014). Queiroz et al. (2016) suggested a decrease in witches' broom diseases of lime transmission by insects under cooler climates in Oman. Countries with variable environmental conditions can therefore focus lime production in areas that are less conducive for phytoplasma symptom development and transmission.

7. Host resistance and/or tolerance?

Host resistance and tolerance are terms associated with different plant responses to the pathogens. Resistance has been described as an ability of the plant to prevent infection and/or development of the disease resulting in the absence of symptoms and a low pathogen titer in the infected plants (Jarausch et al., 2013; Maramorosch and Loebenstein, 2009). Tolerance refers to a mild plant disease allowing commercially acceptable yields to be gained. However, tolerant plant varieties are potential reservoirs of the phytoplasmas because of high titer of the pathogen in the infected plants showing low disease symptoms (Maramorosch and Loebenstein, 2009; Jarausch et al., 2013).

Use of resistant plants seems to be the best approach for preventing phytoplasma diseases. Phytoplasma-resistant genotypes provide an effective and environmentally safe tool for the protection of crops from these pathogens. Resistance against insect vectors based on nonpreference for a plant host was also suggested as a good phytoplasma management tool (Beech, 1981; Laimer et al., 2009; Maramorosch, 2009). Over last decades, plant genotypes with different intra- and interspecific response to various phytoplasmas were defined in natural or experimental conditions (Marcone and Rao, 2019; Marcone et al., 2022). Under natural conditions, the genotype resistance is investigated in the infected fields by scoring the main and distinguishing features of both symptomatic and asymptomatic plants (Akhtar et al., 2009; Ulubas Serçe et al., 2006; Ulubas Serçe et al., 2007). In experimental studies to evaluate host resistance to phytoplasmas, the plants are inoculated via phytoplasma-infected insect vectors (Ustun et al., 2017) or graft transmission (Liu et al., 2004; Marcone et al., 2010) methods. In woody plants, both grafting of diseased bark onto healthy rootstocks or healthy germplasm onto seriously diseased plants are successfully implemented (Liu et al., 2004; Marcone et al., 2010). However, graft transmission methods were not easily applicable to some plant species (Wallace 2002). For these, plants resistance screening is usually performed by planting different varieties under natural infection conditions (Baudouin et al., 2009; Marcone and Rao, 2019).

The traditional evaluation of phytoplasma disease resistance based on the differences in symptoms between genotypes may vary to some degree in scoring by different researchers (Huang et al., 2014). Moreover, in woody plants, the perennial nature of tree species complicates rating of symptom due to different year-to-year overwintering conditions, tree phenology, stage of disease progression, and confusing symptoms with other abiotic or biotic stress factors (Huang et al., 2014). Thus, in some instances, the apparent resistance of genotype may be linked not with genetic traits but with different disease severity or mixed infection with biotic or abiotic agents (Mpunami et al., 2002; Baudouin et al., 2009).

For identification of the pathogen associated with symptoms, more sensitive diagnostic methods are required in addition to disease phenotyping. Quantitative PCR (qPCR) assay provides opportunity not only to detect the pathogen but also to quantify the phytoplasma presence in the plants (Huang et al., 2014). This method used to quantify phytoplasmas in tolerant and resistant genotypes in different crops and has also some potential in plant breeding (Huang et al., 2014; Ikten et al., 2016; Ustun et al., 2017). Besides inoculation experiments (Sinclair et al., 2000a, 2000b) and qPCR, other molecular DNA methods such as the detection of random amplified polymorphic DNA markers associated with the resistance (Cardena et al., 2003) were used for an adequate determination of genotype resistance level and true understanding of phytoplasma diseases. Simple sequence repeats, amplified fragment length polymorphism, and long terminal repeats markers were utilized in phytoplasma resistance mapping (Lenz and Dai, 2017).

The molecular, anatomical, and physiological basis of phytoplasma resistance is still poorly understood (Marcone and Rao, 2019) and additional studies are necessary. The resistance may be associated with morphological and physiological differences in the plants which are designated by the genetic background of each cultivar and their responses to the environment (Oliveira et al., 2019). It may also be linked to their reaction to specific phytoplasma effectors, playing a role in the host–pathogen interaction (Eveillard et al., 2016; Oliveira et al., 2019). Phytoplasmas at metabolic level interrupt the activities of plant enzymes involved in carbohydrate metabolism for their own energy and growth requirements, also they interfere the contents of soluble proteins, pigments, soluble sugars, auxin, and activities of antioxidant enzymes in leaves and spread by using the host plant's phloem system (Lepka et al., 1999; Zafari et al., 2012; Liu et al., 2014). The molecular mechanisms involved in the phytoplasma–plant interaction were examined in limited studies (Francis et al., 2010). For a better understanding the interactions, both the host as well as the pathogen are essential to be studied at a genomic, transcriptomic, proteomic, and metabolomic level (Oliveira et al., 2019).

In some host plants like Chinese jujube, genes associated with phytoplasma resistance through suppressive subtraction hybridization were identified (Liu et al., 2014).

Many studies underlined that the titer of phytoplasma in resistant genotypes is lower than in susceptible genotypes. Huang et al. (2014) found a significant correlation between the copy number of phytoplasmas in qPCR assay and visual phenotypic rating scores of X-disease resistance ('*Ca.* P. pruni') in chokecherry plants. Disease-resistant chokecherries had a significantly lower titer of X-disease phytoplasmas than susceptible plants.

Genotype resistance to phytoplasma is influenced by environmental conditions, aggressiveness of the phytoplasma strains, rootstocks, Atkinson and Urwin (2012) argued that genotypes selected as disease resistant under field conditions showed diverse response of disease resistance under various environmental conditions. This can be due to the fact that the environmental conditions, particularly the interaction between temperature and moisture in the host act directly upon the phytoplasma disease severity (Krishnareddy, 2013) and the population and distribution of insect vectors (Gurr et al., 2016). The genetic diverse phytoplasma populations and appearance of new phytoplasma strains in the different environments may also reduce the level of resistance or may cause the total breakdown of resistance (Quaicoe et al., 2009; Gurr et al., 2016).

Rootstocks have different level of resistance to phytoplasmas, and they significantly affect the susceptibility of the trees (Giunchedi et al., 1994, 1995). Some rootstocks are more resistant or less affected than others. The same scion cultivar depending also on the aggressiveness of the strain may show severe symptoms in susceptible rootstock or mild symptoms on less affected rootstocks. An appropriate rootstock may prolong the productivity of infected trees (Marcone et al., 2010; Marcone and Rao, 2019.). In some cases, when phytoplasmas as '*Ca*. P. prunorum' (ESFY) are able to overwinter in the above-ground parts of the trees, both rootstock and scion cultivars are necessary to be resistant for a successful disease management (Marcone and Rao, 2019). The detailed study performed by graft inoculation, PCR assays, and DAPI fluorescence tests clearly demonstrated the effect of the rootstocks on resistance of stone fruits cultivars to ESFY (Kison and Seemüller, 2001). For instance, ESFY-diseased apricot, Japanese plum, and peach trees grafted on plum rootstocks (e.g., *Prunus Marianna GF8/1* (*Prunus cerasifera* × *Prunus munsoniana*), myrobalan (*P. cerasifera*) seedling, *Prunus insititia*, and *Prunus domestica* stocks) were observed to survive longer than those grafted on peach and apricot rootstocks (Seemüller and Foster, 1995; Kison and Seemüller, 2001). The mortality in the same rootstock and scion combination was found to be strongly dependent on the virulence of phytoplasma strains (Kison and Seemüller, 2001). Although satisfactory ESFY resistance was not found, the scions grafted on *P. domestica* stocks Achermann's, Brompton, P. 2175, and *P. cerasifera* stock Myrabi (P 2032) showed very mild symptoms compared to other rootstocks. On these rootstocks, the phloem necrosis was rarely observed, and even in the presence of severe strains.

Peach Rutgers Red Leaf, Montclair, Rubira, peach and apricot seedlings, and *P. insititia* stock St. Julien 2 were among the most ESFY- susceptible rootstocks (Marcone and Rao, 2019) The grafting of apricot scion on apricot rootstock resulted in most severely damaged trees particularly when the infecting strain was also from apricot. However, the susceptibility of genotypes on peach rootstocks was not related to scion nor inoculum source. Rootstocks have an effect also on host resistance to other ESFY closely related phytoplasmas. Pear trees grafted onto quince rootstocks were demonstrated to suffer less from PD infection in Germany (Seemüller et al., 1986) and England (Davies et al., 1992). Symptoms on diseased apple trees grafted on *Malus sieboldii*-based AP-resistant apomictic rootstocks have been never observed or only rarely, temporary mild symptoms have appeared (Marcone and Rao, 2019).

Rootstocks may affect also vector response to plants and consequently phytoplasma disease. For example, in Israel, during annual mapping of phytoplasma infections in a vineyard, it was discovered that plants on Richter 110 rootstock had less phytoplasma symptoms incidence than on Castel (Weintraub and Beanland, 2006). DAPI test and PCR assays show that colonization of phytoplasma in the resistant rootstocks is lower than in those susceptible (Kison and Seemüller, 2001). Also, phytoplasmas ('*Ca*. P. mali') are more concentrated in roots than in trunks, and in some cases, they are detected only in the roots and not in the trunk. For ESFY, a persistent colonization during the 5—8 years of observation was reported even in resistant rootstocks (Kison and Seemüller, 2001; Marcone and Rao, 2019). The low concentration of phytoplasmas in resistant rootstocks is associated with less recolonization of the upper parts of the tree in spring, leading to the lack of symptoms or mild symptoms (Seemüller et al., 1986).

The scientific efforts on phytoplasma resistance have been ended up with the identification of some resistant genotypes and progenies. The high resistance to AP was determined in apomictic *M. sieboldii* progenies and in some selections having *M. sieboldii* in their ancestors (Seemüller et al., 2008). The AP phytoplasma never colonized in the upper parts of the tree harboring extremely low phytoplasma titer in the roots (Marcone and Rao, 2019). The genotype "Xingguang" (clone of the cultivar Junzao) crossed and released in 2005 in China (Zhao et al., 2009; Liu et al., 2004) has very high resistance to jujube witches' broom phytoplasma, one of the most destructive diseases of Chinese jujube (*Ziziphus jujuba* Mill.) in Asia (Liu et al., 2014; Babaei et al., 2020) showing low disease rate and a late symptom appearance in the field (Zhao et al., 2009; Liu et al., 2004) and keeping always healthy growth after grafting onto different diseased cultivars (Liu et al., 2014). Resistant genotype was developed against brinjal little leaf phytoplasma inducing over 40% yield losses in *Solanum melongena* L. in subtropic and tropic areas using a resistance source from wild relatives of the brinjal, *Solanum gilo* and *Solanum integrifolium* Chakrabarti and Choudhury (1975). Genotypes with different level of resistance against sesame phyllody responsible for the most destructive disease of *Sesamum indicum* L. in some Asian countries (Akhtar et al., 2009; Rao et al., 2015) were described in India, Pakistan, and Turkey (Singh et al., 2007a; Akhtar et al., 2013; Ustun et al., 2017) and inheritance of phyllody resistance was reported (Singh et al., 2007a; Shindle et al., 2011). Natural resistance to X-disease ('*Ca.* P. pruni') infecting many *Prunus* spp. was documented mainly for chokecherry (*Prunus virginiana* L.) (Peterson, 1984; Lenz and Dai, 2017), and the genetic linkage map for identifying genetic regions related to X-disease resistance was developed (Lenz and Dai, 2017). For some phytoplasma diseases, only tolerant genotypes were identified. Apart from resistance meaning an absence of symptoms with a low pathogen titer in the plants, tolerance is understood as the absence or occurrence of mild symptoms under a high pathogen titer (Jarausch et al., 2013). Using of tolerant plants seems to be also promising approach for disease control (Seemüller and Harries, 2010; Jarausch et al., 2013); however, asymptomatic tolerant plants as silent reservoir of high phytoplasma titers may play a role in the disease dispersal (Eveillard et al., 2016). Oliveira et al. (2019) emphasized the importance of phytoplasma absence in plant tissues to prevent these plants of being phytoplasma reservoirs. Tolerance to elm yellows (EY) phytoplasma infections was observed in (*U. glabra* × *U. wallichiana*) × U. minor open-pollinated (= clone 808) and (*U. glabra* Exoniensis × *U. wallichiana*) × U. × hollandica Bea Schwarz selfed (= clone Lobel) clones on the basis of field observations and PCR assays in Italy (Lee et al., 1995; Mittempergher, 2000). Some '*Ca.* P. fraxini', (Griffiths et al., 1999; Davis et al., 2013) tolerant green ash (Sinclair et al., 1994, 1997a, 1997b) and *Syringa vulgaris* cultivars (Hibben et al., 1986; Hibben and Franzen, 1989) were identified. PD-tolerant seedling progenies of open-pollinated genotypes were determined (Seemüller et al., 2009). A stably recovered Bulida apricot trees showed durable tolerance to ESFY (Osler et al., 2014, 2016).

Many economically important phytoplasma diseases are very difficult to manage through resistant plants. For example, resistance screening to lethal yellowing disease associated with '*Ca.* P. palmae' posing a significant threat to global coconut (*Cocos nucifera* L) production is not so easy because of difficulties in genetic improvement of coconut (Cardeña et al., 2003) and the elongation of symptom expression of phytoplasma in the genotype. Up to now, some coconut palm varieties and hybrids less susceptible to lethal yellowing are known; however, no genotype of coconut was declared as resistant (Baudouin et al., 2009). Also, resistance or tolerance to phytoplasmas associated with GYs has not been found in the *Vitis* species examined up to now (Laimer et al., 2009). Unfortunately, most cultivars have been reported as highly susceptible or intermediately susceptible with severe or medium severe symptoms, and only a very limited number were referred to as low susceptible cultivars. Nonetheless, many economically relevant cultivars have not yet been studied for tolerance/susceptibility to FD (Oleveira et al., 2019). No fully resistant cultivars are known to '*Ca* P. phoenicium' (subgroup 16SrIX-B), associated with almond witches' broom, a severe disease affecting almond, peach, and nectarine trees in Lebanon and Iran (Ghayeb Zamharir et al., 2011).

Development of genotypes with durable resistance is required for successful phytoplasma management. For durability of resistance over time, breeding of disease-resistant cultivars carrying not single but multiple sources of resistance is preferable (Luo et al., 2020). So far, some identified resistant genotypes have an unsatisfactory and low agronomic value. However, they can be further used as parentals in breeding programs aiming to develop plants with sufficient agronomical properties and stable inherited phytoplasma resistance (Marcone and Rao, 2019). By using backcrossing models or DNA-based markers, the undesirable features can be thrown off and the quality of plants can be improved to commercially acceptable level (Ustun et al., 2017; Luo et al., 2020). For example, the productivity of apple tree on *M. sieboldii*-based AP-resistant apomictic rootstocks was enhanced by crossing and backcrossing *M. sieboldii*, and its apomictic hybrids with M9 and trees with high resistance and good properties were developed (Marcone and Rao, 2019).

Clustered regularly interspaced short palindromic repeats—based new technology for gene editing has a great potential and can be used for progress and better understanding and exploitation of phytoplasma resistance (Belhaj et al., 2013).

8. Recovery

In the field, "recovery" is an interesting unclear phenomenon of phytoplasma-infected plants which is defined as the disappearance of phytoplasma symptoms. This phenomenon was reported in BN-infected grapevines (Osler et al., 1993), AP-diseased apple trees (Osler et al., 2003; Carraro et al., 2004) and ESFY- infected apricot trees (Marcone et al., 2010). In recovered apricot trees, permanent disappearance of symptoms in the above-ground parts was observed, but the mechanism of recovery or reduction of strain virulence is still poorly understood (Marcone et al., 2010). Musetti et al. (2005) investigated the effect of reactive oxygen species on the recovery phenomenon in ESFY-affected apricot trees and concluded that the overproduction of hydrogen peroxide in recovered plants may be responsible for the reduction of strain virulence and the disease severity leading to the recovery of ESFY-infected apricot trees.

9. Induced resistance

Another strategy for the control of phytoplasma disease is spraying the plant canopy with resistance inducers. Induced resistance is one type of defense system in plants which is activated by biotic and abiotic factors and allows plants resist against attack from a pathogen or parasite (Devendra et al., 2008). Resistance inducers are usually compounds able to start a resistance reaction of the plant. These molecules, called also elicitors, can be of abiotic or biotic nature and challenge the plant, leading to a reaction, often linked to the production of antimicrobial compounds and/or the elicitation of plant defense mechanisms. The most common elicitors used to control phytoplasma diseases are: benzothiadiazole (BTH), Phosetyl-Al, prohexadione calcium, indole-3-butyric acid, indole-3-acetic acid (IAA), chitosan, salicylic acid (SA), and mixture of glutathione and oligosaccharides (GOs) (Romanazzi et al., 2009). One of the commercial forms of phosetyl-Al is Previcur EnergyTm (31% fosetyl-Al plus 53% propamocarb), which is a new systemic synthesized molecule with protective action. It is easy to use (drip application, drench, and spray) and shows root growth stimulant effect. In addition, it strengthens the protective. Most studies were carried out on experimental hosts infected with a phytoplasma, for example, *Arabidopsis thaliana* challenged with X-disease phytoplasma (Bressan and Purcell, 2005), *C. roseus* inoculated with chrysanthemum yellows (CY) or EY phytoplasmas (Prati et al., 2004; Chiesa et al., 2007), or with aster yellows (AY), EY, or "stolbur" (Leljak-Levanic et al., 2010).

SA was applied in crops as tomatoes infected potato purple top phytoplasma (Wu et al., 2012) and lime witches' broom phytoplasma (Ghayeb Zamharir et al., 2020). Trials on woody crops infected by phytoplasmas can be more difficult because it is not easy to find a high number of plants that after application allow getting significant differences in induction of resistance. First evidence of resistance inducer application was obtained in Sardinia on cvs Chardonnay and Vermentino infected with BN that were sprayed with two commercial formulations based on active ingredients Phosetyl-Al and GOs (Garau et al., 2008). Application of humic and fulvic acids, and algae extracts gave promising results, but they were not able to contain BN infections on two different cultivars (Mazio et al., 2008). When five commercial resistance inducers, based on chitosan, Phosetyl-Al, BTH, and two GOs formulations were applied weekly in a vineyard cv Chardonnay naturally infected by BN, BTH, and the two GOs provided a significant reduction of number of symptomatic plants (Romanazzi et al., 2009).

10. Case study 1: witches' broom diseases of lime

WBDL is one of the most destructive diseases of Mexican lime in Iran (Bové et al., 2000), India (Mardi et al., 2011), Sultanate of Oman (Garnier et al., 1991), and the United Arab Emirates (Bové, 1986). WBDL is associated with the presence of '*Ca.* P. aurantifolia' (Zreik et al., 1995). The phytoplasma is transmitted by a leafhopper vector *Hishimonus phycitis* in the field (Salehi et al., 2007). The main management strategies for WBDL control in commercial orchards are based on the control of vectors using systemic insecticides, reducing the inoculum by removing infected trees and transplanting WBDL phytoplasma-free (Ghayeb Zamharir and Salekdeh, 2014). These management techniques have displayed a little role in preventing WBDL severe outbreak. Recently, different techniques including the application of resistance inducer compounds played an important role in plant diseases control (Romanazzi et al., 2009; Schmidt et al., 2015; Romanazzi, 2013). In greenhouse condition, the WBDL symptoms severity was decreased using ascorbic acid (AA) (300 μM); AA (600 μM), AA (300 μM) plus $MgCl_2$ (200 μM), SA (150 μM); SA (300 μM), SA (300 μM) plus $MgCl_2$ (200 μM), Aliette (2/000), nano silver (10 μM) and STARNER (3 g/L), 31% Phosetyl-Al plus 53% propamocarb (Privicore EnergyTM) (17.5 μM), Hymexazol 70% (30 μM) and NORDOX (30 μM) at the end of the experiment, compared with the negative control. In the field condition, the 31% Phosetyl-Al plus 53% propamocarb (Privicore EnergyTM) (17.5 μM), Hymexazol 70% (30 μM), and NORDOX (30 μM)—treated trees was recovered at the end of the experiment, compared with the negative control (Ghayeb Zamharir et al., 2020).

The fruit yield data were collected for the two harvest seasons in Minab, Hormozgan province of Iran, showed that in both harvest seasons, the fruit yield in treated plants with 31% Phosetyl-Al plus 53% propamocarb (Privicore Energy) (17.5 μM) effects positively fruit yield. The average weight of fruit per tree was 2300.1 fruit/tree, while that of the negative control was only 1800.5 (fruit/tree). The same result repeated in field yield trail with slightly higher yield in treated experimental units (plant under treatments). Thus, plants under the treatment of 31% Phosetyl-Al plus 53% propamocarb (Privicore Energy) showed consistently yield increase across 2 years of trials (Ghayeb Zamharir et al., 2020).

11. Case study 2: grapevine yellows

GY is widespread in all viticulture that can have important economic impacts (Ghayeb Zamharir et al., 2017; Salehi et al., 2014; Mirchenari et al., 2015; Karimi et al., 2009). The presence of phytoplasmas associated with yellows disease in grapevine in Iran was reported (Ghayeb Zamharir et al., 2017; Salehi et al., 2014; Mirchenari et al., 2015; Karimi et al., 2009). This disease can decrease yields of different grapevine cultivars in country. In an experiment, two new components 31% Phosetyl-Al plus 53% propamocarb (Privicore Energy) and Hymexazol 70% have been applied with one application per year. The application of Privicore Energy and Hymexazol 70% causes grapevine plants recover from phytoplasmas symptoms for next year without need to spray treatments in new year. Privicore Energy has a complex mechanism of action. Fosetyl aluminum has a complete ascending and descending distribution and systemically in plants, protects especially newly grown parts (Chiesa et al., 2007). It shows a particular and unique mode of action, stimulating the defense system of plants. Propamocarb hydrochloride also has a systemic activity. In case of foliar application, propamocarb hydrochloride is taken by the leaves and is distributed acropetally, being taken into the plant in less than one hour after application; it has a multisite preventive action. Plants treated with Privicore Energy are exhibiting more vigorous growth due to the special development of root system that decreases in phytoplasma infection. On the other hand, Iran is a dry country, and if phytoplasma-infected plant is infected by grape wood fungi, it would control both diseases simultaneously. Hymexazol 70% stimulates the defense system of plants development of root system that decreases in phytoplasma infections (Ghayeb Zamharir and Taheri, 2019).

The application of Aliette and Chito plant also showed the tendency to increase recovery rates as compared to the control in BN-affected grapevines on cvs Lambrusco Salamino and Ancellotta in Italy (Romanazzi et al., 2009). Several studies have shown that Bion can successfully induce resistance to various pathogens, by increasing the content of pathogenesis-related (PR) proteins in many plant species (Vallad and Goodman, 2004). Bion has also been shown to provide some protection and a delay in symptom appearance in chrysanthemums infected by 'Ca. P. asteris' (D'Amelio et al., 2007), a protectant activity against X-disease phytoplasma, and reduced leafhopper survival when applied to *A. thaliana* (Bressan and Purcell, 2005). Both Kendal and Olivis are based on a mixture of glutathione and oligosaccharides. In particular, functions proposed for glutathione in higher plants include storage and transport of reduced sulfur, a protein reductant, a protective role in cellular metabolism, and removal of free radicals. In *Phaseolus vulgaris*, treatment with exogenous glutathione causes a massive and selective induction of transcription of defense genes encoding enzymes of phytoalexin and lignin biosynthesis, as well as the stimulation of genes encoding cell wall components (Wingate et al., 1988). Chito Plant is based on chitosan, a natural biopolymer with antimicrobial properties and an elicitor of plant resistance against fungi, bacteria, and viruses (Kulikov et al., 2006; Romanazzi et al., 2009). In grapevines, chitosan was shown to induce the production of phenylalanine ammonia lyase and to prime catechin and transresveratrol content (Romanazzi et al., 2009). In our trial, Chito Plant showed the tendency to reduce the number of symptomatic plants compared to the control. Similar results were reported by D'Amelio et al. (2007), who observed a delay in symptom development on chrysanthemums infected by '*Ca.* P. asteris' after chitosan application. Treatments with resistance inducers carried out on phytoplasma-infected *C. roseus* showed different responses of the plants sprayed with Phosetyl-Al, contained in Aliette, and chitosan (Chiesa et al., 2007), the first being more effective. Although the extant pathosystem is different, in our trials, chitosan and Phosetyl-Al were not significantly different from each other in both years monitored. Recovery appears to be induced by different factors (Kunze, 1976), among which there are differential varietal propensities (Bellomo et al., 2007; Garau et al., 2007; Romanazzi et al., 2007), rootstock combinations (Romanazzi and Murolo, 2008), and climatic conditions.

12. Biotic resistance inducers

The role of arbuscular mycorrhizal (AM) fungi in phytoplasma infection has been investigated in several pathogenic systems. In "stolbur" infection of tomato, agglutinations and degeneration of phytoplasma cells, coupled to reduced symptom expression, were seen in plants treated with AM fungi. Inoculation with *Glomus intraradices* increased tolerance to PD in infected pear trees. However, the evidence that *Glomus mosseae* BEG 12 inoculation does not decrease periwinkle tolerance to mild and severe '*Ca.* P. asteris' strains has indicated that the effects of AM fungi on phytoplasma infection are complex and probably dependent on a combination of host plant, AM fungus, and phytoplasma strain (Romanazzi, 2013).

Among beneficial microorganisms interacting with plants, there are also fungal endophytes that can result extremely diverse in the different plants, colonizing all or part of the host. In spite of the fact that their biology is still incomplete, it is recognized that fungal endophytes are important source of secondary metabolites and compound of biotechnological value as antibiotics or antitumor drugs (Gimenez et al., 2007). Fungal endophytes establish mutualistic relationships with plants and induce physiological modifications in their hosts making them more resistant against biotic or environmental stresses. Fungal endophyte strains have been identified from grapevines and apple plants grown in areas where recovery phenomenon was recurrent. For fungal community description, a combination of culture-dependent and culture-independent methods has been set up (Grisan et al., 2011). The combined use of the two methods allowed to discover 56 different fungal endophytes grouped in operational taxonomic units (OTUs) on the bases of PCR/RFLP analyses of internal transcribed spacer (ITS) region. 27% of OTUs were obtained by culture-dependent method, 48% by culture-independent method, and 25% by both methods. Furthermore, the collected data revealed that fungal endophytes belonging to genera *Alternaria* sp., *Phoma* sp., *Epicoccum* sp., *Aureobasidium* sp., *Cladosporium* sp., *Pestalotiopsis* sp., and *Pestalotia* sp. constituted, respectively, about 89% of total isolates obtained by the culture-dependent method, and 79% of total clones obtained from the culture-independent method. Strains of *Epicoccum nigrum* and *Aureobasidium pullulans* were chosen for further research activities because they are extensively reported as biocontrol agents or resistance inducers. Using the model plant *C. roseus* infected with '*Ca.* P. mali', it was observed that reduction in symptom severity and lower phytoplasma titer in host tissues occurred when the plants were previously inoculated with endophytic strain of *E. nigrum* (Musetti et al., 2011).

13. Biocontrol

Endophytic bacteria are plant-associated microorganisms that live inside plant without inducing symptoms of disease. Endophytic bacteria seem to positively influence plant host growth through mechanisms similar to those described for plant growth−promoting rhizobacteria. Moreover, they can promote plant growth by reducing the deleterious effects of plant pathogens through direct or indirect mechanisms (biocontrol) (Lugtemberg and Kamilova, 2009). In detail, they are able to suppress pathogens through the competition for an ecological niche, the production of allelochemicals (*e.g.*, antibiotic, siderophores, and lytic enzymes), or through the induction of a systemic plant defense response. The use of plant-associated bacteria in plant protection is related to the understanding of bacteria−host interactions and to the ability to formulate and spread the bacteria under field conditions (Hallmann et al., 1998). A basic point for the success of sustainable management of plant diseases based on biocontrol agents is the study of endophytic bacterial community living inside plants.

AP, associated with the presence of '*Ca.* P. mali', is one of the most important phytoplasma diseases in apple (Frisinghelli et al., 2000; Tedeschi et al., 2002). Recently, endophytic bacterial community associated with healthy and phytoplasma-infected plants has been screened (Bulgari et al., 2009; Martini et al., 2009) to find putative biocontrol agents. In other study, some isolates with belonging to the genus *Pseudomonas* and *Lysinibacillus* showed a minimum of five beneficial traits related to mineral nutrition (phosphate solubilization, siderophores, and nitrogen fixation), development (indolacetic acid synthesis), stress relief (catalase activity), and disease control (siderophores). These strains should be tested for in vivo suppression of phytoplasma-related diseases. The major part of isolates showed the ability to solubilize phosphate and detoxify hydrogen peroxide. Interestingly, some strain, for example, *Lysinibacillus fusiformis* produced high amount of IAA (Bulgari et al., 2014). Interestingly, bacterial strains, here isolated, belong to genera widely studied for developing biocontrol strategies to contain plant pathogens (Trivedi et al., 2010).

Endophytic bacteria promote plant growth and protect them against pathogen infections. Endophytes−plant and endophytes−pathogens interactions are poorly explored. In particular, the ability of endophytes to control pathogens that are not managed directly is still new. Recently, *Pseudomonas putida* S1Pf1Rif was tested, alone or in combination with the mycorrhizal fungus G. mosseae BEG12, against CY phytoplasma infection of chrysanthemum.

14. Conclusions

Phytoplasma disease management is complicated and can be achieved only via multiple approach combining different control methods. Several tools for exclusion of the pathogens from production sites, eliminating or reducing the inoculum sources and prevention of disease and insect vectors spread have to be integrated. Detailed epidemiological data are always necessary for development of the best management strategy. The deep knowledge on pathogen strains and diversities, host resistance and wild hosts, insect species and their behavior, environmental conditions and their effect, and existing control options are essential to take appropriate control measures. Moreover, host−phytoplasma interactions are important to be clearly explained. Over last decades, attempts to understand the phytoplasma diseases have resulted in accumulation of a lot of knowledge on these pathogens. Unfortunately, there are still gaps in many aspects regarding pathogens, host−pathogen interactions, host resistance, insect vectors, etc. For better understanding, host—pathogen interactions and host resistance mechanisms have to be studied at genomic, transcriptomic, proteomic, and metabolomic level, and the function of membrane or secreted proteins and effectors in the phytoplasma genome and also phytoplasma defense-related proteins in the plant have to be clarified. The research should concentrate on the identification of more resistant/tolerant genotypes proving the most environmentally safe tool for the protection of crops from these pathogens. Insect control strategies based on semiochemicals, physical prevention, or another innovative approaches have to be improved, and vector-monitoring systems to be developed for more efficient control. Application of plant resistance inducers, elicitors, or microorganisms triggering the defense mechanisms of the plants can be another promising alternative for phytoplasma containments. However, the studies indicating the potential of inducers for phytoplasma disease protection are still limited; more research is needed to optimize and expand their implementation. In the future, exploitation of new gene-modulating technologies is expected to pioneer better understanding of phytoplasma diseases and development of new managing methodologies.

References

Akhtar, K.P., Sarwar, G., Dickinson, M., Ahmad, M., Haq, M.A., Hameed, S., Iqbal, M.J., 2009. Sesame phyllody disease: symptomatology, etiology and transmission in Pakistan. Turk. J. Agric. For. 33, 477—486.

Akhtar, K.P., Sarwar, G., Sarwar, N., Elahi, M.T., 2013. Field evaluation of sesame germplasm against sesame phyllody disease. Pakistan J. Bot. 45, 1085—1090.

Al-Ghaithi, A.G., Al-Sadi, A.M., Al-Hammadi, M.S., Al-Shariqi, R.M., Al-Yahyai, R.A., Al-Mahmooli, I.H., Carvalho, C.M., Elliot, S.L., Hogenhout, S., 2017. Expression of phytoplasma-induced witches' broom disease symptoms in acid lime (*Citrus aurantifolia*) trees is affected by climatic conditions. Plant Pathol. 66, 1380—1388.

Al-Sadi, A.M., Al-Moqbali, H.S., Al-Yahyai, R.A., Al-Said, F.A., 2012. AFLP data suggest a potential role for the low genetic diversity of acid lime (*Citrus aurantifolia* Swingle) in Oman in the outbreak of witches' broom disease of lime. Euphytica 188, 285—297.

Al-Subhi, A.M., Al-Sadi, A.M., Al-Yahyai, R.A., Chen, Y., Mathers, T., Orlovskis, Z., Moro, G., Mugford, S., Al-Hashmi, K.S., Hogenhout, S.A., 2021. Witches' broom disease of lime contributes to phytoplasma epidemics and attracts insect vectors. Plant Dis. 105 (9), 2637—2648.

Ammar, E., Hogenhout, S., 2006. Mollicutes associated with arthropods and plants. Insect Symb. 2, 97—118.

Anonymous, 2020. Yılı Bitki Sağlığı Uygulamaları. TC Tarım Ve Orman Bakanlığı Gıda Ve Kontrol Genel Müdürlüğü, Eğitim Ve Yayın Dairesi Başkanlığı, Ankara, 297 Sayfa.

Atkinson, N.J., Urvin, P.E., 2012. The interaction of plant biotic and abiotic stresses: from genes to the field. J. Exp. Bot. 63, 3523—3543.

Babaei, G., Esmaeilzadeh-Hosseini, S.A., Zandian, M., Nikbakht, V., 2020. Identification of phytoplasma strains associated with witches' broom and yellowing in *Ziziphus jujube* nurseries in Iran. Phytopath. Medit. 59, 55—61.

Baudouin, L., Philippe, R., Quaicoe, R., Dery, S., Dollet, M., 2009. General overview of genetic research and experimentation on coconut varieties tolerant/resistant to lethal yellowing. Oleagineux 16, 127—131.

Beech, D.F., 1981. Phyllody—its impact on yield and possible control measures. In: Ashri, A., Poetiary, P. (Eds.), Sesame: Status and Improvement. FAO, Rome, Italy, pp. 73—80. FAO Plant Production and Protection Paper No. 29.

Been, B.O., 1995. Integrated pest management for the control of lethal yellowing: quarantine, cultural practices and optimal use of hybrids. In: Oropeza, C., Howard, F.W., Ashburner, G.R. (Eds.), Lethal Yellowing: Research and Practical Aspects, vol. 5. Kluwer Academic Publishers, Dordrecht, pp. 101–109.

Belhaj, K., Chaparro-Garcia, A., Kamoun, S., Nekrasov, V., 2013. Plant genome editing made easy: targeted mutagenesis in model and crop plants using the CRISPR/Cas system. Plant Methods 9, 1–10.

Bellardi, M.G., Bertaccini, A., Madhupriya Rao, G.P., 2018. Phytoplasma Diseases in Ornamental Crops. Phytoplasmas: Plant Pathogenic Bacteria—I: Characterisation and Epidemiology of Phytoplasma—Associated Diseases. Springer Singapore, pp. 191–233.

Bellomo, C., Carraro, L., Ermacora, P., Pavan, F., Osler, R., Frausin, C., Governatori, G., 2007. Recovery phenomena in grapevine affected by grapevine yellows in Friuli Venezia Giulia. Bull. Insectol. 60, 235–236.

Bertaccini, A., Duduk, B., Paltrinieri, S., Contaldo, N., 2014. Phytoplasmas and phytoplasma diseases: a severe threat to agriculture. Am. J. Plant Sci. 5, 1763–1788.

Bertaccini, A., 2019. Plant pathogens, minor (phytoplasmas). In: Encyclopedia of Microbiology, fourth ed., pp. 627–638.

Bertaccini, A., Duduk, B., 2009. Phytoplasma and phytoplasma diseases: a review of recent research. Phytopath. Medit. 48, 355–378.

Bigliardi, E., Sacchi, L., Genchi, M., Alma, A., Pajoro, M., Daffonchio, D., Marzorati, M., Avanzati, A.M., 2006. Ultrastructure of a novel *Cardinium* sp. symbiont in *Scaphoideus titanus* (Hemiptera: Cicadellidae). Tissue Cell 38 (4), 257–261.

Bogoutdinov, D.Z., Kastalyeva, T.B., Girsova, N.V., Samsonova, L.N., 2019. Phytoplasma diseases: a review of 50year history and current advances. Sel'skokhozyaĭstvennaya Biol. 54, 3–18.

Borgi, A.A., Palma, M.S.A., 2014. Tetracycline: production, waste treatment and environmental impact assessment. Brazilian J. Pharmaceut Sci 50 (n. 1). https://doi.org/10.1590/S1984-82502011000100003 jan./mar.

Bové, J-M., 1986. Witches' broom of lime. FAO Plant Prot. Bull. 34, 217–218.

Bové, J-M., Garnier, M., 2002. Phloem-and xylem-restricted plant pathogenic bacteria. Plant Sci. 163, 1083–1098.

Bové, J-M., Danet, J-L., Bananej, K., Hassanzadeh, N., Taghizadeh, M., Salehi, M., Garnier, M., 2000. Witches' broom disease of lime (WBDL) in Iran. Fourteenth IOCV Conf. 207, 212.

Bressan, A., Purcell, A.H., 2005. Effect of benzothiadiazole on transmission of X-disease phytoplasma by the vector *Colladonus montanus* to *Arabidopsis thaliana*, a new experimental host plant. Plant Dis. 89, 1121–1124.

Bulgari, D., Casati, P., Brusetti, L., Quaglino, F., Brasca, M., Daffonchio, D., Bianco, P.A., 2009. Endophytic bacterial diversity in grapevine (*Vitis vinifera* L.) leaves described by 16S rRNA gene sequence analysis and length heterogeneity-PCR. J. Microbiol. 47, 393–401.

Bulgari, D., Casati, P., Quaglino, F., Bianco, P.A., 2014. Isolation of potential biocontrol agents of '*Candidatus* Phytoplasma mali'. In: Phytoplasmas and Phytoplasma Disease Management: How to Reduce Their Economic Impact. IPWG—International Phytoplasmologist Working Group.

CABI, 2021a. Invasive Species Compendium. Grapevine Flavescence dorée Phytoplasma. https://www.cabi.org/isc/datasheet/26184.

CABI, 2021b. Invasive Species Compendium. Phytoplasma mali (Apple Proliferation). https://www.cabi.org/isc/datasheet/6502.

Çağlar, B.K., Satar, S., Bertaccini, A., Elbeaino, T., 2019. Detection and seed transmission of Bermudagrass phytoplasma in maize in Turkey. J. Phytopathol. 167, 248–255.

Cardena, R., Ashburner, G.R., Oropeza, C., 2003. Identification of RAPDs associated with resistance to lethal yellow-ing of the coconut (*Cocos nucifera* L.) palm. Sci. Hortic. 257–263.

References

Carraro, L., Ermacora, P., Loi, N., Osler, R., 2004. The recovery phenomenon in apple proliferation infected apple trees. J. Plant Pathol. 86, 141–146.

Casanova, R., Llacer, G., Sanchez-Capuchino, J.A., 1980. Remission of symptoms of apple proliferation, after injection of concentrated tetracycline solutions. Acta Phytopath. Acad. Sci. Hung. 15, 273–277.

Chakrabarti, A.K., Choudhury, B., 1975. Breeding brinjal resistant to little leaf disease. Proc. Indian Nat. Sci. Academy (Section B) 41, 379–385.

Chalak, L., Choueiri, E., 2015. Contribution to the production scheme of local certified propagating material of almond: *in vitro* sanitation and micropropagation. Acta Hortic. 1083, 163–168.

Chalak, L., Elbitar, A., Rizk, R., Choueiri, E., Salar, P., Bové, J.M., 2005. Attempts to eliminate 'Candidatus Phytoplasma phoenicium' from infected Lebanese almond varieties by tissue culture techniques combined or not with thermotherapy. Eur. J. Plant Pathol. 112 (1), 85–89. http://springerlink.metapress.com/link.asp?id=100265.

Chiesa, S., Prati, S., Assante, G., Maffi, D., Bianco, P.A., 2007. Activity of synthetic and natural compounds for phytoplasma control. Bull. Insectol. 60, 313–314.

Da Silva, F.N., Queiroz, R.B., de Souza, A.N., Al-Sadi, A.M., de Siqueira, D.L., Elliot, S.L., Carvalho, C.M., 2014. First report of a 16SrII-C phytoplasma associated with asymptomatic acid lime (*Citrus aurantifolia*) in Brazil. Plant Dis. 98, 1577.

Davies, D.L., Guise, C.M., Clark, M.F., Adams, A.N., 1992. Parry's disease of pears is similar to pear decline and is associated with mycoplasma-like organisms transmitted by *Cacopsylla pyricola*. Plant Pathol. 41, 195–203.

Davis, R.E., Zhao, Y., Dally, E.L., Lee, I-M., Jomantiene, R., Douglas, S.M., 2013. 'Candidatus Phytoplasma pruni', a novel taxon associated with X-disease of stone fruits, *Prunus* spp.: multilocus characterization based on 16S rRNA, secY, and ribosomal protein genes. Int. J. Syst. Evol. Microbiol. 63, 766–776.

Devendra, K.C., Prakash, A., Johri, B.N., 2008. Induced systemic resistance (ISR) in plants: mechanism of action. Indian J. Microbiol. 47 (4), 289–297.

D'Amelio, R., Massa, N., Gamalero, E., 2007. Preliminary results on the evaluation of the effects of elicitors of plant resistance on chrysanthemum yellows phytoplasma infection. Bull. Insectol. 60, 317–318.

EFSA Panel on Plant Health (PLH), 2020. Pest categorisation of the non-EU phytoplasmas of *Cydonia* Mill., *Fragaria* L., *Malus* Mill., *Prunus* L., *Pyrus* L., *Ribes* L., *Rubus* L. and *Vitis* L. EFSA J. 18 (1), 5929.

EPA, 1993. R.E.D. FACTS: Hydroxytetracycline Monohydrochloride and Oxytetracycline Calcium. EPA831 738-F-93-001. Office of Prevention, Pesticides and Toxic Substances. Retrieved March 8, 2011 from 832. http://www.epa.gov/oppsrrd1/REDs/factsheets/0655fact.pdf.

Eveillard, S., Jollard, C., Labroussaa, F., Khalil, D., Perrin, M., Desqué, D., Salar, P., Razan, F., Hevin, C., Bordenave, L., Foissac, X., Masson, J., Malembic-Maher, S., 2016. Contrasting susceptibilities to "flavescence dorée" in *Vitis vinifera*, rootstocks and wild *Vitis* species. Front. Plant Sci. 7.

Eziashi, E., Omamor, I., 2010. Lethal yellowing disease of the coconut palms (*Cocos nucifera* L.): an overview of the crises. Afr. J. Biotechnol. 9, 9122–9127.

Francis, I., Holsters, M., Vereerke, D., 2010. The Gram-positive side of plant-microbe intercations. Environ. Microbiol. 10, 1–12.

Frisinghelli, C., Delaiti, L., Grando, M.S., Forti, D., Vindimian, M.E., 2000. *Cacopsylla costalis* (Flor1861), as a vector of apple proliferation in Trentino. J. Phytopathol. 148, 425–431.

Galetto, L., Bosco, D., Balestrini, R., Genre, A., Fletcher, J., Marzachì, C., 2011. The major antigenic membrane protein of 'Candidatus Phytoplasma asteris' selectively interacts with ATP synthase and actin of leafhopper vectors. PLoS One 6, e22571.

Garau, R., Sechi, S., Prota, V.A., Moro, G., 2007. Productive parameters in Chardonnay and Vermentino grapevines infected with "bois noir" and recovered in Sardinia. Bull. Insectol. 60, 233–234.

Garau, R., Prota, V.A., Sechi, S., Moro, G., 2008. Biostimulants distribution to plants affected by "bois noir": results regarding recovery. Petria 18, 366–368.

Garnier, M., Zreik, L., Bové, J-M., 1991. Witches' broom, a lethal mycoplasmal disease of lime trees in the sultanate of Oman and the United Arab Emirates. Plant Dis. 75 (6), 546–551.

Ghayeb Zamharir, M., Mardi, M., NHh, Z., Zamanizadeh, H.R., Alizadeh, A., Salekdeh, G.H., 2011. Identification of genes differentially expressed during interaction of acid lime infected with 'Candidatus Phytoplasma aurantifolia'. BMC Microbiol. 11, 1.

Ghayeb Zamharir, M., Hosseini Salekdeh, G., 2014. Molecular response of Mexican lime tree to 'Candidatus Phytoplasma aurantifolia' infection. Afr. J. Microbiol. Res. 7 (51), 5766–5770.

Ghayeb Zamharir, M., Taheri, A., 2019. Effect of new resistance inducers on grapevine phytoplasma disease. Arch. Phytopathol. Plant Protect. 52, 1207–1214.

Ghayeb Zamharir, M., Paltrinieri, S., Hajivand, S., Taheri, M., Bertaccini, A., 2017. Molecular identification of diverse 'Candidatus phytoplasma'species associated with grapevine decline inIran. J. Phytopathol. 165 (7–8), 407–413.

Ghayeb Zamharir, M., Askari Seyahooei, M., Pirseyedi, M., 2020. Witches' broom disease of lime suppressed by some resistance inducers. Indian Phytopathol. 73, 517–525.

Gimenez, C., Cabrera, R., Reina, M., Gonzalez-Coloma, A., 2007. Fungal endophytes and their role in plant protection. Curr. Org. Chem. 11, 707–720.

Giunchedi, L., Poggi Pollini, C., Bissani, R., Vicchi, V., Babini, A.R., 1994. Studi sul deperimento dei peri nell'Italia Centro-settentrionale. Rivista di Frutticoltura 12, 79–82.

Giunchedi, L., Poggi Pollini, C., Bissani, R., Babini, A.R., Vicchi, V., 1995. Etiology of a pear decline disease in Italy and susceptibility of pear variety and rootstock to phytoplasma-associated pear decline. Acta Hortic. 386, 489–495.

Griffiths, H.M., Sinclair, W.A., Smart, C.D., Davis, R.E., 1999. The phytoplasma associated with ash yellows and lilac witches' broom: 'Candidatus Phytoplasma fraxini'. Int. J. Syst. Evol. Microbiol. 49, 1605–1614.

Grisan, S., Martini, M., Musetti, R., Osler, R., 2011. Development of a molecular approach to describe the diversity of fungal endophytes in either phytoplasma-infected, recovered or healthy grapevines. Bull. Insectol. 64 (Supplement), S207–S208.

Gross, J., 2017. New strategies for phytoplasma vector control by semiochemicals. Pheromones and other semiochemicals in integrated production. IOBC-WPRS Bull. 126, 12–17.

Gurr, G.M., Johnson, A.C., Ash, G.J., Wilson, B.A.L., Ero, M.M., Pilotti, C.A., Dewhurst, C.F., You, M.S., 2016. Coconut lethal yellowing diseases: a phytoplasma threat to palms of global economic and social significance. Front. Plant Sci. 7, 1521.

Hallmann, J., Quadt-Hallmann, A., Rodriguez-Kabana, R., Kloepper, J.W., 1998. Interactions between *Meloidogyne incognita* and endophytic bacteria in cotton and cucumber. Soil Biol. Biochem. 30, 925–937.

Hemmati, C., Nikooei, M., 2019. Austroagallia sinuata transmission of 'Candidatus Phytoplasma aurantifolia' to *Zinnia elegans*. J. Plant Pathol. 101, 1223.

Hemmati, C., Nikooei, M., Al-Sadi, A.M., 2021. Five decades of research on phytoplasma-induced witches' broom diseases. CAB Rev. 16, 1–16.

Hibben, C.R., Franzen, L.M., 1989. Susceptibility of lilacs to mycoplasmalike organisms. J. Environ. Hortic. 7, 163–167.

Hibben, C.R., Lewis, C.A., Castello, J.D., 1986. Mycoplasmalike organisms, cause of lilac witches' broom. Plant Dis. 70, 342–345.

Hill, G., Sinclair, W., 2000. Taxa of leafhoppers carrying phytoplasmas at sites of ash yellows occurrence in New York State. Plant Dis. 84, 134–138.

Hill, R.J., Wilson, B.A., Rookes, J.E., Cahill, D.M., 2009. Use of high resolution digital multi-spectral imagery to assess the distribution of disease caused by *Phytophthora cinnamomi* on heathland at Anglesea, Victoria. Australas. Plant Pathol. 38, 110–119.

Hosseini, S.A.E., Salehi, M., Khodakaramian, G., Yazdi, H.B., Salehi, M., Nodooshan, A.J., Jadidi, O., Bertaccini, A., 2015. Status of sesame phyllody and its control methods in Yazd, Iran. Phytopath. Mollic. 5 (1-Supplement), S119−S120.

Huang, D., Walla, J.A., Dai, W., 2014. Quantitative phenotyping of X-disease resistance in chokecherry using real-time PCR. J. Microbiol. Methods 98, 1−7.

Ikten, C., Ustun, R., Catal, M., Yol, E., 2016. Multiplex real-time qPCR assay for simultaneous and sensitive detection of phytoplasmas in sesame plants and insect vectors. PLoS One 11, e0155891.

IPPC, 1997. ISPM 6. Guidelines for Surveillance. http://faperta.ugm.ac.id/perlintan2005/puta_files/attach/ISPM%206%20Guidelines%20for%20Surveillance.pdf.

IPPC, 2017. ISPM 4. Requirements for the Establishment of Pest Free Areas. https://assets.ippc.int/static/media/files/publication/en/2017/05/ISPM_04_1995_En_2017-05-23_PostCPM12_InkAm.pdf.

Ishii, Y., Kakizawa, S., Hoshi, A., Maejima, K., Kagiwada, S., Yamaji, Y., Oshima, K., Namba, S., 2009. In the non-insect-transmissible line of onion yellows phytoplasma (OY-NIM), the plasmid-encoded transmembrane protein ORF3 lacks the major promoter region. Microbiology 155, 2058−2067.

Jakovljević, M., Jović, J., Krstić, O., Mitrović, M., Marinković, S., Toševski, I., Cvrković, T., 2020. Diversity of phytoplasmas identified in the polyphagous leafhopper *Euscelis incisus* (Cicadellidae, Deltocephalinae) in Serbia: pathogen inventory, epidemiological significance and vectoring potential. Eur. J. Plant Pathol. 156, 201−221.

Jarausch, W., Angelini, E., Eveillard, S., Malembic-Maher, S., 2013. Management of fruit tree and grapevine phytoplasma diseases through genetic resistance. In: New Perspectives in Phytoplasma Disease Management—Cost Action FAO807: 56−63.

Karimi, M., Contaldo, N., Mahmoudi, B., Duduk, B., Bertaccini, A., 2009. Identification of "stolbur"-related phytoplasmas in grapevine showing decline symptoms in Iran. Le Progrès Agricole et Viticole HS 31, 208−209.

Kison, H., Seemüller, E., 2001. Differences in strain virulence of the European stone fruit yellows phytoplasma and susceptibility of stone fruit trees on various rootstocks to this pathogen. J. Phytopathol. 149, 533−541.

Krishnareddy, M., 2013. Impact of climate change on insect vectors and vector-borne plant viruses and phytoplasma. In: Singh, H.C.P., Rao, N.K.S., Shivashankar, K.S. (Eds.), Climate-Resilient Horticulture: Adaptation and Mitigation Strategies. Springer), New Delhi, pp. 255−277.

Kulikov, S.N., Chikov, S.N., Il'ina, A.V., Lopatin, S.A., Varlamov, V.P., 2006. Effect of the molecular weight of chitosan on its antiviral activity in plants. Appl. Biochem. Microbiol. 42, 200−203.

Kumari, S., Nagendran, K., Rai, B.A., Singh, B., Rao, G.P., Bertaccini, A., 2019. Global status of phytoplasma diseases in vegetable crops. Front. Microbiol. 10, 1349.

Kunze, L., 1976. The effect of different strains of apple proliferation on the growth and crop of infected trees. Mitteilungen Biologische Bundesamt Land-Forstwirtsch BerlinDahlem 170, 107−115.

Laimer, M., 2003. Detection and elimination of viruses and phytoplasmas from pome and stone fruit trees. Hortic. Rev. 187−236.

Laimer, M., Bertaccini, A., 2008. European stone fruit yellows. In: Harrison, N.A., Rao, G.P., Marcone, C. (Eds.), Characterization, Diagnosis and Management of Phytoplasmas. Studium Press, LLC, Houston, Texas, USA, pp. 73−92.

Laimer, M., Lemaire, O., Herrbach, E., Goldschmidt, V., Minafra, A., Bianco, P., Wetzel, T., 2009. Resistance to viruses, phytoplasmas and their vectors in the grapevine in Europe: a review. J. Plant Pathol. 91 (1), 7−23.

Lee, I-M., Bertaccini, A., Vibio, M., Gundersen, D.E., Davis, R.E., Mittempergher, L., Conti, M., Gennari, F., 1995. Detection and characterization of phytoplasmas associated with disease in *Ulmus* and *Rubus* in northern and Central Italy. Phytopath. Medit. 34, 174−183.

Leljak-Levanič, D., Jesič, M., Cesar, V., Ludwig-Müller, J., Lepeduš, H., Mladinič, M., Katič, M., Curkovič Perica, M., 2010. Biochemical and epigenetic changes in phytoplasmarecovered periwinkle after indole-3-butyric acid treatment. J. Appl. Microbiol. 109, 2069–2078.

Lenz, R.R., Dai, W., 2017. Mapping X-disease phytoplasma resistance in *Prunus virginiana*. Front. Plant Sci. 8.

Lepka, P.M., Moll, E., Seemüller, E., 1999. Effect of phytoplasmal infection on concentration and translocation of carbohydrates and amino acids in periwinkle and tobacco. Physiol. Mol. Plant Pathol. 55 (1), 59–68. July 1999.

Lessio, F., Portaluri, A., Paparella, F., Alma, A., 2015. A mathematical model of "flavescence dorée" epidemiology. Ecol. Model. 312, 41–43.

Linck, H., Reineke, A., 2019. Preliminary survey on putative insect vectors for rubus stunt phytoplasmas. J. Appl. Entomol. 143, 328–332.

Liu, M.J., Zhou, J.Y., Zhao, J., 2004. Screening of Chinese jujube germplasm with high resistance to witches' broom disease. Acta Hortic. 663, 575–580.

Liu, Z., Wang, Y., Xiao, J., Zhao, J., Liu, M., 2014. Identification of genes associated with phytoplasma resistance through suppressive subtraction hybridization in Chinese jujube. Physiol. Mol. Plant Pathol. 86, 43–48.

Lova, M.M., Abou-Jawdah, Y., ChoueiriE, Beyrouthy, M., Fakhr, R., Bianco, P.A., Alma, A., SobhH, Jawhari, M., Mortada, C., et al., 2014. Almond witches' broom phytoplasma: disease monitoring and preliminary control measures in Lebanon. In: Bertaccini, A. (Ed.), Phytoplasmas and Phytoplasma Disease Management: How to Reduce Their Economic Impact. International Phytoplasmologist Working Group, Bolonga, Italy, pp. 71–74.

Lugtemberg, B., Kamilova, F., 2009. Plant-growth-promoting rhizobacteria. Annu. Rev. Microbiol. 63, 541–556.

Luo, F., Evans, K., Norelli, J.L., et al., 2020. Prospects for achieving durable disease resistance with elite fruit quality in apple breeding. Tree Genet. Genomes 16, 21.

MacLean, A.M., Sugio, A., Makarova, O.V., Findlay, K.C., Grieve, V.M., Tóth, R., Nicolaisen, M., Hogenhout, S.A., 2011. Phytoplasma effector SAP54 induces indeterminate leaf-like flower development in Arabidopsis plants. Plant Physiol. 157 (2), 831–841.

Manimekalai, R., Nair, S., Soumya, V.P., 2014. Evidence of 16SrXI group phytoplasma DNA in embryos of root wilt diseased coconut palms. Australas. Plant Pathol. 43, 93–96.

Maramorosch, K., Loebenstein, G., 2009. Plant disease resistance: natural, non-host innate or inducible. . Encyclop. Microbiol. 589, 596.

Marcone, C., Rao, G.P., 2019. Control of phytoplasma diseases through resistant plants. In: Phytoplasmas: Plant Pathogenic Bacteria—II, pp. 165–184.

Marcone, C., Jarausch, B., Jarausch, W., 2010. '*Candidatus* Phytoplasma prunorum', the causal agent of European stone fruit yellows: an overview. J. Plant Pathol. 92, 19–34.

Marcone, C., Pierro, R., Tiwari, A.K., Rao, G.P., 2022. Phytoplasma diseases of temperate fruit trees. Agrica 11 (1), 19–33.

Mardi, M., Khayam Nekouei, S.M., Karimi Farsad, L., Ehya, F., Shabani, M., Shafiee, M., Tabatabaei, M., Safarnejad, M.R., Salehi Jouzani, G., Hosseini Salekdeh, G., 2011. Witches' broom disease of Mexican lime trees: disaster to be addressed before it will be too late. Bull. Insectol. 64 (Supplement), S205–S206.

Martinez, R.T., Baudouin, L., Berger, A., Dollet, M., 2010. Characterization of the genetic diversity of the Tall coconut (*Cocos nucifera* L.) in the Dominican Republic using microsatellite (SSR) markers. Tree Genet. Genomes 6, 73–81.

Martinez, R.T., Narvaez, M., Fabre, S., Harrison, N., Oropeza, C., Dollet, M., et al., 2008. Coconut lethal yellowing on the southern coast of the Dominican Republic is associated with a new 16SrIV group phytoplasma. Plant Pathol. 57, 366.

Martini, M., Musetti, R., Grisan, S., Polizzotto, R., Boselli, S., Favan, F., Osler, R., 2009. DNA dependent detection of the grapevine fungal endophytes *Aureobasidium pullulans* and *Epicoccum nigrum*. Plant Dis. 93, 993–998.

Marzorati, M., Alma, A., Sacchi, L., Pajoro, M., Palermo, S., Brusetti, L., Raddadi, N., Balloi, A., Tedeschi, R., Clementi, E., Corona, S., Quaglino, F., Bianco, P.A., Beninati, T., Bandi, C., Daffonchio, D., 2006. A novel Baceroidetes symbiont is localized in *Scaphoideus titanus*, the insect vector of "flavescence dorée" in *Vitis vinifera*. Appl. Environ. Microbiol. 72 (2), 1467–1475.

Mazio P, Montermini A, Brignoli P (2008) Preliminary trials to test the effectiveness of biological promoters for the control of grapevine yellows symptoms. Giornate Fitopatologiche, 2: 593–600.

McCoy, R.E., 1982. Use of tetracycline antibiotics to control yellows dis-eases. Plant Dis. 66, 539–542.

McCoy, R.E., Thomas, D.L., Tsai, D.H., 1976. Lethal yellowing: a potential danger to date production. In: Date Growers Institute Report, pp. 4–8, 53.

Mirchenari, S.M., Massah, A., Zirak, L., 2015. "Bois noir": new phytoplasma disease of grapevine in Iran. J. Plant Protect. Res. 55 (1), 88–93.

Mittempergher, L., 2000. Elm yellows in Europe. In: Dunn, C.P. (Ed.), The Elms: Breeding, Conservation, and Disease Management. Kluwer Academic Publisher, Dordrecht, The Netherlands, pp. 103–119.

Mondal, S., Sundararaj, R., Yashavantha Rao, H.C., 2020. A critical appraisal on the recurrence of sandalwood spike disease and its management practices. For. Pathol. 50, e12648.

Mpunami, M., Kullaya, A., Mugini, J., 2002. The status of lethal yellowing type diseases in East Africa. In: Proceedings of the Expert Consultation on Sustainable Coconut Production through Control of Lethal Yellowing Disease, pp. 161–168. CFC Technical Paper No. 18. Kingston, Jamaica, Common Fund for Commodities.

Musetti, R., Sanità di Toppi, L., Martini, M., Ferrini, F., Losch, A., Favali, M.A., Osler, R., 2005. Hydrogen peroxide local-ization and antioxidant status in the recovery of apricotplants from European stone fruit yellows. Eur. J. Plant Pathol. 112, 53–61.

Musetti, R., Grisan, S., Polizzotto, R., Martini, M., Paduano, C., Osler, R., 2011. Interactions between 'Candidatus Phytoplasma mali' and the apple endophyte *Epicoccum nigrum* in *Catharanthus roseus* plants. J. Appl. Microbiol. 110, 746–756.

Musetti, R., Ermacora, P., Martini, M., Loi, N., Osler, R., 2013. What can we learn from the phenomenon of "recovery". Phytopath. Mollic. 3, 63–65.

Namba, S., 2019. Molecular and biological properties of phytoplasmas. Proc. Japan. Acad. B Phys. Biol. Sci. 95 (7), 401–418.

Oliveira, M.J.R.A., Roriza, M., Bertaccini, A., Vasconcelosa, M.W., Carvalhoac, S.M.P., 2019. Conventional and novel approaches for managing "flavescence dorée" in grapevine: knowledge gaps and future prospects. Plant Pathol. 68 (15), 3–17.

Orlovskis, Z., 2017. Role of Phytoplasma Effector Proteins in Plant Development and Plant-Insect Interactions. University of East Anglia.

Osler, R., Carraro, L., Loi, N., Refatti, E., 1993. Symptom expression and disease occurrence of a yellows disease of grapevine in northeastern Italy. Plant Dis. 77, 496–498.

Osler, R., Carraro, L., Ermacora, P., Ferrini, F., Loi, N., Loschi, A., Martini, M., Mutton, P.B., Refatti, E., 2003. Roguing: a controversial practice to eradicate grape yellows caused by phytoplasmas. In: Extended Abstracts 14th Meeting of ICVG, Locorotondo, 68.

Osler, R., Borselli, S., Ermacora, P., Loschi, A., Martini, M., Musetti, R., Loi, N., 2014. Acquired tolerance in apricot plants that stably recovered from European stone fruit yellows. Plant Dis. 98, 492–496.

Osler, R., Borselli, S., Ermacora, P., Ferrini, F., Loschi, A., Martini, M., Moruzzi, S., Musetti, R., Giannini, M., Serra, S., Loi, N., 2016. Transmissible tolerance to European stone fruit yellows (ESFY) in apricot: cross-protection or a plant mediated process? Phytoparasitica 44, 203—211.

Peterson, G.W., 1984. 1984 Spread and damage of western X-disease of chokecherry in eastern Nebraska plantings. Plant Dis. 68, 103—104.

Prati, S., Maffi, D., Longoni, C., Chiesa, S., Bianco, P.A., Quaroni, S., 2004. Preliminary study on the effects of two SAR inducers and prohexadione calcium on the development of phytoplasmas in vinca. J. Plant Pathol. 87, 303.

Přibylová, J., Spak, J., 2013. Dodder transmission of phytoplasmas. Methods Mol. Biol. 938, 41—46.

Puterka, G.J., Reinke, M., Luvisi, D., Ciomperlik, M.A., Bartels, D., Wendel, L., Glenn, D.M., 2003. Particle film, surround WP, effects on glassy-winged sharpshooter behaviour and its utility as a barrier to sharpshooter infestations in grapes. Plant Health Prog. https://doi.org/10.1094/PHP2003-0321-01-RS [online].

Quaglino, F., Sanna, F., Moussa, A., Faccincani, M., Passera, A., Casati, P., Bianco, P.A., Mori, N., 2019. Identification and ecology of alternative insect vectors of 'Candidatus Phytoplasma solani' to grapevine. Sci. Rep. 9.

Quaicoe, R.N., Dery, S.K., Philippe, R., Baudouin, L., Nipah, J.O., Nkansah-Poku, J., et al., 2009. Resistance screening trials on coconut varieties to Cape Saint Paul wilt disease in Ghana. Oleagineux 16, 132—136.

Queiroz, R.B., Donkersley, P., Silva, F.N., Al-Mahmmoli, I.H., Al-Sadi, A.M., Carvalho, C.M., Elliot, S.L., 2016. Invasive Mutualisms Between a Plant Pathogen and Insect Vectors in the Middle East and Brazil, vol. 3. Royal Society Open Science, p. 160557.

Rao, G.P., Madhupriya, Tiwari, A.K., Kumar, S., Baranwal, V.K., 2014. Identification of sugarcane grassy shoot-associated phytoplasma and one of its putative vectors in India. Phytoparasitica 42, 349—354.

Rao, G.P., Nabi, S., Madhupriya, 2015. Overview on a century progress in research on sesame phyllody disease. Phytopath. Mollic. 5, 74—83.

Rao, G.P., Madhupriya, Manimekalai R., Tiwari, A.K., Yadav, A., 2017. A century progress of research on phytoplasma diseases in India. Phytopath. Mollic. 7 (1), 1—38.

Raychaudhuri, S.P., Varma, A., Chenulu, V.V., Prakash, N., Singh, S., 1970. Association of Mycoplasma- like Bodies with Little Leaf of *Solanum Melongena* L in Proceedings of the X International Congress of Microbiolpgy Mexico HIV-6.Mexico.

Roddee, J., Kobori, Y., Hanboonsong, Y., 2018. Multiplication and distribution of sugarcane white leaf phytoplasma transmitted by the leafhopper, *Matsumuratettix hiroglyphicus* (Matsumura) (Hemiptera: Cicadellidae), in infected sugarcane. Sugar Tech. 20, 445—453.

Romanazzi, G., 2013. Perspectives for the management of phytoplasma diseases through induced resistance: what can we expect from resistance inducers? Phytopathog. Mollic. 3 (1), 60—62.

Romanazzi, G., Murolo, S., 2008. Partial uprooting and pulling to induce recovery in "bois noir" infected grapevines. J. Phytopathol. 156, 47—750.

Romanazzi, G., Prota, V.A., Casati, P., Murolo, S., Silletti, M.R., Di Giovanni, R., Landi, L., Zorloni, A., D'Ascenzo, D., Virgili, S., Garau, R., Savini, V., Bianco, P.A., 2007. Incidence of recovery in grapevines infected by phytoplasma in different Italian climatic and varietal conditions and attempts to understand and promote the phenomenon. In: Proceedings of the Workshop on "Innovative Strategies to Control Grapevine and Stone Fruit Phytoplasma Based on Recovery, Induced Resistance and Antagonists", pp. 9—11.

Romanazzi, G., D'Ascenzo, D., Murolo, S., 2009. Field treatment with resistance inducers for the control of grapevine bois noir. J. Plant Pathol. (3), 677—682.

Salehi, M., Izadpanah, K., Siampour, M., Bagheri, A., Faghihi, S.M., 2007. Transmission of 'Candidatus Phytoplasma aurantifolia' to bakraee (*Citrus reticulata* hybrid) by feral *Hishimonus phycitis* leafhoppers in Iran. Plant Dis. 91 (4), 466.

Salehi, E., Taghavi, S.M., Salehi, M., Izadpanah, K., 2014. Partial biological and molecular characterization of phytoplasmas associated with grapevine yellows in fars and Lorestan provinces of Iran. Iran. J. Plant Pathol. 50, 55−64.
Schmid, G., 1983. Effects of tetracycline injection against apple proliferation. Acta Hortic. 130, 237−241.
Schmidt, S., Baric, S., Massenz, M., Letschka, T., Vanas, V., Wolf, M., Kerschbamer, C., Zelger, R., Schweigkofler, W., 2015. Resistance inducers and plant growth regulators show only limited and transient effects on infection rates, growth rates and symptom expression of apple trees infected with 'Candidatus Phytoplasma mali'. J. Plant Dis. Prot. 122, 207−214.
Schweigkofler, W., Schmidt, S., Roschatt, C., 2018. Integrated pest management strategies for phytoplasma diseases of woody crop plants: possibilities and limitations. In: Beck, J.J., Rering, C.aitlin C., Duke, S.tephen O. (Eds.), Book: 'Roles of Natural Products for Biorational Pesticides in Agriculture'; ACS Symposium Series #1294. Publisher, ACS Washington DC, USA. Chapter: 7.
Seemüller, E., Foster, A., 1995. European stone fruit yellows. In: Ogawa, J.M., Zehr, E.I., Bird, G.W., Ritchie, D.F., Uriu, K., Uyemoto, J.K. (Eds.), Compendium of Stone Fruit Diseases. American Phytopathological Society Press, St. Paul, Minnesota, United States of America, pp. 59−60.
Seemüller, E., Harries, H., 2010. Plant resistance. In: Weintraub, P.G., Jones, P. (Eds.), Phytoplasmas: Genomes, Plant Hosts and Vectors. CABI, Wallingford, United Kingdom, pp. 147−169.
Seemüller, E., Schaper, U., Kunze, L., 1986. Effect of pear decline on pear trees on Quince A and *Pyrus communis* seedling rootstocks. Z. für Pflanzenkrankh. Pflanzenschutz 93, 44−50.
Seemüller, E., Lorenz, K.H., Lauer, U., 1998a. Pear decline resistance in *Pyrus communis* rootstocks and progenies of wild and ornamental Pyrus taxa. Acta Hortic. 472, 681−691.
Seemüller, E., Stolz, H., Kison, H., 1998b. Persistence of the European stone fruit yellows phytoplasma in aerial parts of *Prunus* taxa during the dormant season. J. Phytopathol. 407−410.
Seemüller, E., Moll, E., Schneider, B., 2008. Apple proliferation resistance of *Malus sieboldii*-based rootstocks in comparison to rootstocks derived from other *Malus* species. Eur. J. Plant Pathol. 121, 109−119.
Seemüller, E., Moll, E., Schneider, B., 2009. Pear decline resistance in progenies of *Pyrus* taxa used as rootstocks. Eur. J. Plant Pathol. 123, 217−223.
Shindle, G.G., Lokesha, R., Naik, M.K., Ranganath, A.G.R., 2011. Inheritance study on phyllody resistance in sesame (*Sesamum indicum* L). Plant Arch. 11, 777−778.
Siampour, M., Izadpanah, K., Salehi, M., Afsharifar, A., 2019. Occurrence and Distribution of Phytoplasma Diseases in Iran. Sustainable Management of Phytoplasma Diseases in Crops Grown in the Tropical Belt. Springer, pp. 47−86.
Sinclair, W.A., Griffiths, H.M., Lee, I-M., 1994. Mycoplasmalike organisms as causes of slow growth and decline of trees and shrubs. J. Arboric. 20, 176−189.
Sinclair, W.A., Whitlow, T.H., Griffiths, H.M., 1997a. Heritable tolerance of ash yellows phytoplasmas in green ash. Can. J. For. Res. 27, 1928−1935.
Sinclair, W.A., Griffiths, H.M., Whitlow, T.H., 1997b. Comparisons of tolerance of ash yellows phytoplasmas in *Fraxinus* species and rootstock-scion combinations. Plant Dis. 81, 395−398.
Sinclair, W.A., Townsend, A.M., Griffiths, H.M., Whitlow, T.H., 2000a. Responses of six Eurasian ulmus cultivars to a North American elm yellows phytoplasma. Plant Dis. 84, 1266−1270.
Sinclair, W.A., Townsend, A.M., Griffiths, H.M., Whitlow, T.H., 2000b. Responses of Six Eurasian *Ulmus*.
Singh, P.K., Akram, M., Vajpeyi, M., Srivastava, R.L., Kumar, K., Naresh, R., 2007a. Screening and development of resistant sesame varieties against phytoplasma. Bull. Insectol. 60, 303−304.
Singh, S.K., Amminudin, Srivastava P., Singh, B.R., Khan, J.A., 2007 b. Production of phytoplasma-free plants from yellow leafdiseased *Catharanthus roseus* L. (G.) Don. J. Plant Dis. Prot. 114, 2−5.

Sugio, A., Hogenhout, S.A., 2012. The genome biology of phytoplasma: modulators of plants and insects. Curr. Opin. Microbiol. 15 (3), 247–254.

Sugio, A., MacLean, A.M., Kingdom, H., Grieve, V.M., Manimekalai, R., Hogenhout, S.A., 2011a. Diverse targets of phytoplasma effectors: from plant development to defense against insects. Annu. Rev. Phytopathol. 49, 175–195.

Sugio, A., Kingdom, H., MacLean, A., Grieve, V., Hogenhout, S., 2011b. Phytoplasma protein effector SAP11 enhances insect vector reproduction by manipulating plant development and defense hormone biosynthesis. Proc. Natl. Acad. Sci. USA 108 (48), E1254–E1263.

Sugio, A., MacLean, A.M., Hogenhout, S.A., 2014. The small phytoplasma virulence effector SAP11 contains distinct domains required for nuclear targeting and CIN-TCP binding and destabilization. New Phytol. 202, 838–848.

Suzuki, S., Oshima, K., Kakizawa, S., Arashida, R., Jung, H.Y., Yamaji, Y., Nishigawa, H., Ugaki, M., Namba, S., 2006. Interaction between the membrane protein of a pathogen and insect microfilament complex determines insect-vector specificity. Proc. Natl. Acad. Sci. USA 103, 4252–4257.

Tedeschi, R., Bosco, D., Alma, A., 2002. Population dynamics of *Cacopsylla melanoneura* (Homoptera: psyllidae), a vector of apple proliferation phytoplasma in northwestern Italy. J. Econ. Entomol. 95, 544–551.

Tiwari, A.K., Tripathi, S., Lal, M., Sharma, M.L., Chiemsombat, P., 2011. Elimination of sugarcane grassy shoot disease through apical meristem culture. Arch. Phytopathol. Plant Protect. 44 (20), 1942–1948.

Tiwari, A.K., Madhupriya, S.rivastava KP., Pandey, B.S., Rao, G.P., 2016. Detection of sugarcane grassy shoot phytoplasma (16SrXI-B subgroup) in *Pyrilla perpusilla* Walker in Uttar Pradesh, India. Phytopath. Mollic. 6 (1), 56–59.

Tiwari, A.K., Kumar, S., Mall, S., Jadon, V., Rao, G.P., 2017. New efficient natural leafhopper vectors of sugarcane grassy shoot phytoplasma in India. Sugar Tech. 19, 191–197.

Tiwari, N.N., Jain, R.K., Tiwari, A.K., 2021. Management of phytoplasma causing little leaf and witches' broom disease in *Cathranthues roseus* through *in vitro* approaches. Agrica 10 (2), 175–181.

Trivedi, P., Duan, Y., Wang, N., 2010. Huanglongbing, a systemic disease, restructures the bacterial community associated with citrus roots. Appl. Environ. Microbiol. 76, 3427–3436.

Ulubaş Serçe, Ç., Gazel, M., Çalayan, K., Başve, M., Son, L., 2006. Phytoplasma diseases of fruit trees in germplasm and commercial orchards in Turkey. J. Plant Pathol. 88, 179–185.

Ulubaş Serçe, C., Gazel, M., Yalcin, S., Çağlayan, K., 2007. Responses of six Turkish apricot cultivars to 'Candidatus Phytoplasma prunorum' under greenhouse conditions. Bull. Insectol. 60 (2), 309–310.

Ustun, R., Yol, E., Ikten, C., Catal, M., Uzun, B., 2017. Screening, selection and real-time qPCR validation for phytoplasma resistance in sesame (*Sesamum indicum* L.). Euphytica 213, 159.

Valad, G.E., Goodman, R.M., 2004. Systemic acquired resistance and induced systemic resistance in conventional agriculture. Crop Sci. 44 (6), 1920–1934.

Viswanathan, R., Rao, G.P., 2011. Disease scenario and management of major sugarcane diseases in India. Sugar Tech. 13, 336–353.

Wallace, M., 2002. Coconut breeding programme for lethal yellowing resistance in Jamaica. In: Proceedings of the Expert Consultation on Sustainable Coconut Production through Control of Lethal Yellowing Disease, pp. 118–127. CFC Technical Paper No. 18, Kingston, Jamaica, Common Fund for Commodities.

Wally, O., Daayf, F., Khadhair, A.H., Adam, L., Elliott, B., Shinners-Carnelley, T., Iranpour, M., Northover, P., Keyworth, S., 2004. Incidence and molecular detection of yellows-type disease in carrots, associated with leafhoppers in southern Manitoba, Canada. J. Indian Dent. Assoc. 26 (4), 498–505.

Walsh, K.B., Guthrie, J.N., White, D.T., 2006. Control of phytoplasma diseases of papaya in Australia using netting. Australas. Plant Pathol. 25 (1), 49–54.

Wang, Q.C., Valkonen, J.P.T., 2008. Efficient elimination of sweetpotato little leaf phytoplasma from sweet potato by cryotherapy of shoot tips. Plant Pathol. 57, 338–347.
Wang, Q.C., Panis, B., Engelmann, F., Lambardi, M., Valkonen, J.P.T., 2009. Cryotherapy of shoot tips: a technique for pathogen eradication to produce healthy planting materials and prepare healthy plant genetic resources for cryopreservation. Ann. Appl. Biol. 351–353.
Weintraub, P.G., 2007. Insect vectors of phytoplasmas and their control—an update. Bull. Insectol. 60 (2), 169–173.
Weintraub, P.G., Beanland, L., 2006. Insect vectors of phytoplasmas. Annu. Rev. Entomol. 51, 91–111.
Weintraub, P.G., Wilson, M.R., 2010. Control of phytoplasmas diseases and vectors. In: Weintraub, P.G., Jones, P. (Eds.), Phytoplasmas: Genomes, Plant Hosts and Vectors. CABI Publishing, pp. 233–249.
Wingate, V.P.M., Lawton, M.A., Lamb, C.J., 1988. Glutathione causes a massive and selective induction of plant defence genes. Plant Physiol. 87, 206–210.
Wu, W., Ding, Y., Wei, W., Davis, R.E., Lee, I-M., Hammond, R.W., Zhao, Y., 2012. Salicylic acid-mediated elicitation of tomato defence against infection by potato purple top phytoplasma. Ann. Appl. Biol. 161, 36–45.
Zafari, S., Niknam, V., Musetti, R., et al., 2012. Effect of phytoplasma infection on metabolite content and antioxidant enzyme activity in lime (*Citrus aurantifolia*). Acta Physiol. Plant. 34, 561–568.
Zhao, J., Liu, M.J., Liu, X.Y., Zhao, Z.H., 2009. Identification of resistant cultivar for jujube witches' broom disease and development of management strategies. Acta Hortic. 840, 409–412.
Zreik, L., Carle, P., Bové, J-M., Garnier, M., 1995. Characterization of the mycoplasmalike organism associated with witches' broom disease of lime and proposition of a Candidatus taxon for the organism, '*Candidatus* Phytoplasma aurantifolia'. Int. J. Syst. Bacteriol. 45, 449–453.

Further reading

McCoy, R.E., Howard, F.W., Tsai, J.H., Donselman, H.M., Thomas, D.L., Basham, R.A., et al., 1983. Lethal yellowing of palms. In: Bulletin 834, vol. 100. University of Florida Institute of Food and Agricultural Sciences), Gainesville, FL.

CHAPTER 8

Management of insect vectors associated with phytoplasma diseases

Chamran Hemmati[1,2], Mehrnoosh Nikooei[1,2], Ajay Kumar Tiwari[3] and Abdullah Mohammed Al-Sadi[4]

[1]*Department of Agriculture, Minab Higher Education Center, University of Hormozgan, Bandar Abbas, Iran;* [2]*Plant Protection Research Group, University of Hormozgan, Bandar Abbas, Iran;* [3]*UPCSR-Sugarcane Research and Seed Multiplication Center, Gola, Uttar Pradesh, India;* [4]*Department of Plant Sciences, College of Agriculture and Marine Sciences, Sultan Qaboos University, Muscat, Oman*

1. Introduction

Phytoplasmas, wall-less and Gram-positive, are the most important pathogens associated with many economically important plants across the globe. Because they are phloem-limited bacteria, only sap-feeder insect species of Auchenorrhyncha and Sternorrhyncha suborders can transmit the phytoplasmas to other plants (Szwedo, 2016). The spread of phytoplasma-associated diseases depends on the inoculum concentration and the presence of insect vectors. The management of the diseases is depending on the plants weather it is annual or perennial crops. Many studies reported the ecology, biology, epidemiology, and control of phytoplasma diseases. However, less attention has been paid to the management of their insect vectors. The chapter presents an up-to-date overview of the vector's management in Asian countries.

2. Vector taxonomy and ecology

The Hemiptera, a diverse and large order of insects, are classified into three suborders: Auchenorrhyncha (planthoppers, leafhoppers, treehoppers, cicadas, and spittlebug), Sternorrhyncha (psyllids, whiteflies, aphids, and scale), and Heteroptera (true bugs). They are vastly distributed in all zoogeographic regions of the world (Dietrich, 2009). They comprise 90,000 species belonging to about 140 families. Leafhoppers and planthoppers are the most known phytoplasma vectors; however, some psyllids and true bugs were reported to vector phytoplasmas (Mitchell, 2004). The number of phytoplasma disease is much more than the identified vectors. All Auchenorrhyncha members except two superfamilies (Cercopoidea and Cicadoidea) and one subfamily (Cicadellinae) feed on phloem tissues. Also, most Typhlocybinae members remove the cell contents from mesophyll cells to feed. Weintraub and Beanland (2006) stated four characters that make leafhoppers and planthoppers successful and efficient vectors of phytoplasma. Feeding selectively and specifically on certain plant tissues, being hemimetabolous that nymphs and adults can feed, having a persistent and propagative relationship with phytoplasmas, and finally having transovarial transmission in some cases are the most effective characters (Hanboonsong et al., 2005). Within the groups of phloem-feeding insects, only a small number, primarily in three ribosomal groups, have been confirmed as vectors of phytoplasmas. Weintraub and Beanland (2006), provide a table of all known vector species, which was recently updated (Wilson and Weintraub, 2007). In addition, Hemmati et al. (2021) updated the vector species status in the Middle East.

The superfamily containing the largest number of vector species is the Membracoidea, within which all known vectors are confined to the family Cicadellidae (leafhoppers). Morphological and molecular evidences indicate that the Membracoidea are a monophyletic superfamily (Dietrich et al., 2001). However, the phylogenetic status and relationships of the families, subfamilies, and tribes remain poorly understood, although progress is being made (Zahniser and Dietrich, 2008).

The subfamily containing the second largest number of confirmed vector species is the Macropsinae. Vector members of the Macropsinae can be monophagous or oligophagous, but most of them feed primarily on woody plants. Vector species are found in four planthopper families (Fulgoromorpha): Cixiidae, Delphacidae, Derbiidae, and one species in the Flatidae (Weintraub and Beanland, 2006). The first three families have at least one species that transmits a phytoplasma in the coconut lethal yellows group. Several species in these families also transmit phytoplasmas from the "stolbur" group. At present, three genera of psyllids include vectors. *Cacopsylla* species transmit apple proliferation group (16SrX) phytoplasmas to pome and stone fruit trees (Jarausch et al., 2019). *Diaphorina citri* Kuwayama transmits the phytoplasmas associated with lime witches' broom disease (Quieroz et al., 2016).

The degree of plasticity and specialization of insect vectors with regards to their host plants enables them to spread phytoplasma diseases. Nickel (2003), depending on vectors' host range, the Hemiptera are divided in five levels: first-degree monophagous, feeding on one species; second-degree monophagous, feeding on one plant genus; first-degree oligophagous, feeding on one plant family; second-degree oligophagous, feeding on two to five plant families, or up to five plant species; and polyphagous, feeding on many plants of diverse families. The open or closed cycle terms were introduced in the epidemiology of phytoplasma fields. When the vector transmits phytoplasmas from one plant species to the other, it is considered as open cycle. However, transmitting from one host plant to the same species is known as closed cycle (Alma et al., 2019). Normally, monophagous vectors lead to a closed epidemiological system of the pathogen, with acquisition access period, latent period (LP), and inoculation access period spent on the same plant species, while polyphagous vectors are occasional vectors, Depending on the susceptibility of each host plant, polyphagous vectors are able to inoculate a broader range of plant species. within the frame of an open epidemiological system (Alma et al., 2019). In addition, expanding or limiting the spread of phytoplasma relies on the host—vector plant interactions. It has been confirmed that leafhoppers that typically do not feed on certain plant species can acquire and vector phytoplasmas to those plants either under laboratory conditions or fields. For example, Salehi et al. (2015) reported that *Orosius albicinctus*, the natural vector of cucumber phyllody, was able to transmit the 16SrII-D phytoplasma to the other plant species including carrot, sunflower, eggplant, tomato, parsley, alfalfa, and pot marigold.

3. Insect vector managements
3.1 Insecticides

Traditionally, the control of phytoplasma-associated plant diseases has relied on the use of insecticides, with or without regard to precisely controlling the vectors. At the time that the symptoms of the phytoplasma disease are observed, the vectors may or may not have been present meaning that the infection may take place weeks or months earlier. Therefore, insecticide spraying may not be useful in these cases as it will result in environmental pollution, killing the parasitoids and predators, and being harmful to nontarget animals (Bianco et al., 2019).

Conservative pesticides cannot control the disease emergence as phytoplasma transmission occurs faster than the insecticide can act, even when regularly used. At best, the insecticides can inhabit the intracrop transmission by controlling the vector populations. It should be noted that the migration of the new population may occur from the surrounding habitat (Weintraub and Wilson, 2010). In the case of *Hyalesthes obsoletus*, the vector of "bois noir" phytoplasma, Mori et al. (2008) applied insecticides to the central canopy of 18 vineyards and revealed that no significant reduction in its population was observed. They stated that it might be because the vector preference was to be on other plants and might incidentally influx on grapevines. However, there are some evidences that the use of insecticides might help control vector populations and consequently decrease intracrop transmission. Reduction in vector population and thus disease incidence was achieved by just spraying the crop borders with systemic insecticides (Pilkington et al., 2004). This method acted only on overwintered leafhoppers that migrated to the plants with a low application. In addition, the use of two insecticides with different modes of action can be useful. Systemic and contact insecticides can be applied to protect plants from carrying migratory vectors and suppress the vector population, respectively. Sarnaik et al. (1986) showed that three sprays of fenvalerate 0.01% gave the most effective control in the sesame field in India. In addition, Reddy et al. (2019) tested some insecticides against *O. albicinctus* in sesame fields in India. They revealed that chemical treatments with pymetrozine showed the highest percent reduction of the nymphal population (91.5%) followed by imidaclopride + ethiprole (80%), acephate (73%), and dinotefuran (70%). The lowest percent age of reduction was recorded in flonicamid (39.3%), thiamethoxam (seed treatment) (25.6%), and imidacloprid (seed treatment) (12.7%), and spray (Tiwari et al., 2016, 2017). There is no information about the other control ways of *O. albicinctus*. The use of insecticides independently or in conjunction with other methods needs more attention to arrive at the appropriate management tactics. Also, the control way applied may change based on the plant species. For example, eradication of symptomatic plants is more effective in orchards or perennials than the use of insecticides.

Kaolin is known as a nonabrasive fine-grained aluminosilicate mineral that is used as a particle film. This compound is a new version of the old type of physical control that inhabits the vector feedings from the plants. It has been shown that *Homalodisca coagulata* (Say), the vector of *Xylella fastidiosa* associated with Pierce's diseases in the grapevine, was less active and fed on the grapevine treated by kaolin (Tubajica et al., 2007). In southern Iran, this compound is widely used to protect the citrus against sunscald. It has been observed that in the orchards sprayed with kaolin the number of lime witches' broom insect vectors was significantly decreased compared with the nonsprayed orchards (C. Hemmati, unpublished; Hemmati, et al., 2021).

3.2 Clean propagation material

Quarantine rules and limitation of the transport from infected areas to new areas are the most important ways of insect vector control. *Hishimonus phycitis* is the vector of lime witches' broom and ornamental plants diseases in some parts of Asia, especially in India (Rao et al., 2017). Based on the population genetic structure, no difference was observed between Oman and Iran populations (Shabani et al., 2013). Therefore, it was thought that the vector could have been introduced from India into Pakistan, then to Iran by infected propagation materials in 1978.

Other hemipteran species including *O. albicinctus*, *Neoaliturus haematoceps*, *N. tenellus*, and *N. fenestratus*, that are the vectors of important diseases, are distributed in some Asian countries. Strict rules should be considered to reduce the phytoplasma disease spreading in the countries. For example, citrus is the major plant in China; however, the leafhopper *H. phycitis* has not been reported from China. The government should apply very strict rules in transporting citrus materials and fruits from where the leafhopper is distributed. Such a scenario was also reported in Europe as *Scaphoideus titanus* Ball was introduced to Europe in the early 1920 and then it become the vector of "flavescence dorée" in (Vidano et al., 1964). In addition, if the materials would transfer to the new area, then using heating or cooling treatment, dipping in hot water, and use of radiation can decrease the chance of new disease introduction to other countries. Heat therapy, hot water treatments, in vitro meristem tip cultures, and tetracycline in combination with in vitro techniques are useful in producing clean materials, and this will be really helpful in reducing the further spread of phytoplasma-associated diseases (Linck et al., 2019; Tiwari et al., 2011, 2021; Singh et al., 2007).

On the other hand, phytoplasma-infected plants should not be transported to areas where the potential or natural vectors are present. The infected *Limonium* sp. were transferred to Israel where the potential vectors were distributed, and consequently, phytoplasma symptoms were observed in other *Limonium* fields within the area (Weintraub et al., 2004).

3.3 Resistant plants

The plants that can deter vectors not to feed on are the most important and the first way of control. Little attention has been paid to find such plant cultivars in Asia. To find the resistant cultivar to sesame phyllody diseases, Selvanarayanan and Selvamuthukumaran (2000) screened four sesame cultivars in India. They revealed that one cultivar namely TMV4 was more resistant, although no significant differences were observed among *O. albicinctus* population feeding on different cultivars. In addition, *H. phycitis* was collected on lime and bakraee plants; however, the leafhoppers could not complete their generation on other citrus species in Iran. This shows that the other citrus species can deter the vectors (Salehi et al., 2017). Also, genetic manipulation of the plants that can secret the defensive compounds against vectors is an effective way. *Sogatella furcifera*, *Nilaparvata lugens* planthoppers, and *Nephotettix virescens* leafhopper are the main phytoplasma vectors on rice plants. Nagadhara et al. (2004) revealed that genetically modified plants, expressing lectin, especially *Galanthus nivalis agglutinin* (GNA), were highly toxic to plant hoppers as the survival, development, and fecundity of *S. furcifera* was significantly reduced. In addition, they revealed that the vectors feeding on GNA rice caused 90% mortality in *S. furcifera*, and 29% and 53% mortality in *N. virescens* and *N. lugens*, respectively (Nagadhara et al., 2004). As a consequence, reduction in planthopper population contributed to a reduction in disease spread and the production of honeydew which provides suitable conditions for mold and fungus growth. The insertion of this gene into other plants could have significant effects which merit more studies. The lectin that can be toxic for vectors is a 25-kDa homodimeric lectin, *Allium sativum* leaf lectin (Dutta et al., 2005). This lectin has a high sequence similarity to GNA that both may be involved in iron homeostasis (Saha et al., 2006).

As it has been confirmed, the rootstock can also play a vital role in attracting vectors to specific plants. For instance, plants grafted on Castel 2016 rootstock had more phytoplasma presence than Richtar 110 ones. Additional research stated that *H. obsoletus* antenna reacted more to the volatiles emitted by Castel 2016 than Richtar 110. Identification of such resistant rootstock can lead to develop a nice control way because most of the woody crops are grafted on rootstock (Weintraube and Wilson, 2010). Recently, the use of plant material as a deterrent was proposed to control the phytoplasma vectors. For instance, hot water and EtOH extracts of spruce could significantly deter *C. picta*, the vector of apple proliferation. In addition, these compounds have been shown to reduce oviposition and nymph developments. However, the authors stated that transmission events were not affected by such treatment and suggested an improvement in such technique before its field application (Gallinger et al., 2020).

Systemic acquired resistance is a response of plants that can be activated when challenged by either a pathogen or arthropods. It is also elicited by a number of chemicals. Studies on *Colladonus montanus* (van Duzee), the vector of X-disease in fruit trees, showed that treatment with benzothiadiazole (BTH) protected plants form the feeding of this vector and consequently phytoplasma infection (74% of incidence reduction). It may be due to that plant phloem has been morphologically modified to prevent phytoplasma from establishing or replicating as well as production of a substance inhibiting vector feeding. In addition, the survival of the leafhopper significantly decreased on plants treated with BTH than nontreated ones (Weintraub and Wilson, 2010).

3.4 Alternative host's and weeds control

It has long been known that alternative hosts can play a vital role in phytoplasma epidemics by being hosts for the phytoplasmas and vectors that overwinter on. *O. albicinctus* is known as the vector of alfalfa witches' broom (AlfWB) phytoplasma in Iran. Esamilzadeh-Hosseini et al. (2011) performed a survey on the alternative host of AlfWB in Iran. They revealed that *Cardaria draba* L. and *Prosopis farcta* (Banks and Sol.) may act as the reservoir for AlfWB-16SrII phytoplasmas in Iran because they captured more leafhoppers on them and act as the preferred hosts for the leafhopper. Other plant species such as *Tamarix aphylla* L. and *Haloxylon aphyllum* (Minkw) which are cultivated nearby the alfalfa fields were identified to be a site for the survival of *O. albicinctus*. These species can also play a role in the epidemiology of AlfWB. They found phytoplasmas in the two species of weeds and recommend disease control by weed removal. Pull strategy has been proposed to control *O. albicinctus* by the use of its preferred host plants, *T. aphylla*, *H. aphyllum*, *C. draba*, and *P. farcta*.

In addition, Salehi et al. (2007) stated that wild lettuce could be a reservoir of lettuce phyllody phytoplasma in Iran, as they were associated with the same strain, and *N. fenestratus*, the confirmed vector, was active in these fields (Salehi et al., 2007). Eradication of the wild lettuce surrounding the lettuce fields may help to control the disease.

'*Ca.* P. australasia' and '*Ca.* P. aurantifolia' strains are known to be associated with many crop diseases such as lime witches' broom, tomato witches' broom, parsley witches' broom, squash and cucumber phyllody, pomegranate little leaf, and ornamental crops such as petunia witches' broom (Hemmati et al., 2019b) and zinnia phyllody (Hemmati and Nikooei, 2017) in Iran. Taking into account the above fact, plants including *Taverniera cuneifolia* and *Aerva javanica* associated with

16SrII-D and 16SrII-B can play a role as an alternative plant of 16SrII-D subgroup phytoplasmas (Hemmati et al., 2021). According to Bagheri et al. (2010) results, the best way to control the lime witches' broom diseases was achieved by the eradication of weeds in citrus orchards in two sequential years in southern Iran. In India, eradication of weeds that harbor phytoplasmas and are the host of their vectors could reduce the phytoplasma diseases in eggplant fields (Rao et al., 2015). More than 30 weed species were confirmed as phytoplasma reservoirs, and there may be chances of transmission from infected plant to healthy or infected weed to healthy plants and vice versa (Rao, 2021). Recently, Maurya et al. (2020) identified *Cannabis sativa* as host of sugarcane grassy shoot phytoplasma (16SrXI-B) from India.

3.5 Habitat management

Habitat management or manipulation is known as an important agroecological approach in integrated pest management (IPM) that employ trap cropping, mulching, intercropping, multicropping, flower strips, or insectary plants, and weed strips techniques. The main goal of this technique is to minimize pest pressure by improving the fitness of natural enemy or unfavorable the habitat for living. Vegetation composition which is adjunct to orchards/fields can have a vital role on the presence and dispersal of phytoplasma vectors (Weintraub and Beanland, 2006). As discussed above, weed species can have a profound effect on vector distribution and infection levels. However, plant density might affect vectors distribution. Weeds on the margin of the fields are the next direction of the vectors when the major crops in the fields are harvested. The fact that the infection is started from the margins showed that the vector overwintering on perennial weeds influx the fields. Lessio and Alma (2004) placed the Malaise traps in ecotonal regions between forests and vineyards. They revealed that the primary direction of *S. titanus* movement was recorded from the wild to the cultivated vegetation. In another experiment, they also stated that the traps in vineyards and those placed at 12 and 24 m beyond the vineyard border captured 3 and 1200 individuals during two years, respectively (Lessio and Alma, 2004).

The application of several synthetic or organic mulches is another means of habitat manipulation. *H. obsoletus* lays eggs at or just below the soil surface. Plastic sheeting as a synthetic mulch is able to prevent the movement of vectors into the soil. Reflective mulches can repel the vector from the plants as maize leafhopper *Dalbulus maidis* was better controlled by plastic reflective mulch (Summers and Stapleton, 2002). The use of aluminum foil mulch could control the aster yellows phytoplasma vector *Macrosteles quadrilineatus* (Forbes) in carrot fields. The type of mulching materials used in controlling is important. For example, the coconut mulched with coarse materials such as pine bark nuggets attracted fewer *Haplaxius crudus*, the vector of lethal yellowing, in comparison with coconut frond and eucalyptus. It can be noted that mulch application is too expensive and not very practical.

3.6 Covering plants

The most consistent way for controlling phytoplasma insect vectors can be achieved by covering crops. However, due to the logistics of large-scale agriculture in major plants such as rice, wheat, sugarcane, and trees, its application is limited that its use cannot even be contemplated. However, the plants that can be grown in the greenhouse are much less affected from the phytoplasma diseases showing the efficacy of this method. Bananas and date fruits are now covered by plastic or papers for controlling other pests and pathogens that can be applied worldwide for other trees.

To compare the effect of covering papayas by insect-exclusion screening (IES), spraying systemic insecticides by Imidacloprid, Elder et al. (2002) and Walsh et al. (2006) performed a survey in Australia. They revealed that covering the trees could control 100% vectors in comparison with insecticide and nontreated control as the phytoplasma symptoms were significantly fewer in covered trees.

This method was also applied in vineyards to control phytoplasma disease (Mannini, 2007). The clonal or mother plants that were continuously covered by IES showed no phytoplasma symptoms. Blua et al. (2005) studied the effect of a 5-m-high barrier screen to prevent the movement of the sharpshooter *Homalodisca vitripennis* (Germar), a vector of *X. fastidiosa*, to high-value vineyards and nursery stock. The results showed that the vector behavior changed as they moved away from the barriers to surrounding plants; only a few flew over the barriers. The results of this demonstrated that a barrier could add significantly to vector control.

3.7 Biological controls

Planthoppers and leafhoppers are fed by a range of predators. Spiders that feed nymphs and adults are very important predators especially in the grassland ecosystem. Bagheri et al. (2017) revealed that two species of predatory spiders namely *Plexippus iranus* Longunov and *Thyene imperialis*, belonging to Salticidae family, voraciously fed *H. phycitis* in citrus orchards of southern Iran. Mirids are known as egg predators in ecosystems.

Auchenorrhyncha are attacked by specialist parasitoids such as Mymaridae (Hymenoptera), Dryinidae (Hymenoptera), and Pipunculidae (Diptera). There are also specialist parasitoids associated with Auchenorrhyncha (Waloff and Jervis, 1987). Mymaridae, known as fairy flies, is very common as egg parasitoids which complete the whole life span in eggs. However, there are little reports on leafhoppers and planthoppers eggs. Bagheri et al. (2017) collected the eggs of *H. phycitis* deposited on the infested leaves of the host tree and maintained inside a Petri dish until parasitoid adult emergence. The parasitoid was then identified as *Polynema* sp. belonging to the Mymaridae family. Dryinidae deposit their eggs into nymphs and adults' planthoppers and leafhoppers that developing larvae are visible externally into the sacs of the hoppers. Their pupal stage completes in the soil or leaf litter. Pipunculidae, the exclusive parasitoids of Auchenorrhyncha, deposit their eggs into the nymphal and adults that the larvae complete their life into leafhoppers' abdomen. Some species of *Strepsiptera* attack some species of delphacids (Denno and Perfect, 1994; Perfect and Cook, 1994).

The study conducted by Jiang and Cheng (2004) revealed that the population of the predators of white-backed planthopper could increase when manure was applied to rice paddies. In addition, the efficiency of the predators enhanced when the supplementary foods (collembolans) were provided when the planthopper populations were low.

Some species of parasitoids that specifically parasite the Auchenorrhyncha might decrease the natural population under economic injury levels and therefore they are not considered pests. However, parasitoids are susceptible to insecticide leading to a decrease in their populations and an increase in the leafhoppers or planthoppers' population (Heinrichs, 1994). A survey of beet leafhopper, *Circulifer tenellus* (Baker), showed that a number of naturally occurring species of egg parasitoids were found, but researchers note that low parasitism rates in winter and spring cannot avoid the need for pesticide treatment of uncultivated land (Bayoun et al., 2008). They postulate that, when beet leafhoppers aggregate in the autumn, parasitoids may play a significant role in reducing populations.

3.8 Symbiotic control

Phytoplasmas are not the only microorganisms living in the vectors' body, and they are involved in complex microbiomes enclosing bacteria, viruses, and fungi, in turn, influencing such associations. Insect symbionts including bacteria, fungi (yeast and yeast-like), and viruses are famous effectors of ecology, life history, and development of their hosts. Similar to all sap feeders, phytoplasma vectors depend on bacterial symbionts supplementing their unbalanced diet (Zchori-Fein and Bourtzis, 2011). Insect microbial symbionts accompanied by supplying the host with nutrients offer protection against pathogens or other stress factors while manipulating reproduction. Such multifaceted interaction could be used for developing microbe-based control approaches against phytoplasma-borne diseases (Dale and Moran, 2006). Vectorial capacity is associated with vector survival, longevity, vector density, duration of LP, probability of the vector feeding on host plants, and vector competence. Therefore, long-lived, very mobile insects and generalist insects are theoretically more efficient vectors, especially those that have an ancient relationship with phytoplasma.

Numerous symbiotic characters could be used for control. Symbiotic control is known as a new biological control technique for plant diseases. In this technique, symbiotic microorganisms are isolated, genetically modified, and then reintroduced to express an antipathogenic agent in the insect vector (Wangkeeree et al., 2012). The identification of insect-associated microorganisms is the first step toward symbiotic control of insect pests. To find the bacterial symbionts of *H. phycitis*, 13 populations of *H. phycitis* were collected from witches' broom disease of lime (WBDL)-contaminated lime orchards from southern Iran. Sequencing and phylogenetic analysis of the 16S rRNA and *wsp* genes uncovered two obligate endosymbionts, 'Ca. Sulcia muelleri' and 'Ca. Nasuia deltocephalinicola', both of which exhibited 100% infection frequencies. Five facultative endosymbionts, *Wolbachia*, *Arsenophonus*, *Pantoea*, *Diplorickettsia*, and *Spiroplasma* exhibited 70%, 90%, 57%, 48%, and 92% infection frequencies, respectively. *Wolbachia* was detected in all tested populations (Hemmati et al., 2021). This bacterium is currently being used to control dengue fever by inducing abnormal reproduction (Ruang-Areerate and Kittayapong, 2006). The level of gene flow observed in *H. phycitis* populations (Hemmati et al., 2018) would assist to introduce a specific gene or a transgenic population to a region and spreading these traits between populations. The technique has been proposed to control "flavescence dorée" vectored by the leafhopper *S. titanus* by cross-colonizing the specific bacteria namely *Asaia* (Crotti et al., 2009). Taking into accounts the above facts, these bacteria can be employed as a new method to control the vectors of WBDL.

Some bacteria found in Auchenorrhyncha were thought to have a pathogenic relationship with their host. *Orosius tenellus*, the vector of diverse phytoplasma groups, is the host of *Rickettsiella* bacterium with a high prevalence in populations and having vertical transmission. Although the role of this bacterium for the vector was not characterized, it is speculated that it can be the pathogens for the vector.

Psyllids are the only groups of Sternorrhyncha that transmit phytoplasmas. Two genus *Cacopsylla* and *Diaphorina* are known as the vector of phytoplasma diseases in the world. The Asian citrus psyllid, *D. citri*, is known as the second vector of lime witches' broom disease in Asia with lower efficiency than the major one. The primary symbiont of the psyllids is 'Ca. Carsonella ruddii' (Gonella et al., 2019) which has a long evolution with the host and provide the essential amino acids. A study conducted by Hosseinzadeh et al. (2019) on endosymbionts of *D. citri* revealed that two obligate symbionts namely 'Ca. Carsonella ruddii' and 'Ca. Proftella armatura' and a reproductive manipulator symbiont namely *Wolbachia pipentis* were present in Asian psyllids. This information can be used in symbiotic controls of the vector.

The evolutionary and ecological success of the insects might rely on yeast and yeast-like symbiont (YLS) (Douglas, 2011). Yeast-like relationships with insects are recognized as obligate and common symbiotic relationships. A novel biological control technique for vector-borne diseases is paratransgenesis. A requirement for the progress of the strategies for the symbiotic control of insect vectors is to identify and characterize the insect-associated YLSs. YLSs living in *H. phycitis* were investigated in insects collected from 13 localities of citrus orchards distributed in southern Iran (Kerman, Hormozgan, Sistan and Baluchestan, and Fars provinces). Results revealed that this vector harbored two YLSs namely YLS of *H. phycitis* and *Candida pimensis* with an identity of (98%–99%) to those reported from the other cicadellids (Hemmati et al., 2017). These complex symbionts (bacterial and fungal symbionts) living in *H. phycitis* may help finding a novel way for controlling WBDL.

4. Conclusion

Numerous tactics have been proposed for controlling phytoplasma insect vectors and consequently phytoplasma diseases. However, to achieve the best control, application and combination of tactics in an IPM manner are required. Although chemical control has been used as the first way of control and will be applied in the predictable future, vector management gradually should move to numerous genetic manipulations of plants to produce truthfully resistant plants. At least, the researchers can produce transgenic plants expressing some chemicals to allow them to be tolerant to phytoplasma infection. Less attention has been paid to apply biological control, although some specific parasitoids have been reported. The awareness of microbial relationships in phytoplasma insect vectors has considerably increased in the last few years. More work is still needed to fully understand the mechanisms and interactions between the endosymbiont (both fungi and bacteria) and insect vectors. This way can be the most reliable way to achieve the best control way in the near future.

References

Alma, A., Lessio, F., Nickel, H., 2019. Insects as Phytoplasma Vectors: Ecological and Epidemiological Aspects. In: Phytoplasmas: Plant Pathogenic Bacteria-II. Springer, Singapore, pp. 1–25.

Bagheri, A., Hemmati, C., Askari Seyahooei, M., Modarres Najafabadi, S.S., Nikooei, M., 2017. Preliminary study on natural enemies of *Hishimonus phycitis* (Distant, 1908) (Hemi., Cicadellidae), the vector of lime witches' broom phytoplasma in Hormozgan province. In: 2nd Iranian International Congress of Entomology. University of Tehran, Karaj, Iran, 2–4 Sep.

Bagheri, A., Askari, M., Faghihi, M., 2010. The Effect of Four Management Ways to Achive Lime Witches Broom Control. Jihad Keshavarzi, pp. 1–58 (In Farsi).

Bayoun, I.M., Walker, G.P., Triapitsyn, S.V., 2008. Parasitization of beet leafhopper eggs, *Circulifer tenellus*, in California. J. Appl. Entomol. 132 (5), 412–424.

Bianco, P.A., Romanazzi, G., Mori, N., Myrie, W., Bertaccini, A., 2019. Integrated Management of Phytoplasma Diseases. In: Phytoplasmas: Plant Pathogenic Bacteria-II. Springer, Singapore, pp. 237–258.

Blua, M.J., Campbell, K., Morgan, D.J.W., Redak, R.A., 2005. Impact of a screen barrier on dispersion behavior of *Homalodisca coagulata* (Hemiptera: Cicadellidae). J. Econ. Entomol. 98 (5), 1664–1668.

Crotti, E., Damiani, C., Pajoro, M., Gonella, E., Rizzi, A., Ricci, I., Negri, I., Scuppa, P., Rossi, P., Ballarini, P., Raddadi, N., 2009. *Asaia*, a versatile acetic acid bacterial symbiont, capable of cross-colonizing insects of phylogenetically distant genera and orders. Environ. Microbiol. 11 (12), 3252−3264.

Dale, C., Moran, N.A., 2006. Molecular interactions between bacterial symbionts and their hosts. Cell 126 (3), 453−465.

Denno, R.F., 1994. The evolution of dispersal polymorphisms in insects: the influence of habitats, host plants and mates. Popul. Ecol. 36 (2), 27−135.

Dietrich, C.H., 2009. Auchenorrhyncha: (cicadas, spittlebugs, leafhoppers, treehoppers, and planthoppers). In: Encyclopedia of Insects. Academic Press, pp. 56−64.

Dietrich, C.H., McKamey, S.H., Deitz, L.L., 2001. Morphology-based phylogeny of the treehopper family Membracidae (Hemiptera: cicadomorpha: Membracoidea). System. Entomology 26 (2), 213−239.

Douglas, A.E., 2011. Lessons from studying insect symbioses. Cell Host Microbe 10 (4), 359−367.

Dutta, I., Saha, P., Majumder, P., Sarkar, A., Chakraborti, D., Banerjee, S., Das, S., 2005. The efficacy of a novel insecticidal protein, *Allium sativum* leaf lectin (ASAL), against homopteran insects monitored in transgenic tobacco. Plant Biotechnol. J. 3 (6), 601−611.

Elder, R.J., Milne, J.R., Reid, D.J., Guthrie, J.N., Persley, D.M., 2002. Temporal incidence of three phytoplasma-associated diseases of *Carica papaya* and their potential hemipteran vectors in central and south-east Queensland. Australas. Plant Pathol 31 (2), 165−176.

Esmailzadeh Hosseini, S.A., Salehi, M., Mirzaie, A., 2011. Alternate hosts of alfalfa witches' broom phytoplasma and winter hosts of its vector *Orosius albicinctus* in Yazd-Iran. Bull. Insectol. 64 (Suppl.), S247−S248.

Gallinger, J., Jarausch, B., Jarausch, W., Gross, J., 2020. Host plant preferences and detection of host plant volatiles of the migrating psyllid species *Cacopsylla pruni*, the vector of European Stone Fruit Yellows. J. Pest Sci. 93 (1), 461−475.

Gonella, E., Musetti, R., Crotti, E., Martini, M., Casati, P., Zchori-Fein, E., 2019. Microbe Relationships With Phytoplasmas in Plants and Insects. In: Phytoplasmas: Plant Pathogenic Bacteria-II. Springer, Singapore, pp. 207−235.

Hanboonsong, Y., Choosai, C., Panyim, S., Damak, S., 2005. Transovarial transmission of sugarcane white leaf phytoplasma in the insect vector *Matsumuratettix hiroglyphicus* (Matsumura). Insect Mol. Biol. 11 (1), 97−103.

Heinrichs, E.A., 1994. Impact of Insecticides on the Resistance and Resurgence of Rice Planthoppers. In: Planthoppers. Springer, Boston, MA, pp. 571−598.

Hemmati, C., Nikooei, M., 2017. Molecular characterization of a 'Candidatus Phytoplasma aurantifolia'-related strain associated with *Zinnia elegans* phyllody disease in Iran. Australas. Plant Dis. Notes 12 (1), 11.

Hemmati, C., Moharramipour, S., Askari Seyahooei, M., Bagheri, A., Mehrabadi, M., 2018. Population genetic structure of *Hishimonus phycitis* (hem.: Cicadellidae), vector of lime witches' broom phytoplasma. J. Agri. Sci. Technol. 20 (5), 999−1012.

Hemmati, C., Moharramipour, S., Askari Siahooei, M., Bagheri, A., Mehrabadi, M., 2017. Identification of yeast and yeast-like symbionts associated with *Hishimonus phycitis* (Hemiptera: Cicadellidae), the insect vector of lime witches' broom phytoplasma. J. Crop Protect. 6 (4), 439−446.

Hemmati, C., Nikooei, M., Al-Sadi, A.M., 2020. Four decades of research on phytoplasma diseases of palms: a review. Int. J. Agri. Biol. 24 (3), 631−644.

Hemmati, C., Nikooei, M., Bertaccini, A., 2019. Identification and transmission of phytoplasmas and their impact on essential oil composition in *Aerva javanica*. 3 Biotech. 9 (8), 1−7.

Hemmati, C., Nikooei, M., Al-Subhi, A.M., Al-Sadi, A.M., 2021. History and current status of phytoplasma diseases in the Middle East. Biology 10, 226.

Hosseinzadeh, S., Shams-Bakhsh, M., Mann, M., Fattah-Hosseini, S., Bagheri, A., Mehrabadi, M., Heck, M., 2019. Distribution and variation of bacterial endosymbiont and 'Candidatus Liberibacter asiaticus' titer in the huanglongbing insect vector, *Diaphorina citri* Kuwayama. Microb. Ecol. 78 (1), 206−222.

Jarausch, B., Tedeschi, R., Sauvion, N., Gross, J., Jarausch, W., 2019. Psyllid Vectors. In: Phytoplasmas: Plant Pathogenic Bacteria-II. Springer, Singapore, pp. 53–78.

Jiang, M.X., Cheng, J.A., 2004. Effects of manure use on seasonal patterns of arthropods in rice with special reference to modified biological control of whitebacked planthopper, *Sogatella furcifera* Horvath (Homoptera: Delphacidae). J. Pest Sci. 77 (4), 185–189.

Lessio, F., Alma, A., 2004. Dispersal patterns and chromatic response of *Scaphoideus titanus* Ball (Homoptera Cicadellidae), vector of the phytoplasma agent of grapevine "flavescence dorée". Agri. Forest Entomol. 6 (2), 121–128.

Linck, H., Lankes, C., Krüger, E., Reineke, A., 2019. Elimination of phytoplasmas in *Rubus* mother plants by tissue culture coupled with heat therapy. Plant Dis. 103 (6), 1252–1255.

Mannini, F., 2007. Hot water treatment and field coverage of mother plant vineyards to prevent propagation material from phytoplasma infections. Bull. Insectol. 60 (2), 311.

Maurya, R., Mall, S., Tiwari, A.K., Jadon, V., Marcone, C., Rao, G.P., 2020. *Cannabis sativa* L.: a potential natural reservoir of sugarcane grassy shoot phytoplasmas in India. Sugar Tech 1–5.

Mitchell, P.L., 2004. Heteroptera as vectors of plant pathogens. Neotrop. Entomol. 33 (5), 519–545.

Mori, N., Pavan, F., Bondavalli, R., Reggiani, N., Paltrinieri, S., Bertaccini, A., 2008. Factors affecting the spread of "bois noir" disease in north Italy vineyards. Vitis 47 (1), 65.

Nagadhara, D., Ramesh, S., Pasalu, I.C., Rao, Y.K., Sarma, N.P., Reddy, V.D., Rao, K.V., 2004. Transgenic rice plants expressing the snowdrop lectin gene (gna) exhibit high-level resistance to the whitebacked planthopper (*Sogatella furcifera*). Theo. App. Genetics 109 (7), 1399–1405.

Nickel, H., 2003. Leafhoppers and Planthoppers of Germany (Hemiptera, Auchenorrhyncha). Goecke & Evers.

Perfect, T.J., Cook, A.G., 1994. Rice Planthopper Population Dynamics: A Comparison Between Temperate and Tropical Regions. In: Planthoppers. Springer, Boston, MA, pp. 282–301.

Pilkington, L.J., Gurr, G.M., Fletcher, M.J., Elliott, E., Nikandrow, A., Nicol, H.I., 2004. Reducing the immigration of suspected leafhopper vectors and severity of Australian lucerne yellows disease. Austra. J. Exp. Agri. 44 (10), 983–992.

Queiroz, R.B., Donkersley, P., Silva, F.N., Al-Mahmmoli, I.H., Al-Sadi, A.M., Carvalho, C.M., Elliot, S.L., 2016. Invasive mutualisms between a plant pathogen and insect vectors in the Middle East and Brazil. Royal Soc.Open Sci. 3 (12), 160557.

Rao, G.P., Nabi, S.U., 2015. Overview on a century progress in research on sesame phyllody disease. Phytopath. Moll. 5 (2), 74–83.

Rao, G.P., 2021. Our understanding about phytoplasma research scenario in India. Indian Phytopathol. 74, 371–401.

Rao, G.P., Madhupriya, T.V., Manimekalai, R., Tiwari, A.K., Yadav, A., 2017. A century progress of research on phytoplasma diseases in India. Phytopath. Moll. 7 (1), 1–38.

Reddy, T.V., Prasad, K.H., Chalam, M., Viswanath, K., 2019. Management of Leafhopper (*Orosius albicintus*) of sesamum with certain insecticides. Andhra Pradesh J Agri. Sci. 5 (3), 181–186.

Ruang-Areerate, T., Kittayapong, P., 2006. *Wolbachia* transinfection in *Aedes aegypti*: a potential gene driver of dengue vectors. Proc. Nat. Acad. Sci. 103 (33), 12534–12539.

Saha, P., Majumder, P., Dutta, I., Ray, T., Roy, S.C., Das, S., 2006. Transgenic rice expressing *Allium sativum* leaf lectin with enhanced resistance against sap-sucking insect pests. Planta 223 (6), 1329–1343.

Salehi, M., Bagheri, A., Faghihi, M.M., Izadpanah, K., 2017. Study of partial biological and behavioral traits of Hishimonus phycitis, vector of lime witches' broom, for management of the disease. Iranian J. Plant Pathol. 53 (1), 75–96.

Salehi, M., Izadpanah, K., Nejat, N., Siampour, M., 2007. Partial characterization of phytoplasmas associated with lettuce and wild lettuce phyllodies in Iran. Plant Pathol 56 (4), 669–676.

Salehi, M., Siampour, M., Esmailzadeh Hosseini, S.A., Bertaccini, A., 2015. Characterization and vector identification of phytoplasmas associated with cucumber and squash phyllody in Iran. Bull. Insectol. 68 (2), 311–319.

Sarnaik, D.N., Ghode, R.N., Peshkar, L.N., Satpute, U.S., 1986. Insecticidal control of sesamum jassids *Orosius albicinctus* (Distant). PKV Res. J. 10 (1), 41–43.

Selvanarayanan, V., Selvamuthukumaran, T., 2000. Field resistance of sesame cultivars against phyllody disease transmitted by *Orosius albicinctus* Distant. Sesame Safflower Newsl. (15), 71–74.

Shabani, M., Bertheau, C., Zeinalabedini, M., Sarafrazi, A., Mardi, M., Naraghi, S.M., Rahimian, H., Shojaee, M., 2013. Population genetic structure and ecological niche modelling of the leafhopper *Hishimonus phycitis*. J. Pest Sci 86 (2), 173–183.

Singh, S.K., Srivastava, P., Singh, B.R., Khan, J.A., 2007. Production of phytoplasma-free plants from yellow leaf diseased *Catharanthus roseus* L.(G.) Don. J. Plant Dis. Prot. 114 (1), 2–5.

Summers, C.G., Stapleton, J.J., 2002. Management of corn leafhopper (Homoptera: Cicadellidae) and corn stunt disease in sweet corn using reflective mulch. J. Econ. Entomol. 95 (2), 325–330.

Szwedo, J., 2016. The unity, diversity and conformity of bugs (Hemiptera) through time. Earth Environ. Sci. Trans. Royal Soc. Edinburgh 107 (2–3), 109–128.

Tiwari, A.K., Kumar, S., Mall, S., Jadon, V., Rao, G.P., 2017. New efficient natural leafhopper vectors of sugarcane grassy shoot phytoplasma in India. Sugar Tech. 19 (2), 191–197.

Tiwari, A.K., Madhupriya, S.V.K., Pandey, K.P., Sharma, B.L., Rao, G.P., 2016. Detection of sugarcane grassy shoot phytoplasma (16SrXI-B subgroup) in *Pyrilla perpusilla* Walker in Uttar Pradesh, India. Phytopath. Moll. 6, 56–59.

Tiwari, A.K., Tripathi, S., Lal, M., Sharma, M.L., Chiemsombat, P., 2011. Elimination of sugarcane grassy shoot disease through apical meristem culture. Arch. Phytopathol. Plant Protect. 44 (20), 1942–1948.

Tiwari, N.N., Jain, R.K., Tiwari, A.K., 2021. Management of phytoplasma causing little leaf and witches broom disease in *Catahranthus roseus* through *in vitro* approaches. Agrica 10 (2), 175–181.

Tubajika, K.M., Civerolo, E.L., Puterka, G.J., Hashim, J.M., Luvisi, D.A., 2007. The effects of kaolin, harpin, and imidacloprid on development of Pierce's disease in grape. Crop Protect 26 (2), 92–99.

Vidano, C., 1964. Scoperta in Italia dello *Scaphoideus littoralis* Ball cicalina americana collegata alla "flavescence dorée" della Vite. L'Italia Agricola 101 (10), 1031–1049.

Waloff, N., Jervis, M.A., 1987. Communities of parasitoids associated with leafhoppers and planthoppers in Europe. Adv. Ecolog. Res 17, 281–376 (Academic Press).

Walsh, K.B., Guthrie, J.N., White, D.T., 2006. Control of phytoplasma diseases of papaya in Australia using netting. Australas. Plant Pathol 35 (1), 49–54.

Wangkeeree, J., Miller, T.A., Hanboonsong, Y., 2012. Candidates for symbiotic control of sugarcane white leaf disease. Appl. Environ. Microbiol. 78 (19), 6804–6811.

Weintraub, P.G., Beanland, L., 2006. Insect vectors of phytoplasmas. Annu. Rev. Entomol. 51.

Weintraub, P.G., Pivonia, S., Rosner, A., Gera, A., 2004. A new disease in *Limonium latifolium* hybrids. II. Investigating insect vectors. HortScience 39 (5), 1060–1061.

Weintraub, P.G., Wilson, M.R., Jones, P., 2010. Control of phytoplasma diseases and vectors. Phytoplasmas: Genomes, Plant Hosts and Vectors 233–249.

Wilson, M.R., Weintraub, P.G., 2007. An introduction to Auchenorrhyncha phytoplasma vectors. Bull. Insectol. 60 (2), 177.

Zahniser, J.N., Dietrich, C.H., 2008. Phylogeny of the leafhopper subfamily Deltocephalinae (Insecta: Auchenorrhyncha: Cicadellidae) and related subfamilies based on morphology. Syst. Biodivers. 6 (1), 1–24.

Zchori-Fein, E., Bourtzis, K. (Eds.), 2011. Manipulative Tenants: Bacteria Associated with Arthropods. CRC press.

Further reading

Chuche, J., Thiéry, D., 2014. Biology and ecology of the "flavescence dorée" vector *Scaphoideus titanus*: a review. Agron. Sustain. Dev. 34 (2), 381–403.

CHAPTER 9

Elimination of phytoplasmas: an effective control perspective

Chamran Hemmati[1,2], Mehrnoosh Nikooei[1,2] and Ajay Kumar Tiwari[3]

[1]*Department of Agriculture, Minab Higher Education Center, University of Hormozgan, Bandar Abbas, Iran;* [2]*Plant Protection Research Group, University of Hormozgan, Bandar Abbas, Iran;* [3]*UPCSR-Sugarcane Research and Seed Multiplication Center, Gola, Uttar Pradesh, India*

1. Introduction

Phytoplasmas, plant-pathogenic bacteria without a cell wall that cause serious crop losses worldwide, are infecting about hundreds of plant species. Dwarfing, witches' broom, yellowing, purple top, and phyllody are some of the symptoms exhibited by infected plants (Hogenhout et al., 2008; Bertaccini and Duduk, 2009). In spite of their importance for agriculture and their unique biological characteristics, phytoplasmas remain among the least understood plant pathogens. The IRPCM (2004) and Bertaccini et al. (2022) classified '*Candidatus* Phytoplasma' species as 49, and several hundred strains of phytoplasma have been described worldwide (Bertaccini and Lee, 2018). Kakizawa and Yoneda (2015) listed five complete phytoplasma genome sequences as well as several draft genome sequences. Early on, optical and electron microscopy were the predominant methods for detecting phytoplasma presence and evaluating the extent of elimination. However, polymerase chain reaction (PCR)-based analyses are now possible with a variety of primers for general and ribosomal group-specific analysis. The 16S rDNA sequence amplified by PCR was used to detect and group phytoplasmas in the 1990s (Lee et al., 1998). Because of its high sensitivity and ease of use, this is a very powerful tool for detecting phytoplasmas in plants and insect vectors. Furthermore, 16S rDNA is very useful for classifying phytoplasmas due to its conservation among bacteria (IRPCM, 2004). As a result, 16S rDNA PCR amplification has become the standard method of detecting phytoplasma strains worldwide. The sanitation techniques being used today are influenced by hypotheses and approaches proposed several decades ago. Plant tissue culture techniques have been effectively used for the eradication of viruses and phytoplasmas from infected plants by apical meristem culture alone or in combination with thermotherapy (Faccioli, 2001; Tiwari et al., 2011, 2012). Thermotherapy and/or chemotherapy with the combination of shoot-tip culture has been earlier suggested for the management of plant viruses (Sanchez et al., 1991; Faccioli, 2001; Tiwari et al., 2012). For the elimination of phytoplasmas, a few techniques have been reported, that is, hot water treatment in case of grapevine and sugarcane (Bianco et al., 2000) and cryotherapy for the removal of little leaf phytoplasma from sweet potato (Wang and Valkonen, 2008). This chapter describes in vivo (hot water treatment and chemotherapy on trees and buds) and in vitro (in vitro thermotherapy, meristem-tip culture, and chemotherapy in vitro) elimination methods, with their obvious advantages of less space and time consuming and higher efficacy.

2. In vivo methods
2.1 Chemotherapy

In the early days of phytoplasma discovery by Doi et al. (1967), they were referred to as mycoplasma-like organisms. Antibiotics were used as treatments to eliminate phytoplasmas. In 1967, Ishiie et al. reported the first clear evidence of the therapeutic effects of tetracycline antibiotics. Only tetracycline and chloramphenicol reduced the disease symptoms among the many antibiotics tested. There was an essential link between phytoplasmas and disease that could only be proven with tetracycline therapy. In most cases, symptoms recurred once antibiotic treatments were suspended, regardless of the method of application. Foliar sprays, root immersion, soil drench, and trunk injection were the most common methods of application. It has been used in the urban areas of Southern Florida for a high landscape value palm tree, but antibiotic therapy is rarely used as a crop protection method today (McCoy, 1982). The first time solubilized oxytetracycline hydrochloride was injected into diseased trees through plastic tubes connected to six to eight small holes drilled in the trunks, Nyland and Moller (1973) reported that tree decline and leaf curl could be prevented. In the United States of America, tetracycline antibiotics have been successful in controlling pear decline (PD) for many years (McCoy, 1982). In California, tetracycline injections into the trunk of trees affected by X-disease phytoplasmas were used to treat trees with chemotherapy (Kirkpatrick et al., 1975). There were many instances of remission of symptoms as well as environmental risks in trees treated with this method, and symptoms usually reappeared after 1–2 years. In most countries, it is illegal to use antibiotics in agricultural plants, especially in Europe, where such use is prohibited mostly out of concern for safety and to avoid raising antibiotic-resistant microbial strains. A common disease that affects paulownia trees is paulownia witches' broom (PaWB), which has been associated with phytoplasma presence. In vivo antibiotics are not effective in curing PaWB, although they suppress the symptoms of the disease and even temporarily eliminate them (Jin, 1982). Infected hosts of *Euphorbia pulcherrima* and *Catharanthus roseus* were successfully treated with tetracycline, but symptoms in treated plants mostly reappeared after the chemical treatments are discontinued. Chemical treatments must be repeated at least once a year and accompanied by thorough vector control to maintain their efficacy (Raju and Nyland, 1988). Tetracycline has also been reported to have the effect of reversing symptoms in brinjal plants infected with little leaf disease; however, it could not eliminate the pathogen completely (Bindra et al., 1972; Anjaneyulu and Ramakrishnan, 1973). The root dip treatment of infected onion seedlings with tetracycline for 15 weeks at 7 days intervals would only allow to detect phytoplasmas in plants that have not been treated. Various molecules, such as biophenicol, chlorophenicol, enteromycelin, lycercelin, paraxin, roscillin, oxytetracycline, eucalyptus oil, rose oil, clove oil were used on brinjal cultivars infected with phytoplasmas, but not significantly effective in controlling the disease. The antibiotic-treated brinjal cultivars failed to produce flowers and fruits (Upadhyay, 2016)

There has been an attempt in India to eradicate '*Candidatus* Phytoplasma australasia' (16SrII-D) from *Chrysanthemum morifolium* cv. Ajay Orange by the application of oxytetracycline in in vivo treatments. Using two sets of experiments, six-month-old chrysanthemum cuttings were treated with different concentrations of oxytetracycline (20, 40, 60, 80, and 100 g/L) for in vivo screening. The dip method involved dipping the cuttings in different concentrations of oxytetracycline for 16 h. Oxytetracycline was sprayed foliar on plants raised from cuttings of infected plants at different concentrations. It was found that a concentration of 80 mg/L of oxytetracycline was optimal for the in vivo foliar spray with a phytoplasma elimination of 80%, whereas in the dip method, 80 and 100 mg/L were both optimal for phytoplasma elimination of 80% and 100%, respectively. The following methods were used to treat mulberry plants with antibiotics: foliar sprays, root immersions, cutting treatment prior to planting stored and unsprouted shoots, and finally, grafting treated budwoods (Asuyama and Iida, 1973). During the summer, Seidl (1979) observed that apple proliferation (AP) ('C. P. mali') phytoplasma could be eradicated from budwood by exposing foliated apple bud sticks to an oxytetracycline or chlortetracycline solution (100−200 ppm) immediately before grafting. Montasser et al. (2012) injected the tetracycline in palm tree showing yellowing symptoms; the grafted plants show recovery of the symptoms.

Several methods of reducing the impact of grapevine yellows diseases were tested under field conditions using a plasma-activated water (PAW), designed to modify the chemical composition of water, resulting in an increase in nitrites, nitrates, and peroxides, and a decrease in pH (Laurita et al., 2015). The PAW solution has an extremely high reactivity. In order to avoid interference with the external environment, it has been applied by endotherapy directly into the vascular system of grapevines. In the field trials, 120 plants were selected from 17 vineyards and treated in April, June, and July for 3 years. Comparing all three treatments, the April treatment was found to be most effective when performed during the coolest hours of the day (Zambon et al., 2017). Plants treated with this treatment showed a slight reduction in symptoms and a delay in their appearance (from the end of August/beginning of September to the beginning of October), allowing them to carry on with their normal productivity. A phytoplasma detection analysis confirmed that the treatment reduced the number of infected plants. This containment strategy represents an entirely novel and ecosustainable way of managing phytoplasma associated diseases in grapevine; however, it is not yet possible to measure its impact, considering the multiple factors present in the field. According to recent studies on different species and cultivation systems, PAW treatment increases the expression of key genes involved in defense mechanisms against pathogens in plants, confirming results found in the lab in *C. roseus* and in the field in grapevines (Zambon et al., 2018, 2020; Laurita et al., 2021).

2.2 Thermotherapy by hot water or hot air

In the past, hot water therapy (HWT) has been proposed and used to obtain plant material from diseased plants associated with phytoplasmas presence, such as peach yellows (Kunkel, 1936), rubus stunt (Thung, 1952), sugarcane white leaf and sugarcane grassy shoot (Liu, 1963; Viswanathan and Rao, 2011), and grapevine phytoplasmas (Caudwell, 1966; Caudwell et al., 1990). In spite of the fact that Kunkel (1936) did not know the nature of the causal agents, he succeeded in curing a dormant peach tree infected with yellows in 1936. Following this treatment, he compared dormant budwood that was immersed in water for 10 min at 50°C, where the "virus" was inactivated without any damage to the buds, to bud sticks that had been immersed for 4−5 days at 34°C to 35°C, for 11 h at 38°C, for 40 min at 42°C, for 15 min at 46°C, for 14 min at 48°C, for 3−4 min at 50°C, for 1.5 min.

Three days immersion at 32°C for "flavescence dorée" of grapevine (Caudwell, 1966) or 10 min at 50°C for peach rosette (Kunkel, 1936) were effective. The "stolbur" phytoplasmas ("bois noir" disease, 'Ca. P. solani') in grapevine were much more difficult to eradicate using HWT than the phytoplasmas associated with "flavescence dorée" (Mannini and Marzachì, 2007).

An HWT was applied to grapevine propagation material as a prevention and quarantine measure. A long-duration treatment is used to control exogenous and endogenous pests and pathogens as recommended in EPPO standard pathogen-tested material from grapevine varieties and rootstocks (EPPO, 2015). It is important to apply HWT with care since it may affect the vitality of propagation materials (Mannini, 2007). Using HWT in autumn on elm plants infected with elm yellows disease ('*Ca.* P. ulmi'), the technique produced phytoplasma-free planting material that could be traded over long distances (Boudon-Padieu et al., 2004). However, HWT treatment in spring followed by harsh planting conditions in the forest led to a decreased plant survival. Similar observations were made in grapevine plants, where HWT treatment must be applied in the start or at the end of the conservation in cold environment to guarantee the best survival of the material (Tassart-Subirats et al., 2003). The first experiences on the use of HWTs on dormant wood from grapevines have been carried out for the elimination of pathogens such as the agent of Pierce's disease, a dangerous bacteriosis present in California, by Goheen et al. (1973) by immersion in water at 45–55°C for 10–150 min. Treatments of this type were also useful for eliminating *Phytophthora cinnamomi*, *Agrobacterium tumefaciens*, *Xanthomonas ampelina*, and the nematodes *Xiphinema index* and *Meloidogyne* sp. (Goussard, 1977; Offer and Goussard, 1980; Burr and Katz, 1989).

3. In vitro culture

For the elimination of phytoplasmas, tissue culture has been successfully used. The callus culture of *C. roseus* and sugarcane (Möllers and Sarkar, 1989; Parmessur et al., 2002), the meristem-tip culture of sugarcane (Wongkaew and Fletcher, 2004; Tiwari et al., 2011), and the stem culture of mulberry (Dai et al., 1997) were used to regenerate symptomless, phytoplasma-free plants. For many perennial crop plants, regeneration via callus culture is not feasible, as it is also risky due to spontaneous mutations. The fact that real meristem tip (dome), which lacks leaf primordia and are largely pathogen-free (Pierik and Tetteroo, 1987), facilitates the production of phytoplasma-free tissue cultures. The absence of vascular elements in the meristem may also prevent phytoplasma from moving through a developing plant (Kartha and Gamborg, 1975). Although meristem culture is a size-dependent technique, it is applied to produce phytoplasma-free plants. It is possible for larger meristems to regenerate plants, but not smaller ones. It was found that plants produced from larger meristems were not phytoplasma-free (Parmessure et al., 2002). According to Tiwari et al. (2011), the sugarcane grassy shoot diseases with meristems length of 2–4 mm were free from the phytoplasma. However, the frequency of survival of explants during initiation of shoot cultures was higher in larger meristems (60%) in comparison to smaller ones (40%).

Dai et al. (1997) attempted to eliminate the mulberry dwarf phytoplasma in vitro from mulberry plants (*Morus alba*). A solid Murashige and Skoog (MS) nutrient medium without growth regulators was used to culture stem segments of heavily phytoplasma-infected shoots. In addition, PCR and DAPI staining were used to determine phytoplasma presence in the tissue cultures, as well as observation of symptoms of mulberry dwarf in regenerated plants grown under greenhouse conditions for three years. According to Dai et al. (1997), about 70% to 90% of the regenerated plants were free of phytoplasma after three years (PCR-negative, DAPI-negative, and symptom-free).

Different tissue culture methods were used to eliminate almond witches' broom ('*Ca.* P. phoenicium') from two Lebanese almond varieties (Chalak et al., 2005). However, the cultures from stem cuttings that were treated with heat therapy, the shoot-tip cultures that were subjected or not to thermotherapy, and the micrograft utilizing shoot tips all resulted in the correct regeneration of shoots or elimination of phytoplasmas. A combination of stem cutting with thermotherapy also resulted in a highly effective method of regenerating almond plantlets free of phytoplasmas.

3.1 In vitro meristem preparation in combination with in vitro thermotherapy

Many plant species can be eliminated with heat treatment, including flowers, sugarcane (Quack, 1977), and cassava (Kartha and Garnborg, 1975). The combined application of heat treatment and meristem culture increased the possibility of obtaining phytoplasma-free plants in tissue-cultured meristems (Caudwell et al., 1990). Shoot-tip cultures, stem-cutting cultures amended with thermotherapy found most practical way to grown phytoplasma-free plantlets in two Lebanese varieties of almond (Chalak et al., 2005). Small-scale explants can be used for in vitro thermotherapy (Laimer and Barba, 2011), which is followed by meristem preparation and plant regeneration. In vitro thermotherapy followed by the meristem-tip culture techniques employed for the '*Ca.* P. prunorum' elimination from diseased *Prunus* plants resulted successful, and the '*Ca.* P. mali' (associated with AP), '*Ca.* P. pyri' (associated with PD), and '*Ca.* P. prunorum' (associated with European stone fruit yellows) were eliminated by in vitro thermotherapy combined with meristem preparation from several cultivars of apple, pear, apricot, and peach (Laimer and Bertaccini, 2019), and more recently of '*Ca.* P. rubi' from *Rubus idaeus* (Ramkat et al., 2014) and "bois noir" (Chalak et al., 2013).

4. In vitro chemotherapy

Moreover, the chemical compounds already mentioned were applied to tissue cultures for in vitro chemotherapy. The phytoplasma elimination with oxytetracycline was independent of the shoot-tip size and may be helpful in overcoming the difficulties with excising meristems of reduced size and subsequent difficult regeneration. Very earlier, Sears and Klomparens (1989) used tetracycline on phytoplasma-infected *Oenothera* plant through leaf culture and observed no symptoms after several weeks of treatment. Tetracyclines in several cases appear to have just a bacteriostatic effect on phytoplasmas, since in treated plants, the symptoms may reappeared after the transfer of the shoots to antibiotic-free media (Da Vies and Clark, 1994; Wongkaew and Fletcher, 2004).

A pear cultivar (Pine) infected with phytoplasma was kept on MS medium with gibberellic acid, indole butyric acid, and benzyl amino purine for more than three years. The phytoplasma concentration in micropropagated shoots was higher than those in field-grown plants, although the first remained symptom-free. The growth medium was treated with 100 g/ml oxytetracycline, and shoots were maintained in this medium for a period of 4 weeks to eliminate phytoplasmas. These results must be taken into consideration when plant propagation schemes are enclosing symptomless phytoplasma-infected shoots obtained by micropropagation.

When 1 cm cuttings were treated with 50, 100, and 150 g/mL of oxytetracycline to eliminate almond witches' broom, no plants regenerated (Chalak et al., 2005). As well as grapevine, oxytetracycline caused severe phytotoxicity when 100 mg/l oxytetracycline was added to a culture medium (Gribaudo et al., 2007). According to Carvalho et al. (2017), cassava shoot-tip culture combined with tetracycline treatment can eliminate cassava frog skin disease. Treatment of infected plant materials of *C. roseus* (susceptible host of phytoplasma) through tetracycline 75–80 mg/L in MS medium was found an effective technique, and phytoplasma can be removed from infected plant materials through in vitro by using optimum dose of tetracycline (Singh et al., 2007; Tiwari et al., 2021).

'*Ca.* P. mali' was tested for susceptibility to antimicrobial agents with different target activities, for example, nisin, esculetin, pyrithione, and chloramphenicol. A tetracycline and an enrofloxacin antibiotic were used as controls to compare their activity. The substances were used in the micropropagation of 'Ca. P. mali'-infected apple shoots. Nisin and pyrithione were used at concentrations of 10, 100, and 500 ppm, respectively. In the presence of 10 and 100 ppm of pyrithione, phytoplasma levels were not detectable after 1 and 2 months. The phytoplasma concentration was also reduced by some of the other substances after 2 months after treatment. Aldaghi et al. (2012) observed that micropropagated shoots grown with essential oils underwent death or withering.

A variant of cryopreservation, shoot-tip cryotherapy, has been described as a tool for eradicating plant pathogens (Wang et al., 2009; Feng et al., 2012). It successfully eliminated several plant pathogens that were transmitted by grafts, mostly viruses. Phytoplasmas, such as sweet potato little leaf phytoplasma (16SrII-D) from sweet potato (Wang and Valkonen, 2008) and Chinese jujube witches' broom phytoplasma ('Ca. P. ziziphi' 16SrV-B) from Chinese jujube (*Ziziphus jujuba*) (Wang et al., 2015) have also been reported to be eliminated. A high frequency of pathogen-free plants can be achieved with cryotherapy regardless of shoot-tip size and cryogenic methods. Cryotherapy of shoot tips results in pathogen-free plants through six major steps: (1) excision of shoot tips, (2) cryotherapy, (3) plant regeneration using specific media, (5) molecular indices to detect pathogen presence after cryotherapy, and (6) production and maintenance of pathogen-free plant materials as nuclear sources. The steps (2), (3), and (4) are similar to those employed for the cryopreservation process but had fundamental importance to produce the pathogen eradication from a high number of shoots (Wang and Valkonen, 2009).

Since their health status is strictly regulated by international rules, grapevine, apple, and peach species are frequent targets for thermotherapy, chemotherapy, and tissue culture trials (Panattoni et al., 2013). Host—pathogen interaction insights will also allow us to design not only resistance breeding strategies (Hogenhout et al., 2008) but also to better understand how different elimination strategies work. Similarly, to the differences in the ease of elimination within *Rubus* and apple stem grooving viruses (Knapp et al, 1994), major differences also exist between phytoplasmas such as "flavescence dorée" versus "bois noir" (Mannini, 2007). There is an increasing amount of evidence and data available for phytoplasmas. Due to their host-dependent life cycles, phytoplasmas lack essential biosynthetic pathways since they are obligate parasites with small genomes (Oshima et al., 2013).

A study of the disappearance of symptoms previously present in infected plants led to partial abandonment of testing different treatments and techniques against phytoplasmas. In apple, apricot, and grapevine, this phenomenon occurs both with and without the elimination of the pathogen from the host (Carraro et al., 2004). Musetti et al. (2004) found that 'Ca. P. mali' disappeared from aerial parts of apple trees during the recovery, but not from the underground apparatus. Grapevine "flavescence dorée" phytoplasma was not detected in recovered leaves either (Musetti et al., 2007), but the asymptomatic leaves of apricot trees recovered from asymptomatic conditions were always positive for 'Ca. P. prunorum' (Musetti et al., 2005). Despite several studies, the factors that contribute to recovery from symptoms in vineyards and orchards plants are not fully understood. A kind of systemic acquired resistance might be involved in the reduction of the pathogen's virulence and its symptoms, according to Musetti et al. (2004, 2005, 2007).

5. Conclusion

It is known that phytoplasmas are propagated by multiplication of scions and/or cuttings taken from diseased plants, especially when symptoms are not present (latency period). The provision of pathogen-free mother plants and the establishment of commercial nurseries remain of paramount importance for a thriving horticultural and agricultural sector. Combining in vitro thermotherapy and shoot-tip culture remains the most promising approach to phytoplasma infection. It remains a major challenge in this approach to skillfully excise small meristems and to make sure the meristem regeneration medium is available (Howell et al, 1997) as well as to verify phytoplasma elimination (Heinrich et al., 2001).

References

Aldaghi, M., Bertaccini, A., Lepoivre, P., 2012. cDNA-AFLP analysis of gene expression changes in apple trees induced by phytoplasma infection during compatible interaction. Eur. J. Plant Pathol. 134 (1), 117–130.

Anjaneyulu, A., Ramakrishnan, K., 1973. Thermotherapy of eggplant little leaf disease. Curr. Sci. 38, 271–272.

Asuyama, H., Iida, T.T., 1973. Effects of tetracycline compounds on plant diseases caused by mycoplasma-like agents. Ann. N. Y. Acad. Sci. 225 (1), 509–521.

Bertaccini, A., Arocha-Rosete, Y., Contaldo, N., Duduk, B., Fiore, N., Montano, H.G., Kube, M., Kuo, C.-H., Martini, M., Oshima, K., Quaglino, F., Schneider, B., Wei, W., Zamorano, A., 2022. Revision of the 'Candidatus Phytoplasma' species description guidelines. Int. J. Syst. Evol. Microbiol. 72 (4), 005353.

Bertaccini, A., Duduk, B., 2009. Phytoplasma and phytoplasma diseases: a review of recent research. Phytopath. Medit. 48 (3), 355–378.

Bertaccini, A., Lee, I-M., 2018. Phytoplasmas: An update. In: Phytoplasmas: Plant Pathogenic Bacteria-I. Springer, Singapore, pp. 1–29.

Bianco, P.A., Fortusini, A., Scattini, G., Casati, P., Carraro, S., Torresin, G.C., 2000. Prove di risanamento di materiale viticolo affetto da flavescenza dorata mediante termoterapia. Inf. tore. Fitopatol. 50, 43–49.

Bindra, O.S., OS, B., AS, S., HL, K., GS, D., 1972. Effect of achromycin (tetracycline hydrochloride) on brinjal little-leaf pahogen. Curr. Sci. 41, 819–820.

Boudon-Padieu, E., Larrue, J., Clair, D., Hourdel, J., Jeanneau, A., Sforza, R., Collin, E., 2004. Detection and prophylaxis of elm yellows phytoplasma in France. Forest System 13 (1), 71–80.

Burr, T.J., Ophel, K., Katz, B.H., Kerr, A., 1989. Effect of hot water treatment on systemic *Agrobacterium tumefaciens* biovar 3 in dormant grape cuttings. Plant Dis. 73 (3), 242–245.

Carraro, L., Ermacora, P., Loi, N., Osler, R., 2004. The recovery phenomenon in apple proliferation-infected apple trees. J. Plant Pathol. 141–146.

Carvalho, M.J.S., Oliveira, E.J., Souza, A.S., Pereira, J.S., Diamantino, M.S.A.S., Oliveira, S.A.S., 2017. Cleaning cassava genotypes infected with cassava frogskin disease via in vitro shoot tip culture. Gene Conser. 16 (63).

Caudwell, A., 1966. Inibition in vivo du virus de la flavescence dorée par la chaleur. Études de virologie. Ann. Epiphyties 17, 61–66.

Caudwell, A., Larrue, J., Valat, C., Grenan, S., 1990. Les traitements á l'eau chaude des bois de vigne atteints de la Flavescence dorée. Le Progrès agricole et viticole 107 (12), 281–286.

Chalak, L., Elbitar, A., Rizk, R., Choueiri, E., Salar, P., Bové, J-M., 2005. Attempts to eliminate '*Candidatus* Phytoplasma phoenicium' from infected Lebanese almond varieties by tissue culture techniques combined or not with thermotherapy. Eur. J. Plant Pathol. 112 (1), 85–89.

Chalak, L., Elbitar, A., Mourad, N., Mortada, C., Choueiri, E., 2013. Elimination of grapevine "bois noir" phytoplasma by tissue culture coupled or not with heat therapy or hot water treatment. Adv. Crop Sci. Technol. 1 (2), 107.

Da Vies, D.L., Clark, M.F., 1994. Maintenance of mycoplasma-like organisms occurring in *Pyrus* species by micropropagation and their elimination by tetracycline therapy. Plant Pathol. 43 (5), 819–823.

Dai, Q., He, F.T., Liu, P.Y., 1997. Elimination of phytoplasma by stem culture from mulberry plants (*Morus alba*) with dwarf disease. Plant Pathol. 46 (1), 56–61.

Doi, Y., Teranaka, M., Yora, K., Asuyama, H., 1967. Mycoplasma-or PLT group-like microorganisms found in the phloem elements of plants infected with mulberry dwarf, potato witches' broom, aster yellows, or paulownia witches' broom. Japanese J. Phytopathol. 33 (4), 259–266.

EPPO, 2015. Hot water treatment of *Vitis* sp. for *Xylella fastidiosa*. EFSA J. 13 (9), 4225.

Faccioli, G., 2001. Control of potato viruses using meristem and stem-cutting cultures, thermotherapy and chemotherapy. In: Virus and Virus-like Diseases of Potatoes and Production of Seed-Potatoes. Springer, Dordrecht, pp. 365–390.

Feng, C., Wang, R., Li, J., Wang, B., Yin, Z., Cui, Z., Li, B., Bi, W., Zhang, Z., Li, M., Wang, Q., 2012. Production of pathogen-free horticultural crops by cryotherapy of in vitro-grown shoot tips. In: Protocols for Micropropagation of Selected Economically-Important Horticultural Plants. Humana Press, Totowa, NJ, pp. 463–482.

Goheen, A.C., Nyland, G., Lowe, S.K., 1973. Association of a rickettsia-like organism with Pierce's disease of grapevines and alfalfa dwarf and heat therapy of the disease in grapevines. Phytopathology 63 (3), 341–345.

Goussard, P.G., 1977. Effect of hot-water treatments on vine cuttings and one-year-old grafts. Vitis 16 (4), 272–278.

Gribaudo, I., Ruffa, P., Cuozzo, D., Gambino, G., Marzachì, C., 2007. Attempts to eliminate phytoplasmas from grapevine clones by tissue culture techniques. Bull. Insectol. 60 (2), 315.

Heinrich, M., Botti, S., Caprara, L., Arthofer, W., Strommer, S., Hanzer, V., Katinger, H., Bertaccini, A., da Camara Machado, M.L., 2001. Improved detection methods for fruit tree phytoplasmas. Plant Mol. Biol. Rep. 19 (2), 169–179.

Hogenhout, S.A., Oshima, K., Ammar, E.D., Kakizawa, S., Kingdom, H.N., Namba, S., 2008. Phytoplasmas: bacteria that manipulate plants and insects. Mol. Plant Pathol. 9 (4), 403–423.

Howell, W.E., Burgess, J., Mink, G.I., Skrzeczkowski, L.J., Zhang, Y.P., June 1997. Elimination of apple fruit and bark deforming agents by heat therapy. Acta Hortic. 472, 641–648.

IRPCM, 2004. '*Candidatus* Phytoplasma', a taxon for the wall-less, non-helical prokaryotes that colonize plant phloem and insects. Int. J. Syst. Evol. Microbiol. 54 (4), 1243–1255.

Jin, K.X., 1982. Studies on Control of Paulownia Witches' Broom. Chinese Forest Society. Chinese Forestry Publishing Co, Beijing, PR China, pp. 116–126.

Kakizawa, S., Yoneda, Y., 2015. The role of genome sequencing in phytoplasma research. Phytopath. Moll. 5 (1), 19.

Kartha, K.K., Granborg, O.L., 1975. Elimination of cassava-mosaic disease by meristem culture. Phytopathology 65, 826–828.

Kirkpatrick, H.C., Lowe, S.K., Nyland, G., 1975. Peach rosette: the morphology of an associated mycoplasma-like organism and the chemotherapy of the disease. Phytopathology 65 (8), 864–870.

Knapp, E., Hanzer, V., Weiss H, H., da Camara Machado, A., Weiss, B., Wang, Q., Katinger, H., Laimer da Camara Machado, M., 1994. New aspects of virus elimination in fruit trees. Acta Hortic. 386, 409–418.

Kunkel, L.O., 1936. Heat treatments for the cure of yellows and other virus diseases of Peach. Phytopathology 26 (9).

Laimer, M., Barba, M., 2011. Elimination of systemic pathogens by thermotherapy, tissue culture, or in vitro micrografting. In: Virus Virus-like Diseases of Pome Stone Fruits, pp. 389–393.

Laimer, M., Bertaccini, A., 2019. Phytoplasma elimination from perennial horticultural crops. In: Phytoplasmas: Plant Pathogenic Bacteria-II. Springer, Singapore, pp. 185–206.

Laurita, R., Barbieri, D., Gherardi, M., Colombo, V., Lukes, P., 2015. Chemical analysis of reactive species and antimicrobial activity of water treated by nanosecond pulsed DBD air plasma. Clin. Plasma Med. 3 (2), 53–61.

Laurita, R., Contaldo, N., Zambon, Y., Bisag, A., Canel, A., Gherardi, M., Laghi, G., Bertaccini, A., Colombo, V., 2021. On the use of plasma activated water in viticulture: induction of resistance and agronomic performance in greenhouse and open field. Plasma Process. Polym. 18, e2000206.

References

Lee, I-M., Gundersen-Rindal, D.E., Davis, R.E., Bartoszyk, I.M., 1998. Revised classification scheme of phytoplasmas based on RFLP analyses of 16S rRNA and ribosomal protein gene sequences. Int. J. Syst. Evol. Microbiol. 48 (4), 1153–1169.

Liu, H.P., 1963. The nature of the causal agent of white leaf disease of sugarcane. Virology 21 (4), 593–600.

Mannini, F., 2007. Hot water treatment and field coverage of mother plant vineyards to prevent propagation material from phytoplasma infections. Bull. Insectol. 60 (2), 311.

Mannini, F., Marzachì, C., 2007. Termoterapia in acqua contro i fitoplasmi della vite. L'Informatore Agrario 63 (24), 62.

McCoy, R.E., 1982. Use of tetracycline antibiotics to control yellows diseases. Plant Dis. 66 (7), 539–542.

Möllers, C., Sarkar, S., 1989. Regeneration of healthy plants from Catharanthus roseus infected with mycoplasma-like organisms through callus culture. Plant Sci. 60 (1), 83–89.

Montasser, M.S., Hanif, A.M., Al_Awadhi, H.A., Suleman, P., 2012. Tetracycline therapy against phytoplasma causing yellowing disease of date palms. The FASEB J. 26 (51), 800–801.

Musetti, R., di Toppi, L.S., Ermacora, P., Favali, M.A., 2004. Recovery in apple trees infected with the apple proliferation phytoplasma: an ultrastructural and biochemical study. Phytopathology 94 (2), 203–208.

Musetti, R., Di Toppi, L.S., Martini, M., Ferrini, F., Loschi, A., Favali, M.A., Osler, R., 2005. Hydrogen peroxide localization and antioxidant status in the recovery of apricot plants from European stone fruit yellows. Eur. J. Plant Pathol. 112 (1), 53–61.

Musetti, R., Marabottini, R., Badiani, M., Martini, M., di Toppi, L.S., Borselli, S., Borgo, M., Osler, R., 2007. On the role of H_2O_2 in the recovery of grapevine (*Vitis vinifera* cv. Prosecco) from "flavescence dorée" disease. Funct. Plant Biol. 34 (8), 750–758.

Nyland, G., Moller, W.J., 1973. Control of pear decline with a tetracycline. Plant Dis. Rep. 57 (8), 634–637.

Orffer, C.J., Goussard, P.G., 1980. Effect of hot-water treatments on budburst and rooting of grapevine cuttings. Vitis 19 (1), 1–3.

Oshima, K., Maejima, K., Namba, S., 2013. Genomic and evolutionary aspects of phytoplasmas. Front. Microbiol. 4, 230.

Panattoni, A., Luvisi, A., Triolo, E., 2013. Elimination of viruses in plants: twenty years of progress. Spanish J. Agric. Res. 1, 173–188.

Parmessur, Y., Aljanabi, S., Saumtally, S., Dookun-Saumtally, A., 2002. Sugarcane yellow leaf virus and sugarcane yellows phytoplasma: elimination by tissue culture. Plant Pathol. 51 (5), 561–566.

Pierik, R.L.M., Tetteroo, F.A.A., 1987. Vegetative propagation of *Begonia venosa* Skan in vitro from inflorescence explants. Plant Cell Tissue Organ Cult. 10 (2), 135–142.

Quack, F., 1977. Meristem Culture and Virus Free Plants. Applied and Fundamental Aspects of Plant Cell, Tissue and Organ Culture, pp. 610–615.

Raju, B.C., Nyland, G., 1988. Chemotherapy of mycoplasma diseases. In: Hiruki, C. (Ed.), Tree Mycoplasmas and Mycoplasma Diseases. The University of Alberta Press, Edmonton, Canada, pp. 207–216.

Ramkat, R., Ruinelli, M., Maghuly, F., Schoder, L., Laimer, M., 2014. Identification of Phytoplasmas Associated with Rubus Spp. As Prerequisite for Their Successful Elimination. Phytoplasmas and Phytoplasma Disease Management: How to Reduce Their Economic Impact, p. 159.

Sanchez, G.E., Slack, S.A., Dodds, J.H., 1991. Response of selected *Solanum* species to virus eradication therapy. Am. Potato J. 68 (5), 299–315.

Sears, B.B., Klomparens, K.L., 1989. Leaf tip cultures of the evening primrose allow stable, aseptic culture of mycoplasma-like organism. Can. J. Plant Pathol. 11 (4), 343–348.

Seidl, V., 1979. Some results of several years' study on apple proliferation disease. Acta Hortic. 94, 241–246.

Singh, S.K., Srivastava, P., Singh, B.R., Khan, J.A., 2007. Production of phytoplasma-free plants from yellow leaf diseased *Catharanthus roseus* L.(G.) Don. J. Plant Dis. Prot. 114 (1), 2–5.

Tassart-Subirats, V., Clair, D., Grenan, S., Boudon-Padieu, E., Larrue, J., 2003. Hot water treatment: curing efficiency for phytoplasma infection and effect on plant multiplication material. In: 14th Meeting of the International Council for the Study of Viruses and Virus Diseases of Grapevine (ICVG), pp. 69–70.

Thung, T.H., 1952. Waarnemingen omtrent de dwergziekte bij framboos en wilde braam II. Tijdschr. Plantenziekten 58 (6), 255–259.

Tiwari, A.K., Tripathi, S., Lal, M., Sharma, M.L., Chiemsombat, P., 2011. Elimination of sugarcane grassy shoot disease through apical meristem culture. Arch. Phytopathol. Plant Protect. 44 (20), 1942–1948.

Tiwari, A.K., Rao, G.P., Khan, M.S., Pandey, N., Raj, S.K., 2012. Detection and elimination of Begomovirus infecting *Trichosanthes dioica* (pointed gourd) plants in Uttar Pradesh, India. Arch. Phytopathol. Plant Protect. 45 (9), 1070–1075.

Tiwari, N.N., Jain, R.K., Tiwari, A.K., 2021. Management of phytoplasma causing little leaf and witches' broom disease in *Catharanthus roseus* by in vitro approaches. Agrica 10, 175–181.

Upadhyay, R., 2016. Varietal susceptibility and effect of antibiotics on little leaf phytoplasma of brinjal (*Solanum melongena L*). Int. J. Emer. Trends Sci. Technol 3, 3911–3914.

Viswanathan, R., Rao, G.P., 2011. Disease scenario and management of major sugarcane diseases in India. Sugar Tech 13 (4), 336–353.

Wang, Q.C., Valkonen, J.P.T., 2008. Elimination of two viruses which interact synergistically from sweet potato by shoot tip culture and cryotherapy. J. Virol. Methods 154 (1–2), 135–145.

Wang, Q., Valkonen, J.P., 2009. Cryotherapy of shoot tips: novel pathogen eradication method. Trends Plant Sci. 14 (3), 119–122.

Wang, B., Wang, Q., Engelmann, F., Lambardi, M., Panis, B., Yin, Z., Feng, C., April 2009. Cryotherapy of shoot tips: a newly emerging technique for efficient elimination of plant pathogens. In: I International Symposium on Cryopreservation in Horticultural Species, 908, pp. 373–384.

Wang, R.R., Mou, H.Q., Gao, X.X., Chen, L., Li, M.F., Wang, Q.C., 2015. Cryopreservation for eradication of jujube witches' broom phytoplasma from Chinese jujube (*Ziziphus jujuba*). Ann. Appl. Biol. 166 (2), 218–228.

Wongkaew, P., Fletcher, J., 2004. Sugarcane white leaf phytoplasma in tissue culture: long-term maintenance, transmission, and oxytetracycline remission. Plant Cell Rep. 23 (6), 426–434.

Zambon, Y., Contaldo, N., Canel, A., Laurita, R., Gherardi, M., Colombo, V., Bertaccini, A., 2017. Plasma atmosferico freddo: energia per una viticoltura eco-sostenibile. Conegliano Valdobbiadene 4, 79–82.

Zambon, Y., Contaldo, N., Canel, A., Laurita, R., Beltrami, M., Gherardi, M., Colombo, V., Bertaccini, A., 2018. Controllo e sostenibilità dei giallumi della vite con il plasma. Vite Vino 2, 66–71.

Zambon, Y., Contaldo, N., Laurita, R., Várallyay, E., Canel, A., Gherardi, M., Colombo, V., Bertaccini, A., 2020. Plasma activated water triggers plant defence responses. Scient. Reports 10, 19211.

Further reading

Faccioli, G., Colalongo, M.C., 2002. Eradication of potato virus Y and potato leafroll virus by chemotherapy of infected potato stem cuttings. Phytopath. Medit. 41 (1), 76–78.

Hadidi, A., Barba, M., Candresse, T., Jelkmann, W. (Eds.), 2011. Virus and Virus-like Diseases of Pome and Stone Fruits. APS Press/American Phytopathological Society.

Ishiie, T., Doi, Y., Yora, K., Asuyama, H., 1967. Suppressive effects of antibiotics of tetracycline group on symptom development of mulberry dwarf disease. Japanese J. Phytopathol. 33 (4), 267–275.

CHAPTER 10

Phytoplasma resistance

Isil Tulum[1,2] and Kadriye Caglayan[3]

[1]Istanbul University, Faculty of Science, Department of Botany, Istanbul, Turkey; [2]Istanbul University, Centre for Plant and Herbal Products Research-Development, Istanbul, Turkey; [3]Plant Protection Department, Faculty of Agriculture, Hatay Mustafa Kemal University, Antakya, Turkey

1. Introduction

Phytoplasmas are cell wall–less plant pathogenic bacteria, belonging to the class *Mollicutes*, associated with more than a 1000 diseases. They are a severe threat to agriculture in many countries causing serious yield losses in a wide range of economically important crops, such as ornamentals, many vegetables, and high-value perennial crops, such as coconut, grapes, and fruit trees (Lee et al., 2000; Hogenhout et al., 2008).

Phytoplasmas contain a minimal genome lacking many genes considered essential for cell metabolism such as genes coding ATP synthases and sugar uptake and use. Therefore, they are dependent on their host and rely on the uptake of nutrients by membrane-transport processes (Oshima et al., 2004).

They exhibit a complex pathosystem, which involves a tritrophic interaction and requires sap-feeding insect herbivores, such as leafhoppers, planthoppers, and psyllids, as vectors for transmission to their host plants (Franco-Lara and Perilla-Henao, 2019). In plants, they remain predominantly restricted to the phloem and are very difficult to cultivate in artificial media (Contaldo et al., 2012). Their unusual life cycle compared with other plant pathogenic bacteria makes them difficult to control.

Integrated management approaches require a deep knowledge of this complex pathosystem, and the disease progress of phytoplasmas is highly variable. In some cases, the complexity of this system is increased by multiple insect vector species, the state of host plant, the genetic variability of both phytoplasma and the vector species, dynamics of the vectors, the titer of the phytoplasma, as well as the environmental conditions (Danet et al., 2011; Jarausch et al., 2019).

The basic control systems of phytoplasma diseases mainly are based on the elimination of sources of phytoplasmas and on the control of insect vectors. The elimination of sources of phytoplasmas can be difficult to implement, costly, and time consuming. Controlling the insect vectors of phytoplasmas relies heavily on the use of insecticides. However, insecticides are not environment-friendly, and they can be unsatisfying in eliminating all insect vectors that are transmitting the disease also because for a relevant number of phytoplasma diseases the insect vectors are still unknown (Weintraub and Wilson, 2009).

In addition, global warming and climate change are urging problems against phytoplasma insect vector management. Vectors that are adapted to warmer climates and warmer winters may facilitate the worldwide spread of phytoplasma diseases. The major changes in geographical distribution and population dynamics of the vectors will affect insect-phytoplasma-plant interaction. Therefore, developing appropriate and efficient control and management strategies for phytoplasma diseases will become more important in the future (Hogenhout et al., 2008; Maejima et al., 2014).

Cultivars that are genetically resistant to phytoplasmas offer the best method of phytoplasma control. However, natural resistance is rare. Although a few resistant varieties of commercially important plant species have been identified as resistant cultivars, the breeding of these plants remains challenging due to the lack of knowledge on the mechanisms of phytoplasma resistance and insufficient resistance screening systems (Jarausch and Torres, 2014).

Although the recent advances on phytoplasma detection techniques and the information gained from the recent studies on phytoplasma epidemiology, physiopathology, and biochemistry on the last decade have helped the basic knowledge on phytoplasmas, there is a huge gap of knowledge on the mechanism of resistance (Nejat and Vadamalai, 2013; Kumari et al., 2019). The practical application of genetic resistance is most advanced in apple. Natural resistance to '*Candidatus* Phytoplasma mali' has been identified in *Malus sieboldii*, where promising apple proliferation (AP)-resistant rootstock genotypes have been selected and are used to develop phytoplasma-resistant apple rootstocks. An alternative way to develop phytoplasma resistance could be using potential biocontrol agents or plant resistance inducers, since the natural resistance is very rare (Jarausch et al., 2013).

This chapter summaries the current knowledge on natural (genetic), induced, and transgenic resistance against phytoplasma diseases on resistant plant view and discuss the current status of phytoplasma resistance against major phytoplasma diseases affecting Asian countries particularly India, Iran, Japan, and Turkey.

2. Natural (genetic) resistance

The most promising strategy for controlling phytoplasma diseases is based on selection of natural resistant or tolerant plant varieties. Differently from the general definitions, the term "resistance" for phytoplasma diseases has been described as the absence of symptoms associated with a low pathogen titer in the infected plants whereas "tolerance" for phytoplasma diseases is mild symptoms under a high pathogen titer (Jarausch et al., 2013).

Although there are several phytoplasma diseases that have a major agronomic impact, a few resistant species or varieties have been determined either by experimental infection and by molecular detection techniques (Jarausch et al., 1999; Cardeña et al., 2002; Seemüller and Harries, 2010). Development of efficient detection methods for phytoplasma diseases such as quantitative PCR enabled researchers to perform a more precise selection of resistant genotypes (Nejat and Vadamalai, 2013; Kumari et al., 2019).

Unfortunately, to date, there is still no single fixed method for identifying the resistant species or varieties due to the lack of an efficient culturing method in artificial media for phytoplasmas, complexity of phytoplasma diseases, and their variability through different plant species. These factors make the resistance screening of a large number of genotypes more difficult. The source of the inoculum, the concentration of the phytoplasmas, and the physiology of the plant are very important points to design a successful and efficient screening method. As an example, natural genetic resistance could be identified in the germplasm of fruit trees and grapevine. Phytoplasmas can be maintained in fruit trees and can efficiently be transmitted by graft inoculation, whereas the phytoplasma itself may become undetectable from the crown of the grapevines, as a result of a phenomenon called "recovery," (Musetti et al., 2013a,b).

Another point to be considered is the physiology of the plant and its behavior during certain periods of the year. The use of resistant stone fruit plants differs from those of pome fruits due to the inconstant distribution of the phytoplasma in the phloem tissues and depends on various parameters such as photoperiod and temperatures. Every year in late winter, pome and stone fruits replace their phloem content by which phytoplasmas can be eliminated from the aerial parts of the plants. The removal of the phloem is complete in pome fruits but partial in stone fruit trees and pathogen persists in the top organs throughout the year (Seemuller et al., 1998). There remains a thin layer of small sieve tubes called "winter phloem," in which phytoplasmas are able to persist. Therefore, the recolonization of phytoplasmas on the aerial part of the trees in springtime can be prevented by developing resistant rootstocks. Unfortunately, this strategy is not applicable to grapevine, which does not exhibit replacement of aerial parts every year.

Intra- and interspecific variation in susceptibility to phytoplasma diseases have been reported for several decades. There are a few extensively studied examples of natural resistance to phytoplasma diseases: three major diseases of temperate fruit trees, which are AP, pear decline (PD), and European stone fruit yellows (ESFY), and phytoplasma diseases of other economically important crops, namely, sesame phyllody, brinjal little leaf, and jujube witches' broom (JWB) (Seemüller and Harries, 2010).

2.1 Natural resistance in temperate fruit trees
2.1.1 Apple proliferation

The search for a resistant species was studied extensively on AP disease. AP is associated with the presence of '*Ca*. P. mali' and is among the most damaging plant disease not only in Europe, but also in Syria and Turkey in Asia (Canik and Ertunc, 2007). It is a quarantine disease in Europe and North America (CABI/EPPO, 2013).

Apple is the main host of '*Ca*. P. mali', and almost all cultivars appear susceptible, as well as wild and ornamental *Malus*. Several genes were identified and a genetic variability of '*Ca*. P. mali' has been reported, and different strains were described (Jarausch et al., 2000). Diseased trees of the cultivated apple (*Malus* x *domestica*) are characterized by witches' broom formation, growth suppression, and undersized, tasteless fruit (EPPO, 2006).

As explained previously, efforts to achieve an efficient control was planned by studying the colonization behavior of the AP pathogen. The pathogen is eliminated in the trunk during winter with degeneration of the current year's phloem, followed by overwintering in the roots. In spring, when fresh phloem is generated, the stem may be recolonized from the roots (Schaper and Seemüller, 1982; Seemüller et al., 1984a,b). The fluctuation in the colonization pattern in apple by the AP phytoplasma has led to the presumption that growing scion cultivars on resistant rootstocks can prevent the disease or at least reduce its impact.

Intensive screening approach of many established and experimental rootstocks of *Malus* x *domestica* resulted with the absence of suitable resistance (Seemüller and Harries, 2010; Seemüller et al., 2011). A suitable resistance was shown by trees on *M. sieboldii* x *M. sieboldii* hybrids. Inoculation and natural infection results indicated that trees on most of the *M. sieboldii*-derived progenies showed a high level of AP resistance expressed by low cumulative disease indices, a high percentage of non- or rarely affected trees. However, these rootstocks were more vigorous and less productive than the major rootstock for commercial apple growing in Europe mainly due to alternate bearing. To solve this problem, a classical breeding program was initiated in 2001(Seemüller et al., 2010). The breeding program was hampered by the presence of latent apple viruses in some of the progeny by *M. sieboldii* (Bisognin et al., 2008). Also, *M. sieboldii* derived clones had to be micropropagated to achieve efficient clonal production (Liebenberg et al., 2010). The breeding program continued for 8 years, and the field trials confirmed that the resistance could be inherited to the breeding progeny. In addition, controlled infections allowed the selection of several resistant genotypes and resistant genotypes were identified showing pomological properties similar to commercial apple (Seemüller et al., 2018).

2.1.2 Pear decline

Pear decline (PD) is a widespread disease, threatening pear trees (*Pyrus communis* L.) in many pear growing countries of Europe, America, Australia, and Asia (Seemüller, 1989; OEPP/EPPO, 2007). The disease is associated with '*Ca.* P. pyri,' phylogenetically closely related to '*Ca.* P. mali' (Seemüller and Schneider, 2004; Marcone et al., 2022).

The PD pathogen exhibits a similar colonization pattern with apple (*Malus* spp.) trees affected by AP disease. Therefore, the same strategy with AP disease, which is growing scion cultivars on resistant rootstocks, was proposed to achieve PD resistance (Seemüller et al., 1984a,b). The disease symptoms largely depend on the rootstock and the severity of PD varies widely depending on pear cultivar and the scion/rootstock combination.

The first natural resistance screening on of *P. communis* genotypes, other *Pyrus* species, and quince (*Cydonia oblonga*) genotypes has been performed in North America (Seemüller, 1992; Seemüller et al., 2011). PD resistance was determined in the open-pollinated seedlings of *P. betulifolia*, *P. elaeagrifolia*, *P. nivalis*, *P. pashia*, *P. syriaca*, and *P. communis*, and clonal Quince A and C (*Cydonia oblonga*). However, *P. amygdaliformis*, *P. caucasica*, *P. cordata*, *P. fauriei* seedlings, which were grown from French seeds, exhibited susceptibility (Giunchedi et al., 1995). Later PD resistance studies in Italy showed similar results with the North America results (Poggi Pollini et al., 1994; Giunchedi et al., 1995).

Studies in Germany evaluated nearly 1200 seedlings of *Pyrus* species covering 39 open-pollinated genotypes in 26 *Pyrus* taxa over 18 years for PD resistance. Seedlings of *P. pyrifolia*, *P. ussuriensis*, *P. amygdaliformis*, *P. caucasica*, *P. cordata*, *P. fauriei,* and unspecified *P. communis* seedlings from France were susceptible to PD, whereas particular progenies of *P. communis* originating from Russia, *P. calleryana* from Bradford and *P. betulifolia* originating from Japan, were more resistant. Considerable variation in susceptibility to PD was also observed between progenies of different accessions of the same taxon. When *P. communis* genotypes were examined, the seedlings from Russia were the most resistant ones, whereas those of the French *P. communis* cultivar Feudière were found to be highly susceptible (Seemüller et al., 1998).

Rootstocks can also affect resistance of the scion. Four pear cultivars were bud-inoculated on different root stocks and evaluated for their host response to PD phytoplasma over a period of 3 years in Hatay province of Turkey. As pear cultivars, Santa Maria, Williams and local cultivars Deveci and Ankara were grafted on *Pyrus communis,* OHF333 and BA29 rootstocks. *P. communis* and OHF 333 were found the most susceptible rootstocks for local pear cv. Deveci and as well as for Santa Maria and Williams cultivars. None of the cultivars grafted on BA 29 were found phytoplasma infected in three years. However, the most tolerant combination was Ankara X *P. communis* (Gazel et al., 2012). These studies showed considerable differences in PD resistance between and within the progenies and revealed that resistance is a segregating feature and cannot be assigned to a certain taxon. Vegetative selection and propagation of resistant genotypes as rootstocks would be more suitable instead of seedling progenies (Seemüller et al., 2009).

2.1.3 European stone fruit yellows

European stone fruit yellow (ESFY) is associated with '*Ca.* P. prunorum' and poses a major threat to stone fruit species. The disease is known to be present in Europe, North Africa, and western parts of Asia. It is referred as apricot chlorotic leaf roll in apricot (*Prunus armeniaca*), "leptonecrosis" in Japanese plum (*P. salicina*) yellows and decline disorders in peach (*P. persica*), European plum (*P. domestica*), almond (*P. dulcis*), ornamental cherry (*P. serrulata*), and several rootstocks used for stone fruits (Seemüller and Schneider, 2004; Marcone et al., 2010; Marcone et al., 2014). ESFY also has been detected in wild plants such as *Rosa canina*, *Celtis australis*, *Fraxinus excelsior* (Jarausch et al., 2001), and even grapevine (Varga et al., 2000; Duduk et al., 2004). Interestingly, ESFY exhibits a diverse severity according to the variety of infected stone fruit. European plum, sweet, and sour cherry (*P. cerasus* and *P. avium*) show less susceptibility whereas apricot, Japanese plum, peach, almond, and ornamental cherry are severely affected.

The strategy to control '*Ca.* P. prunorum' by the use of resistant plants differs from those of pome fruits because the pathogen persists in the above-ground organs of affected trees throughout the year (Seemuller et al., 1998). So, a successful disease control might be possible with both rootstock and scion cultivars due to this stone-fruit-specific colonization conditions.

In a previous study, to examine resistance in stone fruits intensively, 23 clonal or seedling rootstocks for stone fruits belonging to several major rootstock groups were inoculated with the ESFY strains by graft-inoculation and were examined by PCR assays and 4′-6-diamidino-2-phenylindole (DAPI) fluorescence method. Different strains of '*Ca.* P. prunorum' have been detected from ESFY-infected apricot, almond, ornamental cherry, peach, and Japanese plum trees, showing great differences in virulence. A hypervirulent strain is able to kill the host plants very quickly, whereas the milder ones do not induce mortality. *P. domestica*, GF 8/1 (*Prunus cerasifera* x *Prunus munsoniana*) and certain *P. insititia* stocks exhibited low susceptibility, seedlings of Myrobalan (*P. cerasifera*) showed moderate susceptibility whereas trees on peach rootstocks Montclar, Rutgers Red Leaf and Rubira on apricot seedling and St Julien 2 (*P. insititia*) showed high susceptibility. Significant differences in susceptibility were also identified in scion cultivars. Virulence often depended on the pathogen-scion combination, host suitability of the genotypes, and the presence of the original host of the pathogen in the scion (Kison et al., 2001).

In more recent studies, apricot, Japanese plum, and peach trees were shown to be more susceptible than myrobalan and European plum genotypes to ESFY, and significant diversity in susceptibility of new cultivars of European and Japanese plums on rootstock was identified (Landi et al., 2010; Torres et al., 2010). In a study using quantitative PCR analyses, it is shown that the colonization of the '*Ca.* P. prunorum' was lower in the apricot trees infected by hypovirulent strains compared with the plants infected by the hypervirulent strains of the pathogen (Ermacora et al., 2010). A highly efficient, stable, and transmissible tolerance to ESFY in recovered Bulida apricot trees was identified recently (Osler et al., 2013, 2016).

2.2 Natural resistance in economically important crops

2.2.1 Sesame phyllody

Sesame phyllody has been reported as the major disease in sesame, causing a yield loss of up to 80% yield. *Sesamum indicum* L. is one of the most relevant crops for oil production and cultivated in several countries, primarily located in the tropic and subtropic areas of Asia and Africa (Chattopadhyay et al., 2015). Phyllody was first reported in a province of Pakistan (Vasudeva and Sahambi, 1955). Currently, the disease is a significant threat to production in the most important sesame-producing regions of the world, including India, Iran, Taiwan, Turkey, Pakistan, Thailand, South Korea, and Myanmar (Rao and Nabi, 2015).

After rRNA sequencing and restriction analyses, a wide genetic diversity was identified among sesame phyllody-associated phytoplasmas. Although four ribosomal groups 16SrI (aster yellows), 16SrII (peanut witches' broom), 16SrVI (clover proliferation) and 16SrIX (pigeon pea witches' broom) were detected, the majority of identified sesame phytoplasmas (approximately 77%) have been known to belong to the 16SrII and 16SrIX groups (Ikten et al., 2016).

Although the impact of sesame phyllody is worldwide, there are only a few studies focusing on field-based selection and identification of resistant or tolerant sesame genotypes (Kumar et al., 2007; Pervaiz Akhtar et al., 2013). The field experiments and screening showed a diversity in resistance under different environmental conditions, and some cultivars exhibited symptomless infection (Pervaiz Akhtar et al., 2013). Evaluation studies for phyllody resistance using crosses between resistant and susceptible varieties showed that disease resistance in cultivated sesame varieties is governed by a single recessive gene. The wild species *S. alatum* and *S. mulayanum* possess a single dominant gene conferring resistance to sesame phyllody (Kumar et al., 2007; Shindhe et al., 2011; Pervaiz Akhtar et al., 2013).

Recently, a quantitative PCR based method which can accurately and quickly evaluate the resistance level of sesame genotypes against sesame phyllody pathogen in Turkey has been developed (Ikten et al., 2016; Ustun et al., 2017).

The recent advances on reliable detection methods will aid the search for new sources of resistance in order to manage sesame phyllody.

2.2.2 Brinjal little leaf

Eggplant (*Solanum melongena* L.) commonly known as brinjal in South Asia, Southeast Asia, and South Africa is an important vegetable crop cultivated worldwide. Brinjal little leaf (BLL) is a widespread phytoplasma disease in India that causes severe economic losses (Mitra, 1993).

Two phytoplasma groups (16SrI and 16SrVI) were reported to be associated with BLL disease in India (Azadvar and Baranwal, 2012; Kumar et al., 2012). There is still limited knowledge on resistant cultivars of brinjal. Only the wild relatives of brinjal, *Solanum integrifolium* and *S. gilo* showed resistance to the little leaf disease (Chakrabarti and Choudhury, 1975).

2.2.3 Jujube witches' broom

Jujube (*Ziziphus jujuba* Mill.) also called Chinese date or Chinese jujube, an economically important, multipurpose fruit tree of Asia, is one of the most important members in the Rhamnaceae family, and commercial jujube cultivation has developed in China, Japan, South Korea, Iran, Israel, the United States, Italy, Australia, and other 48 countries (Liu et al., 2020).

Jujube witches' broom disease is the most destructive disease of Chinese jujube and causes serious problems to the industry. Analyses of 16S rDNA sequences showed that JWB is associated with 'Ca. P. ziziphi', 16SrV-B (Jung et al., 2003).

In a previous study, 30 JWB-free accessions were collected and subjected to a 5-year test. Liu et al. (2004) developed a protocol grafting healthy germplasm onto seriously diseased plants that was found to be more effective for detecting the highly resistant accessions compared to the method of traditional grafting of diseased bark onto healthy germplasm. In 2005, after 10 years of selection and practice, a highly resistant cultivar, called "Xingguang," which shows a low disease rate and a late symptom appearance, was reported (Liu et al., 2004; Zhao et al., 2009).

Recent analyses showed changes in the transcriptome, proteome, and phytohormone levels in response to grafting a JWB-infected scion onto a susceptible cultivar of jujube. A high number of differentially expressed genes (DEGs) were identified after grafting as early responses to phytoplasma infection (Shao et al., 2016; Fan et al., 2017; Ye et al., 2017).

3. Induced resistance

The recommended use of healthy plants as propagation material is still not sufficient to limit phytoplasma spread. Genetic resistance is very rare, and not many phytoplasma infected plants show resistance or tolerance to disease. Breeding resistant cultivars is time consuming and may not result successfull. In addition, new phytoplasmas are discovered at an increasing pace (Zhao and Davis, 2016).

A promising approach to manage phytoplasma diseases is using resistance inducers. Resistance inducers also called as elicitors are molecules that can have biotic or synthetic origins, which can challenge the plant and cause a reaction by either production of antimicrobials or stimulation of plant defense to trigger induced resistance (IR). Elicitors can have different mode of action. They can cause a fall in the number of infected plants, reduce the severity of the disease symptoms, or delay the appearance of the disease symptoms (Romanazzi, 2013). This strategy was successfully used for the control of grapevine decline by spraying the plant canopy with resistance inducers in Iran (Ghayeb Zamharir and Taheri, 2019). In this study, two molecules (T1: propamocarb/fosetyl 530:310 g/L and T2: hymexazol) were sprayed on the canopy of phytoplasma-infected grapevines (cv. Bidaneh Sefid). The results of this study showed that these resistance-inducers could promote host defence in treated phytoplasma-infected plants and could be useful for control of this grapevine phytoplasma disease in Iran.

3.1 Resistance inducers

Abiotic resistance inducers are chemical compounds that are able to start a reaction in the plant. The most common used to control phytoplasma diseases are: benzothiadiazole (BTH), phosetyl-aluminium, prohexadione calcium, indole-3-butyrric acid (IBA), indole-3-acetic acid (IAA), chitosan, salicylic acid (SA), mixture of glutathione, and oligosaccharines (GOs).

The trials of abiotic elicitors were performed on experimental hosts infected with a phytoplasma, such as *Arabidopsis thaliana* infected with X-disease phytoplasma, *Catharanthus roseus* infected with chrysanthemum yellows (CY), elm yellows (EY), or aster yellows (AY) (Romanazzi, 2013).

One of the first examples of using abiotic resistance inducers to manage phytoplasmas was application of two commercial formulations based on active ingredients of phosetyl-aluminum and GOs on Chardonnay and Vermentino infected with "bois noir" (BN). Unfortunately, the trials showed no significant effects of the treatment on the qualitative and quantitative parameters of healthy and symptomatic plants (Garau et al., 2008). When a combination of five commercial resistance inducers based on chitosan, phosetyl-aluminum, BTH, and two GOs was applied, a significant reduction of number of symptomatic plants was achieved (Romanazzi et al., 2009).

On the other hand, endophytes are also acting as elicitors that have a biotic inducer effect. Endophytes are an endosymbiotic group of microorganisms, mainly fungi and bacteria, that reside inside plants without symptoms of disease. They exhibit complex interactions with host plants to enhance host growth and nutrient gain. They improve the plant's ability to tolerate various types of stresses by inducing the production of phytohormones, other bioactive compounds such as antibiotic molecules, as well as lytic enzymes such as hydrolases, chitinases, laminarinases, and glucanases, and inducing systemic resistance (ISR) (Gouda et al., 2016).

The use of bacterial endophyte, *Pseudomonas putida* S1Pf1Rif, to suppress the symptoms induced by '*Ca.* P. asteris' in *Chrysanthemum carinatum* was reported (Gamalero et al., 2010). A higher level of resistance to phytoplasma infection was reported using *P. putida* S1Pf1Rif in combination with the mycorrhyzal fungus *Glomus mosseae* BEG12, (D'Amelio et al., 2011). Recently, it was shown that bacterial endophyte *Pseudomonas migulae* 8R6 increase resistance to "flavescence dorée" infection in grapevine by synthesizing ACC deaminase, which regulates the level of the stress-related hormone ethylene (Gamalero et al., 2017).

3.2 The "recovery" phenomenon

Elicitors can lead to a phenomenon called as "recovery," which causes a spontaneous remission of the disease symptoms in plants that were previously symptomatic. "Recovery" has been reported in grapevines, apple, and apricot plants infected by phytoplasmas (Musetti et al., 2007, 2010). The mechanism of this spontaneous phenomenon is still not fully understood, but recent studies suggest that the phenomenon can occur due to the expression of acquired resistance by the plant.

It is shown that accumulation of a stable reactive oxygen species (ROS), H_2O_2, in the sieve elements of apple trees is associated with symptom remission from AP (Musetti et al., 2004). The change of the oxidative status of the sieve elements causes biochemical changes in the phloem leading to modifications of phloem protein (P-protein) conformation. The sieve elements of "recovered" apple trees exhibit an abnormal accumulation of callose and protein, associated with the upregulation of callose synthase- and P-protein-coding genes. This phenomenon supports the hypothesis that recovered plants are able to develop resistance mechanisms depending on Ca^{2+} signal activity.

Later, it is shown that, levels of endogenous salicylic acid (SA) were also increased in apple trees with AP (Patui et al., 2013). On the other hand, apple trees "recovered" from AP show up-regulated jasmonate (JA)-related gene expression (Musetti et al., 2013a,b), and high level of JA can be detected in grapevine recovered from BN disease (Paolacci et al., 2017). This "recovery" phenomenon can be common or rare, permanent, or temporary. As an example, grapevines do not acquire permanent immunity from BN after the "recovery". Interestingly, "recovered" plants exhibit similar behavior as healthy plants, with the advantage of less susceptibility to the disease than never infected plants (Osler et al., 2000).

Recently, "molecular memory phenomena" was suggested in grapevine, which are recovering from "flavescence dorée" disease. It is suggested that the "recovery" status associated to symptomless phenotypes is maintained through transcriptional reprogramming processes, switching off biochemical signals previously triggered by the pathogen, and through DNA methylation events (Pagliarani et al., 2020).

To elucidate the molecular mechanisms underlying the differences in susceptibility between diverse varieties of grapevine, the early transcriptional changes were compared the highly susceptible Chardonnay and the scarcely susceptible Tocai friulano by RNA-Seq. Chardonnay plants showed suppression on the gene expression levels of JA-regulated defense pathway (Bertazzon et al., 2019).

A comprehensive overview of transcriptional changes identified in the presence of phytoplasma infections will deepen the knowledge on the molecular regulatory steps affecting the metabolic changes underlying "recovery".

3.3 Effectors and plant immunity

Phytoplasmas produce specific molecules, called effectors that can overcome the plant immune system, alter their hosts' metabolism in order to invade the whole plant. Recently, more than 56 phytoplasma effectors such as SAP11, SAP54, and TENGU have been identified and have been reported to influence plant gene expression leading to modulate plant development, including indeterminate growth of flower organs, morphology alternation, and sterility in *Arabidopsis* plants (Hoshi et al., 2009; Sugio et al., 2011). The results indicate that phytoplasma effectors play an important role in plant immune response and phytoplasma pathogenesis.

Plants detect pathogen attack through a multilayered process consisting of at least two layers. The first layer consists of the perception of conserved microbial signatures, called microbe- or pathogen-associated molecular patterns (MAMPs or PAMPs) or host-derived damage-associated molecular patterns (DAMPs) by plant plasma membrane pattern recognition receptors (PRRs) (Boutrot and Zipfel, 2017). This first layer of immune system is called as PAMP-triggered immunity (PTI). Some pathogens that are adapted to their host plants are capable of inhibiting PTI by delivering virulence effector proteins into host cells. Inhibition of PTI results in effector-triggered susceptibility (ETS) and activates the second layers of the immune system, which is called as effector-triggered immunity (ETI) (Zhang and Zhou, 2010; Héloir et al., 2019).

Interestingly, phytoplasmas have no outer cell wall so that they may escape from the plant immune system. However, phytoplasmas have genes encoding cold shock proteins and the EF-Tu, which may activate PTI. It is still unclear whether phytoplasma effectors can escape from this immune system or phytoplasma effectors may have the ability to inhibit the plant defense response (Sugio et al., 2011).

4. Transgenic resistance

The difficulty of cultivating phytoplasmas in artificial media hampered the efforts of developing efficient management methods. In addition, there are still a few numbers of phytoplasms for which the complete sequences of the genomes were reported. So, the conventional management techniques that are applicable for other plant pathogens are not available for phytoplasma diseases. In order to introduce transgenic resistance against phytoplasma diseases several methods such as enhancing natural defense by promoting cell death at the site of infection by the hypersensitive response, inhibiting pathogen growth by transgenic expression of antimicrobial peptides or single-chain variable-fragment (scFv) antibodies has been explored to generate resistance against phytoplasmas (Laimer et al., 2009).

An antibody-based approach to induce resistance against '*Ca.* P. solani' associated with BN disease in grapevine, has been performed in transgenic tobacco plants, which are expressing an scFv antibody specific for the immunodominant membrane protein of '*Ca.* P. solani' resulting in no significant resistance (Malembic-Maher et al., 2005).

Expression of antibacterial peptides was reported to induce resistance against paulownia witches' broom phytoplasma (16SrI). Trials on transgenic *Paulownia* plants constitutively expressing the gene encoding the cecropin Shiva-1 resulted in fewer phytoplasma and less symptoms (Du et al., 2005).

Later, in order to avoid the unnecessary exposure of antibacterial peptide on nontarget plant tissues, the use of phloem-specific promoters, which can directly deliver antibacterial peptides to phloem sieve elements where phytoplasmas localize and multiply was conducted. The *Arabidopsis* sucrose-H$^+$ symporter AtSUC2, a main phloem-loading transporter, was efficiently transformed in strawberry plants showing directed phloem-specific expression of the GUS reporter gene (Zhao et al., 2004). Recently, the same strategy was applied to cherry (*Prunus avium* L.), which is susceptible to X disease and lethal yellows disease. This strategy can also be applied to engineer phytoplasma resistance in rootstocks rather than individual varieties.

5. Current status of phytoplasma resistance in India, Iran, Japan, and Turkey

5.1 India

In India, 10 different groups of phytoplasmas affecting more than 172 plant species are reported from all parts of the country until now. Aster yellows (16SrI) is the prevalent group where 16SrII, 16SrVI, 16SrXI, and 16SrXIV groups are the widespread groups causing serious economic losses (Rao, 2021).

The search for plant varieties showing resistance to the phytoplasma diseases was successful on sesame and rice. In sesame (*Sesamum indicum*), 150 germplasm, 32 released varieties, and four wild spp. of sesame were evaluated, and resistance to the sesame phyllody has been identified. The results indicated that a single recessive gene governs resistance in cultivated varieties (KMR14 and Pragati); whereas wild species (*S. alatum* and *S. mulayanum*) possess a single dominant gene conferring resistance against sesame phyllody (Kumar et al., 2007). Resistance to the rice yellow dwarf '*Ca.* P. oryzae' has been determined, and disease resistance is reported to be controlled by a single dominant gene (Muniyappa and Raychaudhuri, 1988).

5.2 Iran

Diverse groups of phytoplasma diseases have been reported from most of the provinces in Iran. Over 70 plant species from 26 botanical families are hosts of phytoplasmas and most of the characterized phytoplasmas belonged to 16SrI, 16SrII, 16SrVI, 16SrIX, and 16SrXII groups. Phytoplasmas of the groups 16SrX, 16SrXIV, 16SrXXX, and 16SrVII are also present in Iran (Siampour et al., 2019).

Recent study by Rastegar et al. tested two resistance inducers, fosetyl-Al (Previcur Energy) and propamocarb, in Mexican lime trees infected by lime witches' broom phytoplasma, and the levels of phytohormones in treated plants were compared with those in healthy and infected ones. The results showed that treated plants exhibited a reduction in the presence of '*Ca.* P. aurantifolia' agent of lime witches' broom, and a decrease in disease severity. The phytohormone analyses showed an increase in the contents of jasmonate and its analogs in infected and treated plants (Rastegar et al., 2021). A comprehensive analysis on phytohormones after applying various resistance inducers will help the understanding on how resistance inducers work and will enable to find a universal defense response that is durable and broad-spectrum.

5.3 Japan

More than 50 decades since their discovery, phytoplasmas remain the most poorly characterized phytopathogens (Doi et al., 1967). Japan has been one of the pioneer countries working on the fundamentals of phytoplasmas and phytoplasma-associated diseases. Since there are not efficient resistant cultivars, most of the projects and studies are aiming for the elucidation of the biology and complex life cycle of phytoplasmas and early detection of phytoplasmas.

In Japan, approximately 70 strains of phytoplasmas have been classified as '*Ca*. P. asteris', '*Ca*. P. japonicum', '*Ca*. P. fragariae', '*Ca*. P. aurantifolia', '*Ca*. P. pruni', '*Ca*. P. oryzae', '*Ca*. P. castaneae', '*Ca*. P. luffae', and '*Ca*. P. ziziphi' (Maejima et al., 2014).

'*Ca*. P. asteris' PaWB strain (PaWB) is associated with paulownia witches' broom disease, an important paulownia tree disease. The trees that are once infected with PaWB loose their vigor and often die. Until 1960, more than 1 million PaWB-infected trees were reported in Japan excluding Hokkaido Prefecture and the Tohoku region. But the disease spread was so fast that in 1998, PaWB-infected trees were reported in almost every plantation in the Tohoku region (Ito, 1960; Nakamura et al., 1998; Maejima et al., 2014).

'*Ca*. P. oryzae' RYD strain (RYD) is associated with rice yellow dwarf, a serious rice disease in many regions of Asia, causing a nearly 80% yield loss once infected the plants (Jung et al., 2003). There had been an outbreak of rice yellow dwarf disease affecting 10%−50% of rice growing areas of Japan (Komori, 1966; Sameshima, 1967).

In a recent study, an efficient and reproducible screening system to evaluate the efficacy of antimicrobials against phytoplasmas, using an *in vitro* plant−phytoplasma co-culture system, was developed. This is the first study to predict the genetic factors associated with susceptibility and resistance to antimicrobials (Tanno et al., 2018). In this study, 40 antimicrobials and identified six nontetracyclines (norfloxacin, azithromycin dihydrate, 5-fluorouracil, chloramphenicol, thiamphenicol, and rifampicin) were tested on '*Ca*. P. asteris' micropropagated infected shoots, The results showed a total phytoplasma elimination from infected shoots by the application of both tetracycline and rifampicin. The comparative analyses on nucleotide sequences of rRNAs and amino acid sequences of proteins targeted by antimicrobials between phytoplasmas and other bacteria indicated that the antimicrobial target sequences were conserved among various phytoplasma species. This work will be not only be useful for providing new strategies for phytoplasma disease management but also convenient for monitoring the emergence of antimicrobial-resistant phytoplasma strains.

In another study, Himeno et al. developed an experimental method based on trypan blue staining to detect phloem cell death in presence of phytoplasma infection. The results indicated that phloem cell death impaired phloem loading, and produce a dramatic accumulation of sucrose by about 1000-fold, resulting in activating anthocyanin biosynthesis pathway. Further results suggested that anthocyanin accumulation is involved in the reduction of cell death, and plant immune response pathway is associated with diseased symptoms (Himeno et al., 2014). This study will contribute to development of phytoplasma-tolerant plants.

As explained in the induced resistance section, the fundamental studies on phytoplasma effectors, the mechanism how they modulate plant, and their effect on plant defense mechanisms will contribute to develop strategies to suppress phytoplasma multiplication in crop plants and generate knowledge for the designing of novel management strategies (Oshima et al., 2004; Hoshi et al., 2009; Sugio et al., 2011; Sugawara et al., 2013; Maejima et al., 2015; Kitazawa et al., 2017).

5.4 Turkey

To date, seven groups of phytoplasmas; 16SrI (aster yellows); 16SrII (peanut witches' broom); 16SrVI (clover proliferation); 16SrIX (pigeon pea witches' broom); 16SrX (AP); 16SrXII ("stolbur"); and 16SrXIV (Bermuda grass white leaf) were reported in Turkey associated with devastating losses on various plant species. "Stolbur" (16SrXII-A) and clover proliferation (16SrVI) phytoplasmas are the prevalent and widespread groups in vegetables whereas AP group (16SrX) is dominating in fruit trees (Çağlayan and Gazel, 1999; Çağlayan et al., 2004; Serçe et al., 2006; Serçe et al., 2007; Choudhary et al., 2007; Gazel et al., 2007, 2009, 2016; Çaglayan et al., 2013; Oksal et al., 2017; Usta et al., 2018; Caglayan et al., 2019; Oksal, 2020).

Although the great impact of phytoplasma diseases on agriculture and the efforts of Turkish phytoplasma team, there are still not enough knowledge on resistant cultivars in Turkey. The most promising results were reported for the field and greenhouse experiments of sesame phyllody disease. Ikten et al. developed a diagnostic multiplex qPCR assay using TaqMan chemistry based on detection of the 16S ribosomal RNA gene of phytoplasmas and the 18S ribosomal gene of sesame (Ikten et al., 2016). This accurate quantification method was used for the selection of resistant sesame varieties. Later, a total of 542 sesame genotypes from 29 different countries were screened for the phytoplasma resistance in the field and greenhouse for two consecutive years to ensure high disease pressure. Two sesame accessions (ACS38 and ACS102) were identified as resistant to the phytoplasmas causing phyllody disease, which are also confirmed by qPCR assays (Ustun et al., 2017). These resistant genotypes have potential for understanding the mechanism of resistance to sesame phyllody disease and for sesame farming in the regions where the disease is widespread.

An extensive screening study on different pear cultivars and rootstock-scion combinations to 'Ca. P. pyri' was assessed in order to determine resistant/tolerant variants to PD disease. The Turkish local pear cultivar Deveci resulted the most susceptible as compared to the other local cultivar Ankara and to well-known pear cultivars, Williams and Santa Maria, on Quince rootstock (BA29) than *Pyrus communis* seedlings. The combination of Deveci x *P. communis* plants could not last one year after inoculating with 'Ca. P. pyri' (Gazel et al., 2007, 2020; Caglayan et al., 2019).

In order to cope with 'Ca. P. mali' infections, two different strategies are being applied. One is using *Malus sieboldii* rootstocks to develop phytoplasma resistance and the other one is using endophytic bacteria 'Ca. P. mali' interactions as biocontrol agents (Bulgari et al., 2012). A collaborative study of Turkish phytoplasma group on identifying endophytic bacteria living in the roots of healthy and 'Ca. P. mali'-infected apple trees showed the presence of the groups *Proteobacteria*, *Acidobacteria*, *Bacteroidetes*, *Actinobacteria*, *Chlamydiae*, and *Firmicutes* by 16S rDNA sequence analysis. These bacteria can be used to control 'Ca. P. mali' to develop sustainable approaches for managing AP disease.

6. Conclusion and perspectives

Phytoplasmas are one of the most aggressive phloem-limited pathogens that are obligate parasites of plants transmitted by sap-feeding insects, which also serve as their hosts. They infect more than 1000 plant species including many important crops and are associated with devastating yield losses worldwide.

Despite their agricultural importance and unique features, phytoplasmas remain one of the most poorly characterized plant pathogens mostly due to the lack of a routinely applicable cultivation in artificial media, gene delivery, and mutagenesis systems. It is agriculturally important to identify factors involved in their pathogenicity to discover effective measures to control phytoplasma diseases.

Research on resistant or tolerant phytoplasma host plants remains challenging as the mechanisms are not yet completely understood. However, further research on phytoplasma disease resistance is of high importance as resistant cultivars would be a highly efficient control measure for these difficult to control diseases.

The recent research on genetic and induced resistance to phytoplasmas in fruit trees and grapevine, application of genetic resistance in apple, development of model systems to study the genetic basis of the resistance mechanism, and application of resistance inducer such as bioactive compounds and endophytes are increasing the knowledge on phytoplasma resistance and are offering environmentally friendly strategies to manage them.

References

Akhtar, K.P., Sarwar, G., Sarwarm, N., Tanvir Elani, M., 2013. Field evaluation of sesame germplasm against sesame phyllody disease. Pak. J. Bot. 45 (3), 1085–1090.

Azadvar, M., Baranwal, V.K., 2012. Multilocus sequence analysis of phytoplasma associated with brinjal little leaf disease and its detection in *Hishimonus phycitis* in India. Phytopath. Moll. 2 (1), 15.

Bertazzon, N., Bagnaresi, P., Forte, V., Mazzucotelli, E., Filippin, L., Guerra, D., Zechini, A., Cattivelli, L., Angelini, A., 2019. Grapevine comparative early transcriptomic profiling suggests that "flavescence dorée" phytoplasma represses plant responses induced by vector feeding in susceptible varieties. BMC Genom. 20, 526.

Bisognin, C., Schneider, B., Salm, H., Grando, M.S., Jarausch, W., Moll, E., Seemüller, E., 2008. Apple proliferation resistance in apomictic rootstocks and its relationship to phytoplasma concentration and simple sequence repeat genotypes. Phytopathology 98 (2), 153–158.

Boutrot, F., Zipfel, C., 2017. Function, discovery, and exploitation of plant pattern recognition receptors for broad-spectrum disease resistance. Annu. Rev. Phytopathol. 55, 257–286.

Bulgari, D., Bozkurt, A.I., Casati, P., Caglayan, K., Quaglino, F., Bianco, P.A., 2012. Endophytic bacterial community living in roots of healthy and 'Candidatus Phytoplasma mali'-infected apple (*Malus domestica*, Borkh.) trees. Anton. Leeuw. 102 (4), 677–687.

CABI/EPPO, 2013. *Candidatus* Phytoplasma mali. [Distribution Map]. Distribution Maps of Plant Diseases, (Edition 2). Map 761. Wallingford.

Çağlayan, K., Gazel, M., 1999. Primary studies for viroid and phytoplasma problems of stone fruits in East Mediterranean Area of Turkey. In: XIVth International Plant Protection Congress (IPPC). Jerusalem, p. 16.

Çağlayan, K., et al., 2004. Doğu Akdeniz Bölgesindeki sert çekirdekli meyve ağaçlarında Avrupa Sert Çekirdekli Meyve Sarılığı (ESFY) fitoplazmasının yaygınlık durumunun PCR/RFLP yöntemiyle saptanması. In: Türkiye I. Bitki Koruma Kongresi. Samsun, p. 141.

Çaglayan, K., Gazel, M., Küçükgöl, C., Paltrineri, S., Contaldo, N., Bertaccini, A., 2013. First report of '*Candidatus* Phytoplasma asteris'(group 16SrI-B) infecting sweet cherries in Turkey. J. Plant Pathol. 95 (4), 77.

Caglayan, K., Gazel, M., Škorić, D., 2019. Transmission of phytoplasmas by agronomic practices. In: Phytoplasmas: Plant Pathogenic Bacteria—II. Springer Singapore, Singapore.

Canik, D., Ertunc, F., 2007. Distribution and molecular characterization of apple proliferation phytoplasma in Turkey. Bull. Insectol. 60 (2), 335—336.

Cardeña, R., Ashburner, G.R., Oropeza, C., 2002. Identification of RAPDs associated with resistance to lethal yellowing of the coconut (*Cocos nucifera* L.) palm.

Chakrabarti, A., Choudhury, B., 1975. Breeding brinjal resistant to little leaf disease. Proceedings of the Indian National Science Academy (Section B) 379—385.

Chattopadhyay, C., Kolte, S.J., Waliyar, F., 2015. Diseases of Edible Oilseed Crops. CRC Press, Boca Raton, FL.

Choudhary, D.K., Prakash, A., Johri, B.N., 2007. Induced systemic resistance (ISR) in plants: mechanism of action. Indian J. Microbiol. 289—297.

Contaldo, N., Bertaccini, A., Paltrinieri, S., Windsor, H.M., Windsor, D.G., 2012. Axenic culture of plant pathogenic phytoplasmas. Phytopath. Medit. 51, 607—617.

Danet, J-L., Balakishiyeva, G., Cimerman, A., Sauvion, N., Marie-Jeanne, V., Labonne, G., Lavina, A., Batlle, A., Krizanac, I., Skoric, D., Ermacora, P., Ulubas-Serce, C., Caglayan, K., Jarausch, W., Foissac, X., 2011. Multilocus sequence analysis reveals the genetic diversity of European fruit tree phytoplasmas and supports the existence of inter-species recombination. Microbiology 157 (2), 438—450.

Doi, Y., Teranaka, M., Yora, K., Asuyama, H., 1967. Mycoplasma- or PLT group-like microorganisms found in the phloem elements of plants infected with mulberry dwarf, potato witches' broom, aster yellows, or paulownia witches' broom. Jpn. J. Phytopathol. 33 (4), 259—266.

Du, T., Wang, Y., Hu, Q-X., Chen, J., Liu, S., Huang, W-J., Lin, M-L., 2005. Transgenic paulownia expressing shiva-1 gene has increased resistance to paulownia witches' broom disease. J. Integr. Plant Biol. 47 (12), 1500—1506.

Duduk, B., Botti, S., Ivanović, M., Krstić, B., Dukić, N., Bertaccini, A., 2004. Identification of phytoplasmas associated with grapevine yellows in Serbia. J. Phytopathol. 152 (10), 575—579.

D'Amelio, R., Berta, G., Gamalero, E., Massa, E., Avidano, L., Cantamessa, S., D'Agostino, G., Bosco, D., Marzachì, C., 2011. Increased plant tolerance against chrysanthemum yellows phytoplasma ('*Candidatus* Phytoplasma asteris') following double inoculation with *Glomus mosseae* BEG12 and *Pseudomonas putida* S1Pf1Rif. Plant Pathol. 60 (6), 1014—1022.

EPPO, 2006. '*Candidatus* Phytoplasma mali'. EPPO Bulletin 36. European and Mediterranean Plant Protection Organisation.

Ermacora, P., Loi, N., Ferrini, F., Loschi, A., Martini, M., Osler, R., Carraro, L., 2010. Hypo-and hyper-virulence in apricot trees infected by European stone fruit yellows. Julius-Kühn-Archiv. 427, 198—200.

Fan, X.P., Liu, W., Qiao, Y-S., Shang, Y-J., Wang, G-P., Tian, X., Han, Y-H., Bertaccini, A., 2017. Comparative transcriptome analysis of *Ziziphus jujuba* infected by jujube witches' broom phytoplasmas. Sci. Hortic. 226, 50—58.

Franco-Lara, L., Perilla-Henao, L., 2019. Management of Phytoplasmas in Urban Trees. In: Olivier, C., Dumonceaux, T., Pérez-López, E. (Eds.), Sustainable Management of Phytoplasma Diseases in Crops Grown in the Tropical Belt. Sustainability in Plant and Crop Protection, 12. Springer, Cham.

Gamalero, E., D'Amelio, R., Musso, C., Cantamessa, S., Pivato, B., D'Agostino, G., Duan, J., Bosco, D., Marzachì, C., Berta, G., 2010. Effects of *Pseudomonas putida* S1Pf1Rif against chrysanthemum yellows phytoplasma infection. Phytopathology 100 (8), 805—813.

Gamalero, E., Marzachì, C., Galetto, L., Veratti, F., Massa, N., Bona, E., Novello, G., Glick, B.R., Ali, S., Cantamessa, S., D'Agostino, G., Berta, G., 2017. An 1-Aminocyclopropane-1-carboxylate (ACC) deaminase-expressing endophyte increases plant resistance to flavescence dorée phytoplasma infection. Plant Biosyst. 151 (2), 331–340.

Garau, R., et al., 2008. Biostimulants distribution to plants affected by "bois noir": results regarding recovery. Petria 18, 366–368.

Gazel, M., Ulubas Serçe, C., Çaglayan, K., Öztürk, H., 2007. Detection of 'Candidatus Phytoplasma pyri' in Turkey. Bull. Insectol. 60, 125–126.

Gazel, M., Caglayan, K., Ulubas Serce, C., Son, L., 2009. Evaluations of apricot trees infected by 'Candidatus Phytoplasma prunorum' for horticultural characteristics. Roman. Biotechnol. Lett. 14 (1), 4123–4129.

Gazel, M., et al., 2012. Susceptibility of some apricot and pear cultivars on various rootstocks to 'Candidatus Phytoplasma pruni' and 'Ca. P. pyri'. In: 22nd International Conference on Virus and Other Graft Transmissible Diseases of Fruit Crops, Rome, 119.

Gazel, M., Çağlayan, K., Başpınar, H., Mejia, J.F., Paltrinieri, S., Bertaccini, A., Contaldo, N., 2016. Detection and identification of phytoplasmas in pomegranate trees with yellows symptoms. J. Phytopathol. 164 (2), 136–140.

Gazel, M., Ulubaş Serçe, C., Öztürk, H., Çağlayan, K., 2020. Detection of 'Candidatus Phytoplasma pyri' in different pear tissues and sampling time by PCR-RFLP analyses. Mustafa Kemal Üniversitesi Tarım Bilimleri Dergisi, 25, 406–412.

Ghayeb Zamharir, M., Taheri, M., 2019. Effect of new resistance inducers on grapevine phytoplasma disease. Arch. Phytopathol. Plant Protect. 52 (17–18), 1207–1214.

Giunchedi, L., Poggi Pollini, C., Bissani, R., Babini, A.R., Vicchi, V., 1995. Etiology of a pear decline disease in Italy and susceptibility of pear variety and rootstock to phytoplasma-associated pear decline. Acta Hortic. (386), 489–495.

Gouda, S., Das, G., Sen, S.K., Shin, H-S., Patra, J.K., 2016. Endophytes: a treasure house of bioactive compounds of medicinal importance. Front. Microbiol. 7, 1538.

Héloir, M.C., Adrian, M., Brulé, D., Claverie, J., Cordelier, S., Daire, X., Dorey, S., Gauthier, A., Lemaître-Guillier, C., Negrel, J., Trdá, L., Trouvelot, S., Vandelle, E., Poinssot, B., 2019. Recognition of elicitors in grapevine: from MAMP and DAMP perception to induced resistance. Front. Plant Sci. 10, 1117.

Himeno, M., Kitazawa, Y., Yoshida, T., Maejima, K., Yamaji, Y., Oshima, K., Namba, S., 2014. Purple top symptoms are associated with reduction of leaf cell death in phytoplasma-infected plants. Sci. Rep. 4, 4111.

Hogenhout, S.A., Oshima, K., Ammar, E-D., Kakizawa, S., Kingdom, H.N., Namba, S., 2008. Phytoplasmas: bacteria that manipulate plants and insects. Mol. Plant Pathol. 9, 403–423.

Hoshi, A., Oshima, K., Kakizawa, S., Ishii, Y., Hashimoto, M., Komatsu, K., Kagiwada, S., Yamaaji, Y., Namba, S., 2009. A unique virulence factor for proliferation and dwarfism in plants identified from a phytopathogenic bacterium. PNAS 106, 15.

Ikten, C., Ustun, R., Catal, M., Yol, E., Uzun, B., 2016. Multiplex real-time qPCR assay for simultaneous and sensitive detection of phytoplasmas in sesame plants and insect vectors. PLoS One 11 (5), e0155891.

Ito, K., 1960. Witches' broom of paulownia trees. For. Pests News 9, 221–225.

Jarausch, W., Torres, E., 2014. Management of Phytoplasma-Associated Diseases. Phytoplasmas and Phytoplasma Disease Management: How to Reduce Their Economic Impact. IPGW (International Phytoplasmologist Working Group), pp. 199–208.

Jarausch, W., Eyguard, J.P., Mazy, K., Lansac, M., Dosba, F., 1999. High level of resistance of sweet cherry (Prunus avium L.) towards European stone fruit yellows phytoplasmas. Adv. Hortic. Sci. 13 (3), 108–112.

Jarausch, W., Saillard, C., Helliot, B., Garnier, M., Dosba, F., 2000. Genetic variability of apple proliferation phytoplasmas as determined by PCR-RFLP and sequencing of a non-ribosomal fragment. Mol. Cell. Probes 14 (1), 17–24.

Jarausch, W., Jarausch-Wertheim, B., Danet, J-L., Broquaire, J.M., Dosba, F., Saillard, C., Garnier, M., 2001. Detection and indentification of European stone fruit yellows and other phytoplasmas in wild plants in the surroundings of apricot chlorotic leaf roll-affected orchards in southern France. Eur. J. Plant Pathol. 107 (2), 209−217.

Jarausch, W., Angelini, E., Eveillard, S., Malembic-Maher, S., 2013. Management of European fruit tree and grapevine phytoplasma diseases through genetic resistance. Phytopath. Moll. 3 (1), 16−24.

Jarausch, B., Tedeschi, R., Sauvion, N., Gross, J., Jarausch, W., 2019. Psyllid vectors. In: Phytoplasmas: Plant Pathogenic Bacteria—II. Springer Singapore, Singapore.

Jung, H-Y., Sawayanagi, T., Kakizawa, S., Nishigawa, H., Wei, W., Oshima, K., Miyata, S-i., Ugaki, M., Hibi, T., Namba, S., 2003. '*Candidatus* Phytoplasma ziziphi', a novel phytoplasma taxon associated with jujube witches'-broom disease. Int. J. Syst. Evol. Microbiol. 53 (4), 1037−1041.

Kison, H., Seemüller, E., 2001. Differences in strain virulence of the European stone fruit yellows phytoplasma and susceptibility of stone fruit trees on various rootstocks to this pathogen. J. of Phytopath 149, 533−541.

Kitazawa, Y., Iwabuchi, N., Himeno, M., Sasano, M., Koinuma, H., Nijo, T., Tomomitsu, T., Yoshida, T., Okano, Y., Yoshikawa, N., Maejima, K., Oshima, K., Namba, S., 2017. Phytoplasma-conserved phyllogen proteins induce phyllody across the Plantae by degrading floral MADS domain proteins. J. Exp. Bot. 68 (11), 2799−2811.

Komori, N., 1966. Occurrence and control of rice yellow dwarf disease in Ibaraki Prefecture. Plant Prot. 20, 285−288.

Kumar, S.P., Akram, M., Vajpeyi, M., Srivastava, R.L., Kumar, K., Naresh, R., 2007. Screening and development of resistant sesame varieties against phytoplasma. Bull. Insectol. 60 (2), 303−304.

Kumar, J., Gunapati, S., Singh, S.P., Lalit, A., Sharma, N.C., Tuli, R., 2012. First report of a '*Candidatus* Phytoplasma asteris' (16SrI group) associated with little leaf disease of *Solanum melongena* (brinjal) in India. New Dis. Rep. 26 (1), 21.

Kumari, S., Krishnan, N., Rai, A.B., Singh, B., Rao, G.P., Bertaccini, A., 2019. Global status of phytoplasma diseases in vegetable crops. Front. Microbiol. 10, 1349.

Laimer, M., Lemaire, O., Herrbach, E., Goldschmidt, V., Minafra, A., Bianco, P.A., Wetzel, T., 2009. Resistance to viruses, phytoplasmas and their vectors in the grapevine in Europe: a review. J. Plant Pathol. 91 (1), 7−23.

Landi, F., Prandini, A., Paltrinieri, S., Missere, D., Bertaccini, A., 2010. Assessment of susceptibility to European stone fruit yellows phytoplasma of new plum variety and five rootstock/plum variety combinations. Julius-Kühn-Archiv. 427, 378−382.

Lee, I-M., Davis, R.E., Gundersen-Rindal, D.E., 2000. Phytoplasma: phytopathogenic *Mollicutes*. Ann. Rev. Microbiol. 54, 221−255.

Liebenberg, A., Wetzel, T., Kappis, A., Herdemertens, M., Krczal, G., Jarausch, W., 2010. Influence of apple stem grooving virus on *Malus sieboldii*-derived apple proliferation resistant rootstocks. Julius-Kühn-Archiv 427, 186−188.

Liu, M.J., Zhou, J.Y., Zhao, J., 2004. Screening of Chinese jujube germplasm with high resistance to witches' broom disease. Acta Hortic. 663, 575−580.

Liu, M., Wang, J., Wang, L., Liu, P., Zhao, J., Zhao, Z., Yao, S., Stănică, F., Liu, Z., Wang, L., Ao, C., Dai, L., Li, X., Zhao, X., Jia, C., 2020. The historical and current research progress on jujube—a superfruit for the future. Hortic. Res. 7 (1), 119.

Maejima, K., Oshima, K., Namba, S., 2014. Exploring the phytoplasmas, plant pathogenic bacteria. J. Gen. Plant Pathol. 80, 210−221.

Maejima, K., Kitazawa, Y., Tomomitsu, T., Yusa, A., Neriya, Y., Himeno, M., Yamaji, Y., Oshima, K., Namba, S., 2015. Degradation of class E MADS-domain transcription factors in arabidopsis by a phytoplasmal effector, phyllogen. Plant Signal. Behav. 10 (8), e104263.

Malembic-Maher, S., Le Gall, F., Danet, J-L., Dorlhac de Borne, F., Bové, J-M., Garnier-Semancik, M., 2005. Transformation of tobacco plants for single-chain antibody expression via apoplastic and symplasmic routes, and analysis of their susceptibility to stolbur phytoplasma infection. Plant Sci. 168 (2), 349−358.

Marcone, C., Jarausch, B., Jarausch, W., 2010. '*Candidatus* Phytoplasma prunorum', the causal agent of European stone fruit yellows: an overview. J. Plant Pathol. 19–34.

Marcone, C., Guerra, L.J., Uyemoto, J.K., 2014. Phytoplasmal diseases of peach and associated phytoplasma taxa. J. Plant Pathol. 96 (1), 15–28.

Marcone, C., Pierro, R., Tiwari, A.K., Rao, G.P., 2022. Phytoplasma diseases of temperate fruit trees. Agrica 11 (1), 19–33.

Mitra, D., 1993. Little leaf, a serious disease of eggplant (*Solanum melongena*). In: Raychaudhuri, S. (Ed.), Management of Plant Diseases Caused by Fastidious Prokaryotes. Associated Publishing Co, New Delhi, pp. 73–78.

Muniyappa, V., Raychaudhuri, S.P., 1988. Rice yellow dwarf disease. In: Mycoplasma Diseases of Crops. Springer New York, New York, NY.

Musetti, R., Sanità di Toppi, L., Ermacora, P., Favali, M.A., 2004. Recovery in apple trees infected with the apple proliferation phytoplasma: an ultrastructural and biochemical study. Phytopathology 94 (2), 203–208.

Musetti, R., Marabottini, R., Badiani, M., Martini, M., Sanità di Toppi, L., Borselli, S., Borgo, M., Osler, R., 2007. On the role of H_2O_2 in the recovery of grapevine (*Vitis vinifera* cv. Prosecco) from "flavescence dorée" disease. Funct. Plant Biol. 34 (8), 750–758.

Musetti, R., Paolacci, A., Ciaffi, M., Tanzarella, O.A., Polizzotto, R., Tubaro, F., Mizzau, M., Ermacora, P., Badiani, M., Osler, R., 2010. Phloem cytochemical modification and gene expression following the recovery of apple plants from apple proliferation disease. Phytopathology 100 (4), 390–399.

Musetti, R., Farhan, K., De Marco, F., Polizzotto, R., Paolacci, A., Ciaffi, M., Ermacora, P., Grisan, S., Santi, S., Osler, R., 2013a. Differentially-regulated defence genes in *Malus domestica* during phytoplasma infection and recovery. Eur. J. Plant Pathol. 136 (1), 13–19.

Musetti, R., Ermacora, P., Martini, M., Loi, N., Osler, R., 2013b. What can we learn from the phenomenon of "recovery". Phytopath. Moll. 3 (1), 63–65.

Nakamura, H., Ohgake, S., Sahashi, N., Yoshikawa, N., Kubono, T., Takahash, T., 1998. Seasonal variation of paulownia witches'-broom phytoplasma in paulownia trees and distribution of the disease in the Tohoku district of Japan. J. For. Res. 3 (1), 39–42.

Nejat, N., Vadamalai, G., 2013. Diagnostic techniques for detection of phytoplasma diseases: past and present. J. Plant Dis. Prot. 120 (1), 16–25.

OEPP/EPPO, 2007. Pear Decline Phytoplasma. EPPO A2 List of Pests Recommended for Regulation as Quarantine Pests. No.95 Version 2007–09.

Oksal, H.D., 2020. Natural phytoplasma infections on fruit, vegetable and weed plants at the same agroecosystem and their molecular properties. Not. Bot. Horti Agrobot. Cluj-Napoca 48 (2), 615–625.

Oksal, H., Apak, F.K., Oksal, E., Tursun, N., Sipahioglu, H.M., 2017. Detection and molecular characterization of two '*Candidatus* Phytoplasma trifolii' isolates infecting peppers at the same ecological niche. Int. J. Agric. Biol. 19 (6), 1372–1378.

Oshima, K., Kakizawa, S., Nishigawa, H., Jung, H-Y., Wei, W., Suzuki, S., Arashida, R., Nakata, D., Miyata, S-i., Ugaki, M., Namba, S., 2004. Reductive evolution suggested from the complete genome sequence of a plant-pathogenic phytoplasma. Nat. Genet. 36 (1), 27–29.

Osler, R., et al., 2000. Recovery in plants affected by phytoplasmas. In: 5th Congress of the European Foundation for Plant Pathology. Taormina, pp. 589–592.

Osler, R., Borselli, S., Ermacora, P., Loschi, A., Martini, M., Musetti, R., Loi, N., 2013. Acquired tolerance in apricot plants that stably recovered from European stone fruit yellows. Plant Dis. 98 (4), 492–496.

Osler, R., Borselli, S., Ermacora, P., Ferrini, F., Loschi, A., Martini, M., Moruzzi, S., Musetti, R., Giannini, M., Serra, S., Loi, N., 2016. Transmissible tolerance to European stone fruit yellows (ESFY) in apricot: cross-protection or a plant mediated process? Phytoparasitica 44 (2), 203–211.

Pagliarani, C., Gambino, G., Ferrandino, A., Chitarra, W., Vrhovsek, U., Cantu, D., Palmano, P., Marzachì, C., Schubert, A., 2020. Molecular memory of "flavescence dorée" phytoplasma in recovering grapevines. Hortic. Res. 7 (1), 7, 126.

Paolacci, A.R., Catarcione, G., Ederli, L., Zadra, C., Pasqualini, S., Badiani, M., Musetti, R., Santi, S., Ciaffi, M., 2017. Jasmonate-mediated defence responses, unlike salicylate-mediated responses, are involved in the recovery of grapevine from "bois noir" disease. BMC Plant Biol. 17 (1), 118.

Patui, S., Bertolini, A., Clincon, L., Ermacora, P., Braidot, E., Vianello, A., Zancani, M., 2013. Involvement of plasma membrane peroxidases and oxylipin pathway in the recovery from phytoplasma disease in apple (*Malus domestica*). Physiol. Plantarum 148 (2), 200—213.

Poggi Pollini, C., Bissani, R., Giunchedi, L., 1994. Overwintering of pear decline agent in some quince rootstocks. Acta Hortic 385, 496—499.

Rao, G.P., 2021. Our understanding about phytoplasma research scenario in India. Indian Phytopathology 74, 371—401.

Rao, G., Nabi, S., 2015. Overview on a century progress in research on sesame phyllody disease. Phytopath. Moll. 5 (1), 74—83.

Rastegar, L., Ghayeb Zamharir, M., Cai, W-J., MIghani, H., Ghassempour, A., Feng, Y-Q, 2021. Treatment of lime witches' broom phytoplasma-infected Mexican lime with a resistance inducer and study of its effect on systemic resistance. J. Plant Growth Regul. 40 (4), 1409—1421.

Romanazzi, G., 2013. Perspectives for the management of phytoplasma diseases through induced resistance: what can we expect from resistance inducers. Phytopath. Moll. 3 (1), 60—62.

Romanazzi, G., D'Ascenzo, D., Murolo, S., 2009. Field treatment with resistance inducers for the control of grapevine bois noir. J. Plant Pathol. 91 (3), 677—682.

Sameshima, T., 1967. Occurrence and control of rice yellow dwarf disease in Miyazaki Prefecture. Plant Prot. 21, 47—50.

Schaper, U., Seemüller, E., 1982. Condition of the phloem and the persistence of mycoplasmalike organisms associated with apple proliferation and pear decline. Phytopathology 72, 736—742.

Seemüller, E., Kunze, L., Schaper, U., 1984a. Colonization behavior of MLO, and symptom expression of proliferation-diseased apple trees and decline-diseased pear trees over a period of several years. J. Plant Dis. Protect. 91, 525—532.

Seemüller, E., Schaper, U., Zimbelmann, F., 1984b. Seasonal variation in the colonization patterns of mycoplasmalike organisms associated with apple proliferation and pear decline. J. Plant Dis. Protect. 91 (4), 371—382.

Seemüller, E., 1989. Pear decline. In: Fridlund, P. (Ed.), Virus and Virus-like Diseases of Pome Fruits and Simulating Non-infectious Disorders. Washington State University Press, Pullman, pp. 188—201.

Seemüller, E., 1992. Pear decline. In: Kumar, J., et al. (Eds.), Diseases of Fruit Crops: Plant Diseases of International Importance. New Jersey, pp. 308—334.

Seemüller, E., Lorenz, K.-H., Lauer, U., 1998. Pear decline resistance in pyrus communis rootstocks and progenies of wild and ornamental pyrus taxa. Acta Hortic. 472, 681—692.

Seemuller, E., Stolz, H., Kison, H., 1998. Persistence of the European stone fruit yellows phytoplasma in aerial parts of *Prunus* taxa during the dormant season. J. Phytopathol. 146, 407—417.

Seemüller, E., Schneider, B., 2004. '*Candidatus* phytoplasma mali', '*Candidatus* Phytoplasma pyri' and '*Candidatus* Phytoplasma prunorum', the casual agents of apple proliferation, pear decline and European stone fruit yellows, respectively. Int. J. Syst. Evol. Microbiol. 54 (4), 1217—1226.

Seemüller, E., Moll, E., Schneider, B., 2009. Pear decline resistance in progenies of *Prunus* taxa used as rootstocks. Eur. J. Plant Pathol. 123 (2), 217—223.

Seemüller, E., Bisognin, C., Grando, M.S., Schneider, B., Velasco, R., Jarausch, W., Seemüller, E., 2010. Breeding of rootstocks resistant to apple proliferation disease. Julius-Kühn-Archiv 427, 183—185.

Seemüller, E., Harries, H., 2010. Plant resistance. In: Weintraub, P., Jones, P. (Eds.), Phytoplasmas: Genomes, Plant Hosts and Vectors. CABI, Wallingford.

Seemüller, E., 2011. Apple proliferation phytoplasma. In: Hadidi, A., Barba, M., Candresse, T., Jelkmann, W. (Eds.), Virus and Virus-like Diseases of Pome and Stone Fruits. American Phytopathological Society Press, Minnesota, pp. 67–73.

Seemüller, E., Schneider, B., Jarausch, B., 2011. Pear decline phytoplasma. In: Hadidi, A., Barba, M., Candresse, T., Jelkmann, W. (Eds.), Virus and Virus-like Diseases of Pome and Stone Fruits. American Phytopathological Society Press, Minnesota, pp. 77–84.

Seemüller, E., Galliger, J., Jelkmann, W., Jarausch, W., 2018. Inheritance of apple proliferation resistance by parental lines of apomictic *Malus sieboldii* as donor of resistance in rootstock breeding. Eur. J. Plant Pathol. 151 (3), 767–779.

Shao, F., Zhang, Q., Liu, H., Lu, S., Qiu, D., 2016. Genome-wide identification and analysis of MicroRNAs involved in witches'broom phytoplasma response in *Ziziphus jujuba*. PLoS One 11 (11).

Shindhe, G.G., et al., 2011. Inheritance study on phyllody resistance in sesame (*Sesamum indicum* L.). Plant Archiv. 11 (2), 775–776.

Siampour, M., Izadpanah, K., Salehi, M., Afsharifar, A., 2019. Occurrence and Distribution of Phytoplasma Diseases in Iran, Sustainability in Plant and Crop Protection book series (SUPP,volume 12), pp. 47–86.

Sugawara, K., Honma, Y., Komatsu, K., Himeno, M., Oshima, K., Namba, S., 2013. The alteration of plant morphology by small peptides released from the proteolytic processing of the bacterial peptide TENGU. Plant Physiol. 162 (4), 2005–2014.

Sugio, A., MacLean, A.M., Kingdom, H.N., Grieve, V.M., Manimekalai, R., Hogenhout, S.A., 2011. Diverse targets of phytoplasma effectors: from plant development to defense against insects. Annu. Rev. Phytopathol. 175–195.

Tanno, K., Maejima, K., Miyazaki, A., Koinuma, H., Iwabuchi, N., Kitazawa, Y., Nijo, T., Hashimoto, M., Yamaji, Y., Namba, S., 2018. Comprehensive screening of antimicrobials to control phytoplasma diseases using an in vitro plant–phytoplasma co-culture system. Microbiology 164 (8), 1048–1058.

Torres, E., Laviña, A., Sabaté, J., Bech, J., Batlle, A., 2010. Evaluation of susceptibility of pear and plum trees varieties and root- stocks to '*Candidatus* Phytoplasma prunorum' by means of real-time PCR. Julius-Kühn-Archiv 427, 395–398.

Ulubas Serce, C., Gazel, M., Caglayan, K., Bas, M., Son, L., 2006. Phytoplasma diseases of fruit trees in germplasm and commercial orchards in Turkey. J. Plant Pathol. 179–185.

Ulubas Serçe, C., Gazel, M., Yalçin, S., Çaglayan, K., 2007. Responses of six Turkish apricot cultivars to '*Candidatus* Phytoplasma prunorum' under greenhouse conditions. Bull. Insectol. 60 (2), 309–310.

Usta, M., Güller, A., Sipahioğlu, H.M., 2018. Molecular analysis of '*Candidatus* Phytoplasma trifolii' and '*Candidatus* Phytoplasma solani' associated with phytoplasma diseases of tomato (PDT) in Turkey. Int. J. Agric. Biol. 20 (9), 1991–1996.

Ustun, R., Yol, E., Ikten, C., Catal, M., Uzun, B., 2017. Screening, selection and real-time qPCR validation for phytoplasma resistance in sesame (*Sesamum indicum* L.). Euphytica 213, 159.

Varga, K., Kolber, M., Martini, M., 2000. Phytoplasma identification in Hungarian grapevines by two nested-PCR systems. In: XIIIth Meeting of the International Council for the Study of Viruses and Virus-like Diseases of the Grapevine (ICVG). Adelaide, pp. 113–115.

Vasudeva, R.S., Sahambi, H.S., 1955. Phyllody in sesamum (*Sesamum orientale* L.). Indian Phytopathol. 8, 124–129.

Weintraub, P.G., Wilson, M.R., 2009. Control of phytoplasma diseases and vectors. In: Phytoplasmas: Genomes, Plant Hosts and Vectors. CABI, Wallingford.

Ye, X., Wang, H., Chen, P., Fu, B., Zhang, M., Li, J., Zheng, X., Tan, B., Feng, J., 2017. Combination of iTRAQ proteomics and RNA-seq transcriptomics reveals multiple levels of regulation in phytoplasma-infected *Ziziphus jujuba* Mill. Hortic. Res. 4, 17080.

Zhang, J., Zhou, J.M., 2010. Plant immunity triggered by microbial molecular signatures. Molecular Plant. Oxford University Press, pp. 783–793.

Zhao, Y., Davis, R.E., 2016. Criteria for phytoplasma 16Sr group/subgroup delineation and the need of a platform for proper registration of new groups and subgroups. Int. J. Syst. Evol. Microbiol. 66 (5), 2121–2123.

Zhao, Y., Liu, Q., Davis, R.E., 2004. Transgene expression in strawberries driven by a heterologous phloem-specific promoter. Plant Cell Rep. 23 (4), 224–230.

Zhao, J., Liu, M.J., Liu, X.Y., Zhao, Z.H., 2009. Identification of resistant cultivar for jujube witches' broom disease and development of management strategies. Acta Hortic. 840, 409–412.

CHAPTER 11

microRNAs role in phytoplasma-associated developmental alterations

Sapna Kumari[1], Amrita Singh[2] and Suman Lakhanpaul[1]

[1]Department of Botany, University of Delhi, New Delhi, Delhi, India; [2]Department of Botany, Gargi College, University of Delhi, New Delhi, Delhi, India

1. Introduction

1.1 Role of noncoding RNAs

RNA is a dedicated field of biological research, as it plays diverse roles in cellular processes and molecular mechanisms in living organisms (Jiayan et al., 2014). In the past several years, considerable research has been done in the field of small noncoding RNAs as an effective regulator of gene expression in plants (Yishai et al., 2007). Among these small noncoding RNA family members, microRNAs can be considered key players that regulate various developmental processes in plants during different life cycle stages. microRNAs are 20–24 nucleotides long noncoding RNAs (Rhoades et al., 2006) that have emerged as essential regulators of gene expression. They act as negative regulators of their target genes through translational repression or mRNA cleavage, thus regulating gene expression at transcriptional and posttranscriptional levels (Rhoades et al., 2006). In addition to regulating multiple biological processes in plants, including vegetative and reproductive development (Zhu et al., 2011; Li et al., 2016), their role has also been shown in biotic and abiotic stresses (Sunkar et al., 2007; Ruiz et al., 2009).

1.2 Discovery of microRNAs

The microRNAs were discovered in *Caenorhabditis elegans* by the Ambros and Ruvkun group in 1993 (Lee et al.,1993) and were found to control the temporal development pattern of larval stages in this nematode. It was found that these miRNAs show complementarity to multiple sites in the 3′UTR region of lin-14 RNA, and binding of the miRNAs to the lin-14 mRNA results in a decrease in the expression of LIN-14 protein (Lee et al., 1993; Wightman et al., 1993). The discovery of microRNA in plants is very recent. Its first evidence was reported in *Arabidopsis thaliana* by RNA sequencing method, wherein a large number of microRNAs of nearly 21–24 nucleotides in length were cloned (Llave et al., 2002; Reinhart et al., 2002). In 2003, Aukerman and Sakai described the role of miR172 in plant development, followed by numerous reports of their diverse functions in plants.

1.3 Biogenesis of microRNAs

Multiple steps are involved in forming mature miRNA from their genes (Bartel, 2004; Kurihara and Watanabe, 2006). MIRs, genes coding miRNA, are present either in the intergenic regions or within the introns of other genes (Rhoades, 2006). They are transcribed by RNA polymerase II resulting in the formation of pri-miRNA (Lee et al., 2004). Many transcription factors (TFs) are involved in this transcription process including transcription complexes formed by MED20A, MED17, and MED18. These complexes can recruit pol II to the promoter of miRNA genes (Chadick et al., 2005; Kim et al., 2011). SA1 is another protein that can affect the binding of Pol II to the promoter of MIR genes (Li et al., 2018).

DCL1 (Dicer-like 1) is the protein that plays a crucial role in the processing miRNA as the primary transcript undergoes subsequent cleavage by the DCL1 processing complex. Several other proteins such as HYL1, SERRATE, CBC, TGH, etc., interact with DCL1 and processing efficiency. For example, HYL1 (HYPONASTIC LEAVES 1) interacts with DCL1 to improve the accuracy of price-miRNA cleavage (Kurihara et al., 2006), and SERRATE is a C2H2 zinc finger protein that provides a platform for the interaction between DCL1 and HYL1. After this step, there is the cleavage of pri-miRNA into the stem loop intermediate called as pre-miRNA, which is further stabilized by DAWDLE (DDL), an RNA binding protein (Khairwesh et al., 2012).

The pri-miRNA undergoes modification by the DCL1 processing complex, which is cleaved to form double-stranded miRNA/miRNA* (Song et al., 2007). These processed double-stranded miRNAs undergo further modification and stabilization to form mature miRNAs. The nascent miRNA—miRNA duplex produced by the DCL-mediated processing exhibits two nucleotide 3' overhangs on both strands, and each strand possesses a 5'end phosphate and two 3' end hydroxyl groups. Methylation occurs only at the 2'OH position with the help of small RNA methyltransferase HUA enhancer 1 (Yang et al., 2006). HEN 1 protein, a methyltransferase with a binding site for double-stranded miRNA at its N-terminal and a conserved C-terminal methylation domain (Yu et al., 2005), plays a significant role in this process by improving the stability of double-stranded miRNA through methylation. The pre-miRNAs are transported from the nucleus to the cytoplasm through Exp5 homolog HASTY protein, which is also involved in stabilizing miRNA (Park et al., 2005).

The miRNA can regulate the target gene expression by transcriptional cleavage or by translational repression through the RNA-induced silencing complex (RISC complex). RISC complex shows interaction with the AGO1 protein. In the first step, pre-miRNAs bind to AGO protein to form pre-RISC, after which the miRNA* in double-stranded RNAs is removed from the complex to form a mature RISC protein complex. The AGOU1 component can modulate the silencing activity of the RISC protein complex (Eamens et al., 2009). The AGO proteins contain the PAZ domain, PIWI domain, mid-domain, and cap-binding domain. AGO1 is chiefly involved in the transcriptional suppression and cleavage of miRNA. The double-stranded RNA binding protein, DRBI/HYL1, interacts with AGO1 and modulates the strand selection through directional loading of miRNA duplex into RISC for passenger strand degradation (Eamens et al., 2009). In the RISC complex, miRNA can bind to the target mRNA owing to complementarity and inhibit the gene expression of target mRNA (Bartel, 2004).

2. Role of microRNAs in vegetative and reproductive development in plants

The microRNAs are considered one of the most important posttranscriptional gene regulators in plants as mutations in their genes have a very pronounced impact on the plants. After that, their role in diverse aspects of plant development has also been demonstrated (Chen et al., 2008) and has been highlighted here.

2.1 Leaf development

MiR159/JAW plays a vital role in the regulation of leaf development. Their mechanism of action involves targeting a subset of TCP TF genes (Palatnik et al., 2003). When there is overexpression of miRJAW, it results in low levels of all tested TCP mRNAs and causes jaw-D phenotype, along with uneven leaf shapes and curvature. In contrast, in another case, overexpression of miRJAW-resistant TCP mutants shows that miRJAW-guided mRNA cleavage is sufficient to stop the TCP function (Palatnik et al., 2003).

2.2 Root and shoot development in plants

The miRNAs act as a mobile signal for stem cell maintenance and are also involved in post-transcriptional regulation of SAM-related genes (Bauman, 2013).

NAC-domain TF family plays a critical role in root and shoot development in plants. Members of the NAC-domain gene family in *Arabidopsis* such as CUC1, CUC2, NAC1, At5g07680, and At5g61430 have a complementarity site with miR164 and are thus its targets (Laufs et al., 2004; Mallory et al., 2004). Further, role of miR164 is important because NAC1 TF plays a very important role in the lateral root development in plants and separation of aerial parts (Aida et al., 1997).

2.3 Vascular development

Overexpression of miR166a in plant tissue results in a decreased level of ATHB15 and causes an increase in the vascular cell differentiation from procambial cells. Kim et al. (2005) reported the formation of an altered vascular system with expanded xylem tissue, along with the interfascicular region in such plants.

2.4 Flower and fruit development

The microRNAs like miR319 and miR159 play a pivotal role in flower development by regulating the TCP and MYB TFs. Both these TFs interact with each other for the regulation of miR167 expression and form a regulatory network that facilitates flower development. The miR159 is involved in regulating LFY transcript levels and has an effect on anther development and control of flowering time; it regulates the expression of some GAMYB-related genes (Waheed et al., 2020). Overexpression of miR159 in transgenic *Arabidopsis* has reduced the level of MYB33 and shows many pleiotropic development defects such as small siliques, stunted anther development, reduced fertility (Palatnik et al., 2007). In *Oryza sativa*, miR159 restricts the OsGAMYB expressions to the anthers only (Tsuji et al., 2006), whereas OsGAMYB loss of function mutant results in defective pollen and anther development (Kaneko et al., 2004).

The miR169 is considered one of the most abundant miRNA families in *Arabidopsis*. Scientific evidence suggest that miR159 plays crucial role in fruit development. For example, in the case of strawberries, miR159 is strongly expressed in the receptacle tissue of the fruit and appears to regulate the GAMYB, as fruit development is GA-regulated (Csukai et al., 2012). GAMYB acts as a key regulator of the fruit development in strawberry as the repression of GAMYB through RNA interference results in inhibiting receptacle ripening and color formation (Vallarino et al., 2015).

2.5 Abiotic and biotic stress

It was observed that the level of miR159 increases in response to drought in the crops such as maize, barley, and wheat (Zhang, 2015). In *A. thaliana*, many miRNAs related to drought stress are identified (Liu et al., 2008). In cold stress, miRNAs like miR397 and miR169 are found to be upregulated in the case of *Arabidopsis, Poplar,* and *Brachypodium* (Zhang et al., 2009). In maize, nine miRNAs are identified to be differentially expressed in response to the chronic low nitrogen availability in which six of the miRNAs are downregulated, that is, miR167, miR169, miR395, miR399, miR408, and miR528, whereas three miRNAs are upregulated, that is, miR172, miR164, and miR827 (Xu et al., 2011).

Scientific studies show that microRNAs also play a very important role against biotic stresses such as in response to the fungus attack, *Verticillium dahlia*; in *Arabidopsis*, it results in the accumulation of high level of miR159 which is then exported to the fungal hyphae, where it targets the gene encoding Isotrichodermin C-15 hydroxylase, which is essential for fungal growth (Zhang, 2016).

3. microRNAs associated with phytoplasma-induced diseases

Phytoplasmas are pleomorphic bacteria that are obligate parasites in the phloem tissues of the plants (Doi et al., 1967). They are associated with diseases in several hundred plant species, including many vegetables, fruits, food crops, ornamental plants, shade trees (Bertaccini and Duduk, 2009). Diseases associated with phytoplasmas are spread by the insects belonging to the families Cixiidae, Psyllidae, Delphacidae, Cicadellidae, and Derbidae (Weintraub and Beanland, 2006). The primary symptoms associated with phytoplasma diseases in the plants are shortened internodes, virescence (green pigmentation of flowers), stunting, floral malformation, witches' broom, phyllody (conversion of floral organs into leaves), sterility of flowers, etc. (Bertaccini, 2007, 2022).

Phytoplasmas are chiefly restricted to the phloem tissues. A recent study demonstrated that phytoplasma releases certain virulence proteins known as effectors, which induce morphological alterations in the host plants (Hogenhout et al., 2008; Sugio et al., 2011). Effector molecules like SAP54 and SAP11 are most likely to be released into the phloem. These molecules may remain restricted to the phloem cells or get unloaded into the target cells or the associated organs (Bai et al., 2009; Hoshi et al., 2009).

3.1 Role of primary microRNAs involved in phytoplasma infection

Several microRNAs have been found to be upregulated or downregulated in response to phytoplasma infection in different host plants (Table 11.1). miR156, miR172, and miR159 are the major miRNAs involved in phytoplasma infection, whose biogenesis and regulation have been studied in detail (Table 11.2).

3.1.1 miR156

Genomic studies have shown that miR156 is highly conserved during plant evolution (Zhang et al., 2015). miR156 controls SPL genes at posttranscriptional level, and SPL TF genes contain a miRNA-responsive element, which helps them to bind to miR156 to regulate their expression (Wang et al., 2009).

The level of miR156 is generally higher in the seedling stage, and it gradually decreases with time (Ahsan et al., 2019). The gradual downregulation of miR156 expression results in the upregulation of

3. microRNAs associated with phytoplasma-induced diseases

Table 11.1 Phytoplasma effect on upregulation/downregulation of miRNA in its associated hosts and its relationship with symptoms.

Host	Phytoplasma	Symptoms	Upregulated/Downregulated Novel microRNAs	References
Citrus aurantifolia L.	'Candidatus Phytoplasma aurantifolia'	Witches' broom, many thin secondary shoots each with shortened internodes only, dry leaves	**Upregulated**: miR156, miR169, miR395, miR157a, miR390b **Downregulated**: miR172, miR2911, miR159a **Novel**: miR-11, miR-13, miR-16, miR-17, miR-110miR-111, miR-113, miR-114miR-116, miR-118, miR-119	Ehya et al. (2013)
Ziziphus jujuba	'Candidatus Phytoplasma ziziphi'	Jujube witches' broom, phyllody, stunting, sterile flowers	**Upregulated**: miR156a, miR156b, miR156c, miR156d, miR156e, miR156h, miR159e, miR319a, mir395a, miR395b, zju-miR23, zju-miRn24 **Downregulated**: miR159a, **miR172**, miR2111, miR399, miR477, miR2950, miR858b, zju-miRn2, zju-miRn8, zju-miRn16 **Novel**: zju-miRn23, zju-miRn24, zju-miRn3, zju-miRn4, Zju-miRn7, zju-miRn8, zju-miRn9, zju-miRn10, zju-miRn11	Shao et al. (2016)
Morus alba	'Candidatus Phytoplasma asteris'	Yellow dwarf disease, yellowing,	**Upregulated**: mul-miR156a, mul-miR5813, mul-	Gai et al. (2018)

Continued

Table 11.1 Phytoplasma effect on upregulation/downregulation of miRNA in its associated hosts and its relationship with symptoms.—cont'd

Host	Phytoplasma	Symptoms	Upregulated/Downregulated Novel microRNAs	References
		phyllody, stunting, witches' broom	miR2199, mul-miR166a, mul-miR166h-3p, mul-miR482a-5p **Downregulated**: mul-miR160a, mul-miR167d-5p, mul-miR3630—3p, miR3630—5p, mul-miR396a **Novel**: mul-miRn21—3p, mul-miRn22—5p, mul-miRn23—5p, mul-miRn24—5p, mulmiRn28—3p, mul-miRn29—5P, mul-miRN30—3P, mul-miRn-31—3P, mul-miRN32—5P	
Vitis vinifera	'Candidatus Phytoplasma asteris'	Grapevine yellows, leaf yellowing (chlorosis), discolouration and necrosis of veins and laminae, abnormal leaf shape and size, stunting and necrosis of shoots, flower abortion, berry withering, downward curling of leaves, incomplete lignification	**Upregulated**: vvi-miR159c, vvi-miR160c, d, e, vvi-miR171 acdij, vvi-miR2950—5p, vvi-miR3627—5p, vvi-miR319bcef, miRn027—3p, vvi-miR172d, vvi-miR395a-m **Downregulated**: vvi-miR156 b, c, d, vvi-miR3629, vvi-miR399aheg, vvi-miR479, vvi-miR3638—5p **Novel**: vvi-miRn040—3P, miRn117—5p, miRn027—3p, vvi-miRn137—5p, vvi-miRn140p, vvi-miRn115—3p, vvi-miRn 027—3p	Synman et al. (2017)

Table 11.1 Phytoplasma effect on upregulation/downregulation of miRNA in its associated hosts and its relationship with symptoms.—cont'd

Host	Phytoplasma	Symptoms	Upregulated/Downregulated Novel microRNAs	References
Paulownia tomentosa × *Paulownia fortune*	'Candidatus Phytoplasma asteris'	Paulownia witches' broom, yellow purple discoloration of leaves and shoots, virescence, phyllody (conversion of floral organ into leaf like structures), proliferation of shoots, stunting	**Upregulated**: pau-miR156, pau-miR166 g/h/i/j, pau-miR482e/f, pau-miR397a/b/c/b, pau-miR396c/d, pau-miR61, pau-miR62, pau-miR9 **Downregulated**: pau-miR169e, pau-miR319e/f/g, pau-miR 441 a/b, pau-miR 477, pau-miR90 a/b **Novel**: pau-miR32, pau-miR34, pau-miR 90a, b, pau-miR41, pau-miR46a, pau-mir 1, pau-miR2, pau-miR3, pau-miR4, pau-miR5, pau-miR6, pau-miR7	Fan et al. (2015)

Table 11.2 Biogenesis, target, and symptoms associated with major miRNAs involved in the phytoplasma infection.

miRNA	Biogenesis/regulation	Target	Symptoms	References
miR156	Transcribed by the RNA polymerase II. Precursor is processed into miRNA/miRNA* COMPLEX through DICER LIKE 1. Duplex is transported into cytoplasm, then pre-miRNA is processed by the aid of HEN1	SPL family downregulates SPL gene expression via transcriptional cleavage and translational inhibition	Leaf development, root development; biosynthesis of secondary metabolites such as anthocyanins. Promoting flowering, tillering, flowering time, phase transition from juvenile to reproductive	Xu et al. (2016), Dai et al. (2018), Wang et al. (2008), Chuck et al. (2007), Abe et al. (2010), Wu et al. (2010)

Continued

Table 11.2 Biogenesis, target, and symptoms associated with major miRNAs involved in the phytoplasma infection.—cont'd

miRNA	Biogenesis/ regulation	Target	Symptoms	References
	protein into a mature double stranded miRNA/ miRNA*complex that is 2 methylated at the 3′ end. HASTY protein transports this small RNA duplex into cytoplasm Antisense miRNA is degraded in cytoplasm; the mature miRNA is embedded in the RISC complex			
miR172	POWERDRESS (PWR), gene, promotes the transcription of miR172 a, b, c; enhance pol II occupancy at their promoters LEUNIG (LUG), a transcriptional regulator, act upstream of miR172 and directly represses, its expression in petals	AP2 family Expression is regulated via translation inhibition	Vegetative phase change; stamen and carpel development; flower patterning; flowering repression; shoot meristem identity and branching; trichome formation on leaves	Aukerman et al. (2003), Chen et al. (2004), Li et al. (2021), Lian et al. (2021), Zhu and Helliwell (2011)
miR159 (highly abundant miRNA)	Cleavage of the loop, instead of unusual cut at the base of the stem loop structure. High number of bulges may be responsible for specific type of processing	GAMYB and GAMYB-like genes	Anther and pollen development, regulation of flowering time. Fruit development; seed development; flower development	Allen et al. (2007); Millar et al. (2019)

SPL expression. Overexpression of miR156 results in a longer vegetative phase and late-flowering phenotypes (Chuck et al., 2007; Wang et al., 2009; Jiao et al., 2010). In rice, overexpression of *OsSPL14* promotes panicle branching. It affects the branching of shoots by cleaving SPL transcripts (Miura et al., 2010; Liu et al., 2017).

Overexpression of miR156 in plants results in smaller leaves (Poethig et al., 2009; Zhang et al., 2011). Reduced level of miR156 results in fewer lateral and adventitious roots. It also reduces apical dominance, delays flowering time, causes dwarfism, and increases the total leaf number and biomass (Schwab et al., 2005). Wang et al. (2020) show that overexpression of miR156 has a comprehensive effect on the biosynthesis of anthocyanins.

3.1.2 miR172

The miR172 is encoded mainly by five loci, that is, miR172a–e. During reproductive development, the expression level of miR172a–c increases, while the expression of miR172d–e decreases (Zhu et al., 2011). The promoter of miR172 contains several copies of SPL binding elements which act as a transcription activator. Plants overexpressing miR172 resulted in an accelerated flowering phenotype (Jung et al., 2007).

It is observed that in *Arabidopsis*, miR172 negatively regulates the expression of AP2 and five AP2 genes including TARGET OF EAT1(TOE1), TOE2, TOE3, SCHLAFMUTZE (SMZ), and SCHNARCHZAPFEN (Chen, 2004; Yamaguchi et al., 2012). The miR172 controls its target gene expression mainly by translational inhibition, but in certain cases, transcript cleavages were observed (Fang et al., 2007).

Studies show that accumulation of miR172 takes place in the center of flower primordia, which restricts AP2 expression to the two outer whorls of the flower meristem and decides the boundary between stamens and petals (Zhou et al., 2007).

Overexpression of miR172 affects floral patterning in many cereal crops like barley and rice, where it targets AP2-like TFs Oryza Sativa INTERMINET SPIKELET 1(osIDS1) and SUPERNUMARY BRACT (SNB) in rice, CLEISTOGAMY 1 in barley, to control lodicule development (Nair et al., 2010).

3.1.3 miR159

The miR159 is present in eudicots, monocots, and also in some of the basal angiosperms, ferns, gymnosperms, and lycopods (Chavez et al., 2014).

Arabidopsis has three miR159 genes (MIR159a, MIR159b, and MIR159c) each encoding a different isoform. The isoforms differ from one another by one–two nucleotides (Rajagopalan et al., 2006). It has been demonstrated that miR159 overexpression results in a decline in the activity of LFY and MYB33, and plants showed late flowering phenotypes. miR159a is the major miRNA that is expressed in seed and throughout plant development at a constant high level, but it is absent in male reproductive organs like stamens. The miR159b is expressed at a lower level than miR159a, but its expression is almost similar to miR159a, whereas miR159c expression is mainly confined to stamens and it is weakly expressed (Millar et al., 2019).

The DELLA proteins repress GA response leading to declining in the expression of both miR159 and its target GAMYB. GA treatment degrades DELLA proteins and increases the level of miR159 and GAMYB TFs, which bind to LFY promoter through GA responsive cis-elements to activate flowering (Jin et al., 2013). The miR159 downregulates the expression of some GAMYB-related genes and also regulates LFY transcript levels and has a serious impact on another formation and control of flowering time (Archard et al., 2004).

Transgenic *Arabidopsis* plants overexpressing miR159 have a reduced level of MYB33 and show pleiotropic development effects such as reduced fertility, stunted anther development, small siliques, etc. (Schwab et al., 2005). Overexpression of miR159 also results in delayed flowering phenotypes (Tsuji et al., 2006; Kaneko et al., 2004). It is observed that the application of gibberellin leads to elevated levels of miR159 accumulation, whereas a lower level of miR159 is observed in GA deficient mutants (Allen et al., 2007).

The efficacy of miR159 is attenuated in seeds which enables GAMYB-like gene to express, and it further results in promoting programmed cell death (PCD) of aleurone, but throughout vegetative development, the efficacy of miR159 is very strong and MYB33/65 expression is strongly silenced (Guo et al., 2008). It only occurs via the inhibition of miR159 or mutation of binding sites within MYB65 and MYB33 resulting in strong deleterious effects like curled leaves and stunted growth (Millar et al., 2019). The miR159ab double mutants show pleiotropic morphological defects which include curled leaves and shorter siliques, whereas miR159a and miR159b single mutants are phenotypically normal (Allen et al., 2007).

In the anthers, the activity of miR159 is very low. MYB33 and MYB65 are expressed to promote PCD in the tapetum (Millar et al., 2005). MYB97/120/101 expression is required for pollen function. Thus, miR159 plays a crucial role in fertilization (Zho et al., 2018).

4. Putative mechanism(s) of interaction between miRNAs and their target genes in response to phytoplasma infection

4.1 Defective development of anthers

Under normal conditions, miR159 is very weakly expressed, so it is not able to target MYB33/MYB65 TFs. A high level of MYB33/MYB65 is able to positively regulate programmed cell death in a tapetal layer of anther resulting in the normal development of anthers with fertile pollens (Fig. 11.1A), whereas in case of phytoplasma infection in hosts such as *Vitis vinifera*, miR159 is very highly expressed (Synman et al., 2017). A high expression level of miR159 leads to the inhibition of TFs like MYB33/MYB65 (Palatnik et al., 2007), which perturbs programmed cell death in anther tapetum and further leads to defective pollen formation and anther development (Fig. 11.1B).

4.2 Anthocyanin accumulation in leaves (leaf margins and petioles)

Anthocyanin biosynthesis involves several structural genes like F3H (flavanone 3-hydroxylase), DFR (dihydroflavonol-4-reductase), ANS (anthocyanidin synthase), CHI (chalcone isomerase), CHS (chalcone synthase), F3′H (flavonoid 3 Hydroxylase), AAT (anthocyanin acyltransferase), (Tanaka et al., 2008). Several TFs are involved in the regulation of anthocyanin biosynthesis including MYB, bHLH, and WD40 repeat which form an MYB/bHLH/WDR (MBW) complex that furthers regulates the expression of anthocyanin biosynthetic genes in diverse species (Xu et al., 2015).

In this model of interaction, it is hypothesized that under normal conditions, the expression level of miR156 is low, so it cannot target the SPL9 (Squamosa Promoter-Binding Protein-Like gene; a plant-specific TF). It further impairs the stability of MBW complex via direct protein interactions.

4. Putative mechanism(s) of interaction between miRNAs and their target genes

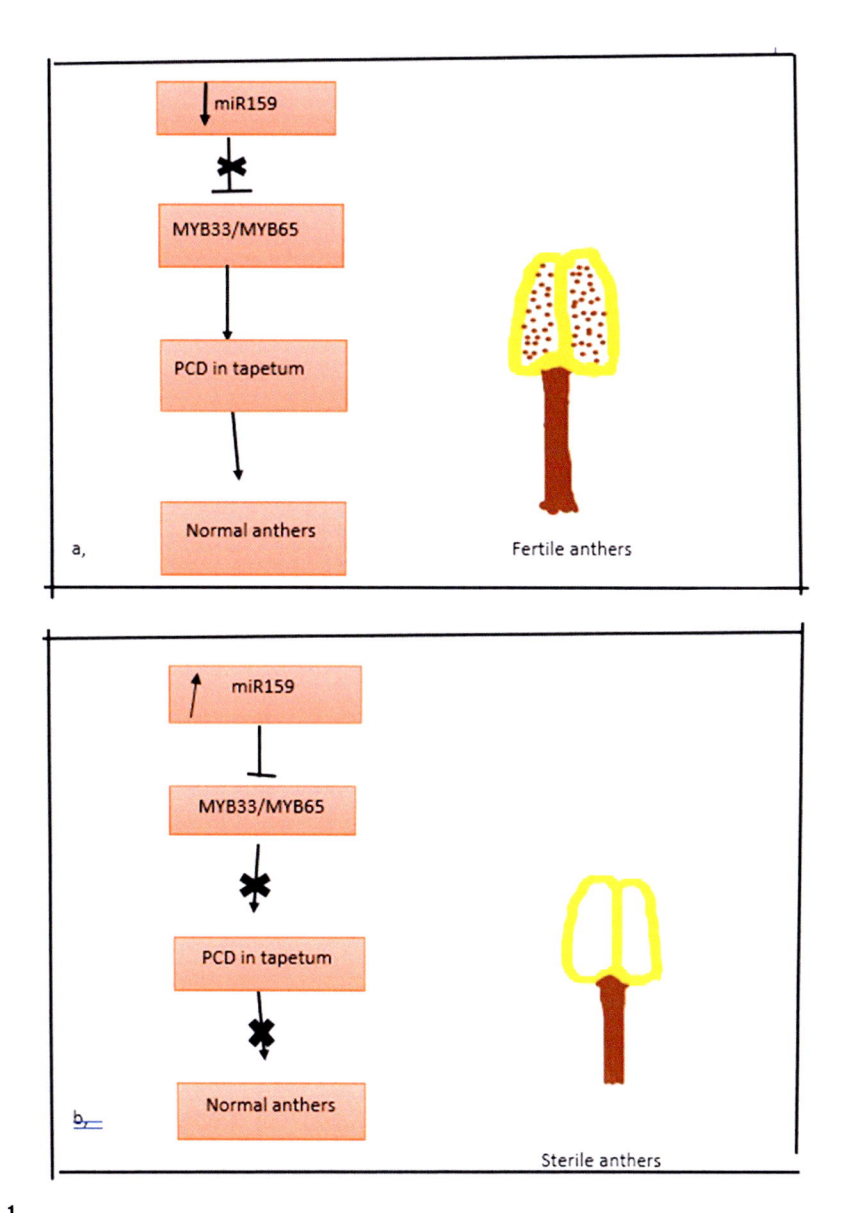

FIGURE 11.1

Schematic model highlights the anther development via the regulation of miR159 (A) under normal condition and (B) under phytoplasma infection.

MBW complex is required to express anthocyanin biosynthetic genes like F3H and DFR. But, when MBW complex is unavailable, F3H and DFR genes are unable to express, and thus, no anthocyanin accumulation occurs (Fig. 11.2A). But in the case of phytoplasma infection, it is observed that the expression of miR156 increases, so it targets SPL9 and inhibits it. Thus, the MBW complex is now

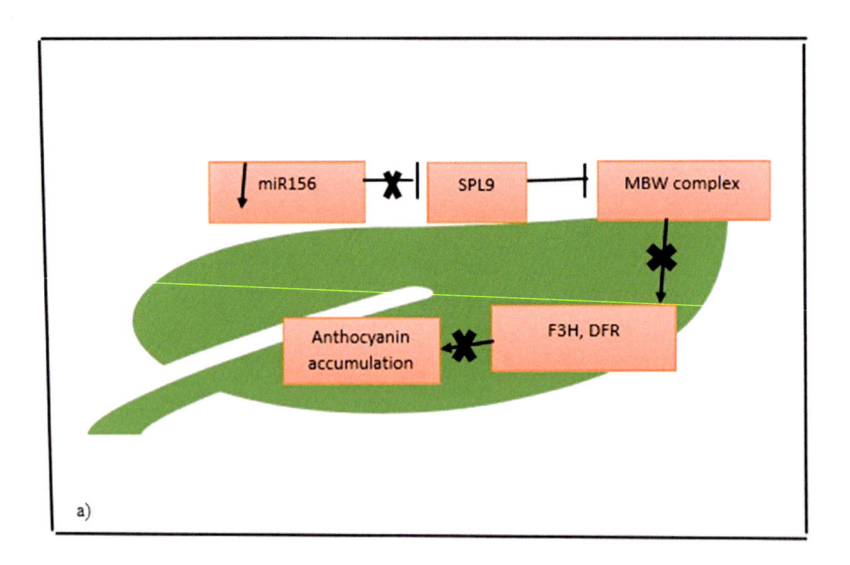

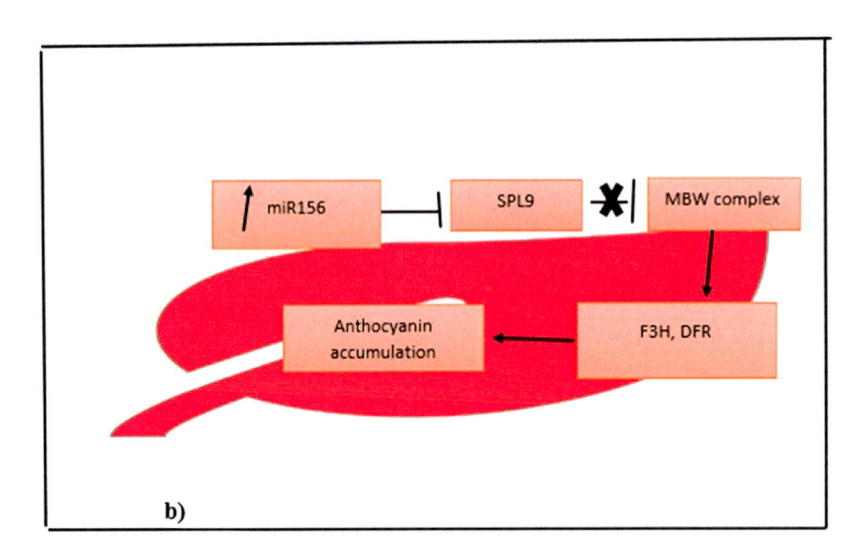

FIGURE 11.2

Schematic model highlights the regulation of anthocyanin accumulation via miR156 (A) under normal condition and (B) under phytoplasma infection.

available to bind to F3H and DFR genes and allow these genes to express and show anthocyanin accumulation in bright red color (Fig. 11.2B).

4.3 Witches' broom and phyllody

The interaction between miRNA and its target genes under normal conditions is illustrated in Fig. 11.3. During phytoplasma infection, virulence effectors are released into the phloem, and are then unloaded

4. Putative mechanism(s) of interaction between miRNAs and their target genes 179

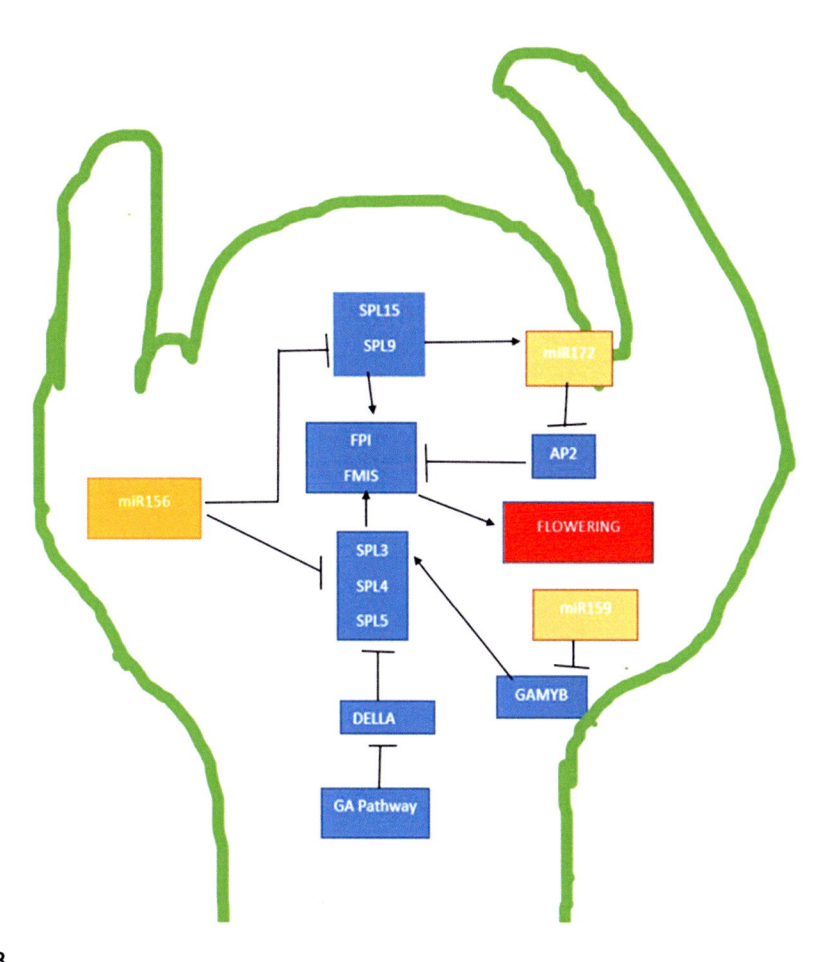

FIGURE 11.3

Model represents the interactions between miRNAs and their target genes, under normal conditions.

from the phloem to access distal tissues and induce changes in the development of the host plant. SAP11 and SAP54 are two major effector molecules. SAP 11 molecules are reported to destabilize class II TB/CYC-TCP TFs, resulting in witches' broom symptoms (Sugio et al., 2014). It is observed that SAP11 molecules also increase the level of miR156 in the host plant, but the exact mechanism is still to be deciphered. Similarly, SAP54 is another effector molecule that is responsible for phyllody in phytoplasma-infected plants by degrading MADS-box TFs (Kitazawa et al., 2017). It modulates the level of miR156 in the host plant and may be responsible for expressed symptoms (Fig. 11.4).

4.4 Dwarfism or stunting

Studies show that in plants like *Arabidopsis* and maize, increase in the expression of miR156 results in a decrease in plant height (Schwab et al., 2005). In *Morus alba* and *Ziziphus jujuba*, the level of

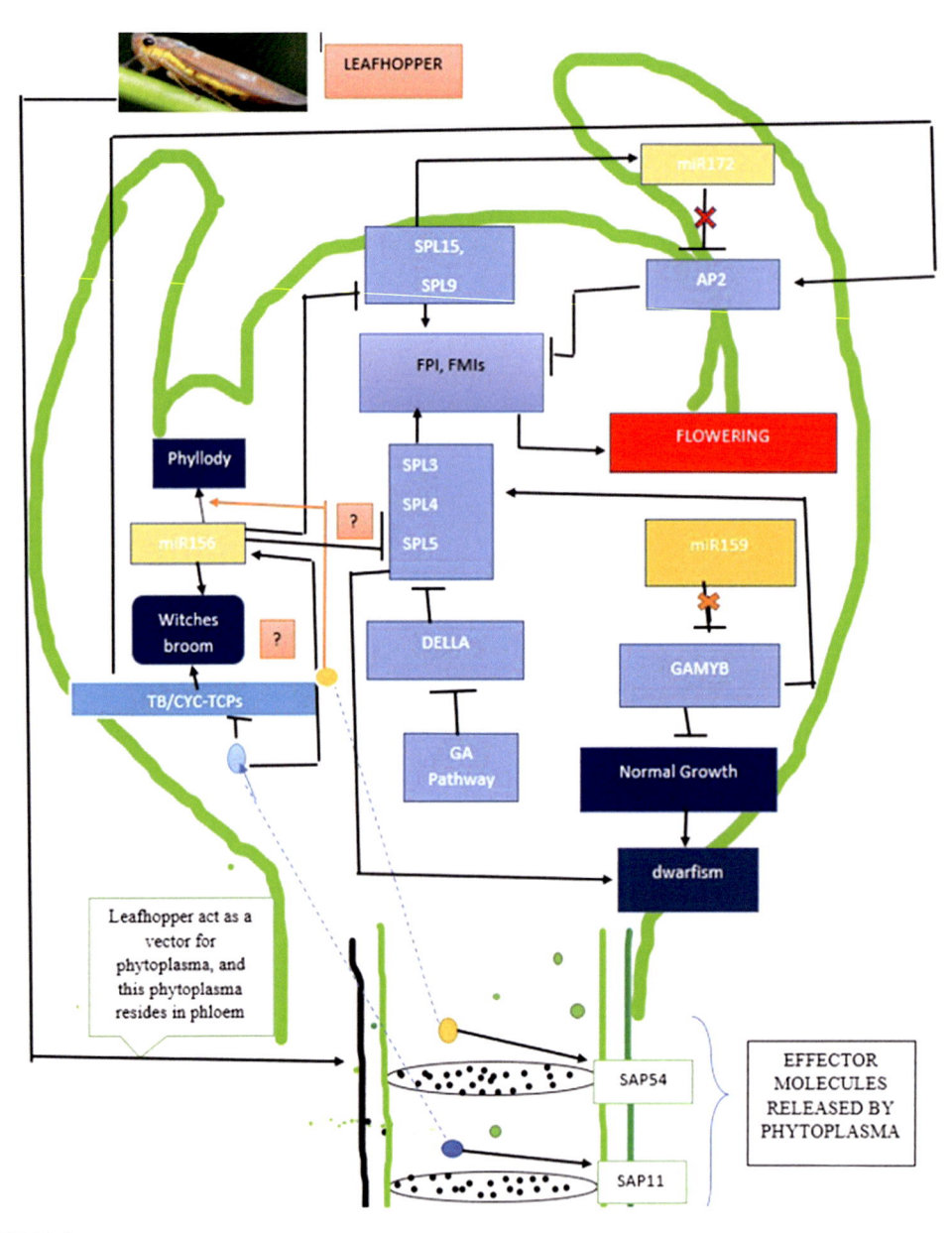

180 Chapter 11 microRNAs role in phytoplasma-associated developmental alterations

FIGURE 11.4

Putative mechanism proposed for the interaction between the miRNA and their target genes, during phytoplasma infection.

miR156 increases in response to phytoplasma infection (Shao et al., 2016; Gai et al., 2018), which represses SPL3, SPL4, and SPL5 and causes dwarfism. The other mechanism that may be involved is that under phytoplasma infection, the level of miR156 increases, which represses GAMYB TF and results in dwarfism. In *V. vinifera*, the level of miR156 decreases and miR159 increases, but dwarfism is still observed in the infected plant. It may be because of the presence of an alternative pathway that is responsible for dwarfism in *V. vinifera* (Fig. 11.4).

There exist certain unanswered questions in the case of phytoplasma-infected *Paulownia*, upregulation of miR156 occurs (Fan et al., 2015) which is responsible for anthocyanin accumulation in plant parts, but in the case of phytoplasma-infected *V. vinifera*, there is downregulation of miR156 (Synman et al., 2017), but still there is anthocyanin accumulation. This discrepancy might be because of some other alternative pathways in *V. vinifera*, which is responsible for anthocyanin accumulation in this plant. But the exact mechanism behind the existence of this variation in the expression of miRNAs still needs to be explored. Another gap exists because SAP54 destabilizes TB/TYC-TCPs (Kitazawa et al., 2017), which results in overexpression of AP2. Still, the exact mechanism behind this observation and its implications needs further scientific exploration. It is necessary to look for the vulnerable points targeted in SAP54 either directly or indirectly.

5. Conclusion

Significant leads suggest the role of flowering genes getting affected by phytoplasma effectors. However, the precise mechanisms or steps having loopholes need detailed investigations using suitable techniques. Multiple vulnerable points have been identified, suggesting a plurality of the mechanism induced by a phytoplasma, thereby revealing the generic nature of this pathogen. The molecular and physiological mechanisms of phytoplasma infections are still unclear. Therefore, further research needs to be done to elucidate the roles of these miRNAs, which can improve the understanding of the level of diversity of specific plant responses to phytoplasmas.

6. Future directions

In the past few years, significant progress has been made in characterization of microRNAs and their role in different developmental processes in plants. This chapter summarizes the role of microRNAs in the development of plants and in phytoplasma-associated developmental alterations. But still, there are unanswered questions which need to be addressed to have a deeper understanding of the miRNAs-regulated molecular mechanisms. The major areas of research in the field of miRNAs such as the diverse effects of miRNA target interactions, the differential expression of MIR loci, the interplay of miRNAs with various components of flowering pathways, upstream processes involved about the biogenesis of miRNAs, which are being triggered by phytoplasma-induced effector can be investigated by further advancements in experimental techniques. The miRNAs can be considered as emerging targets for genetic engineering to improve agronomic properties in plants. Understanding molecular mechanisms regulated by miRNAs in the near future enables agricultural scientists to manipulate specific traits in plants and potentially improve and also enables agricultural scientists to manipulate crop improvement.

Acknowledgments

Authors are grateful to financial support from National Agricultural Science Fund (NASF), Indian Council for Agricultural Research (ICAR), Government of India, and IoE FRP Grant, University of Delhi. University Grants Commission Fellowship awarded to AS by the Ministry of Human Resource Development, Government of India is also gratefully acknowledged (Grant Number 2121330649).

References

Abe, M., Nonaka, M., Sakakibara, H., Sato, Y., Nagato, Y., Itoh, J., 2010. Wavy leaf 1, an ortholog of *Arabidopsis* HEN1, regulates shoot development by maintaining microRNA and transacting small interfering RNA accumulation in rice. Plant Physiol. 154, 1335–1346.

Archard, P., Herr, A., Baulcombe, D.C., Harberd, N.P., 2004. Modulation of floral development by a gibberellin-regulated microRNA. Development 131, 3357–3365.

Ahsan, M.U., Hayward, A., Irihmovitch, V., Fletcher, J., Tanurdzic, M., Pocock, A., Beviridge, C.A., Mitter, N., 2019. Juvenility and vegetative phase transition in tropical/subtropical tree crops. Front. Plant Sci. 10, 729.

Aida, M., Ishida, T., Fukaki, H., Fujisawa, H., Tasaka, M., 1997. Genes involved in organ separation in Arabidopsis: an analysis of the cup- shaped cotyledon mutant. Plant Cell 9, 841–857.

Allen, R.S., Li, J., Stahle, M.I., Dubroue, A., Gubler, F., Millar, A.A., 2007. Genetic analysis reveals functional and the major target genes of the *Arabidopsis* miR159 family. Proc. Natl. Acad. Sci. USA 104, 16371–16376.

Aukerman, M.J., Sakai, H., 2003. Regulation of flowering time and floral organ identity by a microRNA and its APETALA 2 -like target genes. Plant Cell 15, 2730–2741.

Bai, X., Correa, V.R., Toruno, T.Y., Ammar, D., Kamoun, S., Hogenhout, S.A., 2009. Phytoplasma secretes a protein that targets plant cell nuclei. Mol. Plant Microbe Interact. 2, 18–30.

Bartel, D.P., 2004. MicroRNAs: Genomics: biogenesis, mechanism and function. Cell 116 (2), 281–297.

Baumann, K., 2013. Plant cell biology: mobile miRNAs for stem cell maintenance. Nat. Rev. Mol. Cell Biol. 14 (3).

Bertaccini, A., 2007. Phytoplasmas: diversity, taxonomy and epidemiology. Front. Biosci. 12, 673–689.

Bertaccini, A., 2022. Plants and phytoplasmas: when bacteria modify plants. Plants 11, 1425.

Bertaccini, A., Duduk, B., 2009. Phytoplasma and phytoplasma diseases. Phytopathol. Mediterr. 48 (3), 355–278.

Chadick, J.Z., Asturias, F.J., 2005. Structure of eukaryotic mediator complexes. Science 30, 264–271.

Chavez, M., Cardenas, F., Mahalingam, G., 2014. Sample sequencing of vascular plants demonstrates widespread conservation and divergence of microRNAs. Nat. Commun. 5, 3722.

Chen, X., 2004. MicroRNA as a translational repressor of APETALA2 in *Arabidopsis* flower development. Science 303, 2022–2025.

Chen, X., 2008. MicroRNA metabolism in plants. Curr. Top. Microbiol. Immunol. 117–136.

Chuck, G., Cigan, A.M., Saeturn, K., Hake, S., 2007. The heterochromatic maize mutant Corngrass 1 results from overexpression of a tandem microRNA. Nat. Genet. 39, 544–549.

Csukai, F., Donaire, L., Csanal, A., Martinez-Priego, L., Botella, M.A., 2012. Two strawberry miR159 family members display developmental specific expression patterns in the fruit receptacle and cooperatively regulate GAMYB. Phytol. 195, 47–57.

Dai, Z., Wang, J., Yang, X., Lu, H., Miao, X., Shi, Z., 2018. Modulation of plant architecture by the miR156f-OsSPL7-OsGH3.8 pathway in rice. J. Exp. Bot. 69 (21), 5117–5130.

Doi, Y., Teranaka, M., Yora, K., 1967. Asuyama.Mycoplasma or PLT group like microorganisms found in the phloem elements of plants infected with mulberry dwarf, potato witches' broom. Ann. Phytopathol. Soc. Jpn. 33, 259–266.

Eamens, L., Smith, N., Curtin, S., Wang, M., Waterhouse, P.M., 2009. The *Arabidopsis thaliana* double stranded RNA binding protein DRB1 directs guide strand selection from microRNA duplxes. RNA 15 (12), 2219–2235.

Ehya, F., Monavarfes, H., MohseniFard, E., 2013. Phytoplasma-responsive microRNAs modulate hormonal, nutritional, and stress signalling pathways in Mexican lime trees. PLoS One 8 (6).

Fan, G., Niu, S., Xu, T., Deng Zhao, Z., Wang, Y., 2015. Plant-pathogen interaction-related microRNAs an their targets provides indicators of phytoplasma infection in *Paulownia tomentosa* ×*Paulownia fortune*. PLoS One 10 (10).

Fang, Y., Spector, D.L., 2007. Identification of nuclear dicing bodies contain proteins for microRNA biogenesis in living *Arabidopsis p*lants. Curr. Biol. 17, 818–823.

References

Gai, Y.P., Zhao, H.N., Zhao, Y.N., 2018. miRNA-seq-based profiles of miRNAs in mulberry phloem sap provide insight into the pathogenic mechanisms of mulberry yellow dwarf disease. Sci. Rep. 8 (812).

Guo, W.J., Ho, T.H.D., 2008. An abscissic acid induced protein HVA22 inhibits gibberellin mediated programmed cell death in cereal aleurone cells. Plant Physiol 147, 1710–1722.

Hogenhout, S.A., Oshima, K., Anmar, D., Kakizawa, S., Kingdom, H.N., Namba, S., 2008. Phytoplasmas: bacteria that manipulate plants and insects. Mol. Plant Pathol. 9, 403–423.

Hoshio, A., Oshima, K., Kakizawa, S., Ishii, Y., Ozeki, J., Hashimoto, M., Komatsu, K., Kagiwada, S., Namba, S., 2009. A unique virulence factor for proliferation and dwarfism in plants identified from pathogenic bactrerium. Proc. Natl. Acad. Sci. U.S.A. 106 (125), 6416–6421.

Jiao, Y., Wang, Y., Xue, D., Wang, J., Yan, M., Liu, G., Dong, G., 2010. Regulation of OsSPL14 by OsmiR156 defines ideal plant architecture in rice. Nat. Genet. 42, 541–544.

Jiayan, W., Jingfa, X., Zhang, Z., Xumin, W., Songnian, H., Jun, Y., 2014. Ribogenomics:the science and knowledge of RNA. Dev. Reprod. Biol. 12 (2), 57–63.

Jin, D., Wang, Y., Zhao, Y., Chen, M., 2013. MicroRNAs and their cross talk in plant development. J. Genet.Genom. 40, 161–170.

Jung, J., Seo, Y., Seo, J., Reyes, J., Yun, J., Chua, H., Park, C., 2007. The gigantea-regulated microRNA172 mediates photopereiodic flowering independent of constants in *Arabidopsis*. Plant Cell 19 (9), 2736–2748.

Kaneko, M., Inukai, Y., Ueguchi, T.M., Itoh, H., Izawa, T., Kobayashi, Y., Hattori, T., Miyao, A., Ashikari, M., 2004. Loss of function Mutations of the Rice GAMYB impair alpha amylase expression in aleurone and flower development. Plant Cell 16, 33.

Khraiwesh, B., Zhu, J.K., Zhu, J., 2012. Role of miRNAs and siRNAs in biotic and abiotic stress responses of plants. Biochim. Biophys. Acta BBA Bioenerg. 1819, 137–148.

Kim, J., Jung, J., Reyes, J., Kim, Y., Chung, K., Lee, M., Lee, Y., Kim, V., Chua, N., Park, C., 2005. microRNA-directed cleavage of ATHB15 mRNA regulates vascular development in *Arabidopsis* inflorescence stems. Plant J. 42 (1), 84–94.

Kim, Y.J., Zheng, B., Yu, Y., Won, S.Y., Chen, Mo B., 2011. The role of mediator in small and long noncoding RNA production in *Arabidopsis thaliana*. EMBO J. 30, 814–822.

Kitazawa, Y., Iwabuchi, N., Himeno, M., Sasano, M., Koinuma, H., Nijo, T., 2017. Phytoplasma—conserved phyllogen proteins induce phyllody across the plantae by degrading floral MADS domain proteins. J. Exp. Bot. 568 (11), 2799–2811.

Kurihara, Y., Takashi, Y., Watanabe, Y., 2006. The interaction between DCL1 and HYL1 is important for efficient and precise processing of pri-miRNA in plant microRNA biogenesis. RNA 12 (2), 206–212.

Laufs, P., Peaucelle, A., Maorin, H., Tras, J., 2004. MicroRNA regulation of the CUC genes is required for boundary size control in *Arabidopsis* meristems. Development 131, 4311–4322.

Lee, R.C., Feinbaum, R.L., Ambros, V., 1993. The *C. elegans* heterochronic gene lin-4 encodes small .RNAs with antisense complementarity to lin-14. Cell 75, 843–854.

Lee, Y., Kim, M., Han, J., Yeom, K.H., Lee, S., Baek, S.H., Kim, V.N., 2004. MicroRNA genes are transcribed by RNA polymerase II. EMBO J. 23 (20), 4051–4060.

Lian, H., Wang, L., Ma, N., Zhou, C.-M., Han, L., Zhang, 2021. T-Redundant and specific roles of individual MIR172 genes in plant development. PLoS Biol. 19 (2).

Li, C., Zhang, B., 2016. MicroRNAs in control of plant development. J. Cell. Physiol. 231, 303–311.

Li, M., Yu, B., 2021. Recent advances in the regulation of plant miRNA biogenesis. RNA Biol. 18 (12), 2087–2096.

Li, S., Xu, R., Li, A., 2018. SMA1, ahomolog of the splicing factor Prp28,has a multifaceted role in miRNA biogenesis in *Arabidopsis*. Nucleic Acids Res. 46 (17), 9148–9159.

Liu, H.H., Tian, X., Li, Y.J., Wu, C.A., Zheng, C., 2008. Microarray- based analysis of stress regulated microRNAs in *Arabidopsis thaliana*. RNA 14 (5), 836–843.

Liu, J., Cheng, X., Liu, P., Sun, J., 2017. miR156-Targeted SBP-box transcription factors interact with DWARF53 to regulate teosinte branched 1 and Barren stalk1 expression in bread wheat. Plant Physiol. 174, 1931–1948.

Llave, C., Rector, M.A., Carrington, J.C., Kasschau, K.D., 2002. Endogenous and silencing-associated small-RNAs in plants. Plant Cell 14, 1605–1619.

Mallory, A.C., Dugas, D.V., Bartel, D.P., Bartel, B., 2004. MicroRNA regulation of NAC-Domain targets is required for proper formation and separation of adjacent embryonic,vegetative and floral organs. Curr. Biol 14, 1035–1046.

Millar, A.A.,G.F., 2005. The Arabidopsis GAMYB like genes,MYB33 and MYB65,are microRNA regulated genes that redundantly facilitate anther development. Plant Cell 17, 705–721.

Millar, A., Lohe, A., Wong, R., 2019. Biology and function of miR159 in plants. Plants 8 (8), 255.

Miura, K., Ikeda, M., Matsubara, A., Song, X.J., Ito, M., Asano, K., Matsuoka, M., Kitano, H., Ashikari, M., 2010. OsSPL14 promotes panicle branching and higher grain productivity in rice. Nat. Genet. 42, 545–549.

Nair, S.K., Wang, N., Turuspekov, Y., Pourkheirandish, M., Sinsuwongwat, S., Chen, G., Sameri, M., Tageri, A., Honda, I., Watabe, Y., 2010. Cleistogamous flowering in barley arises from the suppression of microRNA guided Hv APR Mrna cleavage. Proc.Natl. Acad.Sci.USA 107, 490–494.

Palatnik, J.F., Allen, E., Wu, X., Schommer, C., Schwab, R., Carrington, J.C., Weigel, D., 2003. Control of leaf morphogenesis by microRNAs. Nature 425 (6955), 257–263, 18.

Palatnik, J.F., Wollmann, H., Schommer, C., Schwab, R., Boisbouvier, J., Rodriguez, R., Warthamann, N., Allen, E., Dezulian, T., Huson, D., 2007. Sequences and expression differences underlie functional specialisation of *Arabidopsis* microRNAs miR159 and miR319. Dev. Cell 132, 115–125.

Park, M., Gang, W., Gonzalez, A., Vaucheret, H., Poething, S., 2005. Nuclear processing and export of microRNAs in *Arabidopsis*. Proc. Natl. Acad. Sci. USA 102 (10), 3691–3696.

Poethig, R.S., 2009. Small mRNAs and developmental timings in plants. Curr. Opin. Genet. Dev. 19374–19378.

Rajagopalan, R., Vaucheret, H., Trejo, J., Bartewl, D.P., 2006. A diverse and evolutionarily fluid set of microRNAs in *Arabidopsis thaliana*. Genes Dev 20, 3407–3425.

Reinhart, B.J., Rhoades, M.W., Bartel, D.P., Bartel, B., 2002. MicroRNAs in plants. Genes Dev. 16, 1616–1626.

Rhoades, M.W., Bartel, D.P., Bartel, B., 2006. MicroRNAs and their regulatory roles in plants. Annu. Rev. Biol. 57, 19–53.

Ruiz-Ferrer, V., Voinnet, O., 2009. Roles of plant small RNAs in biotic stress responses. Annu. Rev. Plant Biol. 60, 485–510.

Schwab, R., Palatnik, J.F., Riester, M., Schommer, C., Schmid, M., Weigel, D., 2005. Specific effects of microRNA on the plant transcriptome. Dev. Cell 8, 517–527.

Shao, F., Zhang, Q., Liu, H., Liu, S., Qiu, D., 2016. Genome—wide identification and analysis of microRNAs involved in witches' broom phytoplasma response in *Ziziphus jujuba*. PLoS One 11 (11).

Song, L., Han, M., Lesicka, J., Federoff, N., 2007. *Arabidopsis* primary microRNA processing proteins processing proteins HYL1 and DCL1 define a nuclear body distinct from the cajal body. Proc. Natl. Acad. Sci. U.S.A. 104 (13), 5437–5442.

Sugio, A., Maclean, A.M., Hogenhout, S.A., 2014. The small phytoplasma virulence effector SAP11 contains distinct domains required for nuclear targeting and CIN-TCP binding and destabilization. New Phytol. 202 (3), 838–848.

Sugio, A., MacLean, A.M., Kingdom, H.N., Grieve, V.M., Manimekalai, R., Hogenhout, S.A., 2011. Diverse targets of phytoplasma effectors: from plant development to defense against insects. Annu. Rev. Phytopathol. 49 (1), 175–195.

Sunkar, R., Chinnuswamy, V., Zhu, J.H., Zhu, J., 2007. Small RNAs as big players in plant abiotic stress responses and nutrient deprivation. Trends Plant Sci. 12, 301–309.

Synman, M.C., Solofoharivelo, M.C., Souza-Richards, R., Stephan, D., Murray, S., Burger, J.T., 2017. The use of high throughput small RNA sequencing reveals differentially expressed microRNAs in response to aster yellows Phytoplasma-infection in Vitis vinifera cv.Chardonnay. PLoS ONE 12.

Tanaka, Y., Sasaki, N., Ohmiya, A., 2008. Biosynthesis of plant pigments: anthocyanins, betalains and carotenoids. Plant J. 54, 733–749.

References

Tsuji, H., Aya, K., Ueguchi-Tanaka, M., Shimada, Y., Nakazono, M., Watanabe, R., Nishizawa, N.K., Gomi, K., Shimada, A., Kitano, H., 2006. GAMYB control different sets of genes and is differentially regulated by microRNA in aleurone cells and anthers. Plant J. 47, 427−444.

Vallarino, J.G., Osorio, S., Bombarely, A., Cruz, E., Amaya, I., 2015. Central role of Fa GAMYB in the transition of the strawberry receptacle from development to ripening. New Phytol. 208, 482−496.

Waheed, S., Zeng, L., 2020. The critical role of miRNAs in regulation of flowering time and flower development. Gene 11, 31917.

Wang, J.W., Czech, B., Weigel, D., 2009. miR156-Regulated SPL transcription factors define an endogenous flowering pathway in *Arabidopsis thaliana*. Cell 138, 738−774.

Wang, J., Schwab, R., Czech, B., Mica, E., Weigel, d, 2008. Targeted SPL genes and CYP78A5/KLUH on plastrochon length and organ size in *Arabidopsis thaliana*. Plant Cell 20 (5), 1213−1243.

Wang, Y., Liu, W., Wang, X., Yang, R., Wu, Z., Wang, H., Wang, L., Hu, Z., Guo, S., Zhang, H., Lin, J., Fu, C., 2020. miR156 regulates anthocyanin biosynthesis through SPL targets and other microRNAs in poplar. Hortic. Res. 7 (18).

Weintraub, P.G., Beanland, L.A., 2006. Insect vectors of phytoplasmas. Annu. Rev. Entomol. 51, 91−111.

Wightman, B., Ruvkun, G., 1993. Posttranscriptional regulation of the heterochronic gene lin -14 by lin 4 mediates temporal pattern formation in *C. elegans*. Cell 75, 855−862.

Wu, L., Zhou, H., Zhang, Q., Zhang, J., Ni, f., Qi, Y., 2010. DNA methylation mediated by a microRNA pathway. Mol. Cell. 38, 465−475.

Xu, M., Hu, T., Zhao, J., Park, M.Y., Earley, K.W., Wu, G., 2016. Developmental functions of miR156-regulated SQUAMOSA promoter binding proteing-like (SPL) genes in *Arabidopsis thaliana*. PLoS Genet. 12 (8).

Xu, W.J., Dubos, C., Lepiniec, L., 2015. Transcriptionsl control of flavonoid biosynthesis by MYB-bHLH-WDR complexes. Trends Plant Sci. 20, 176−185.

Xu, Z., Zhong, S., Li, X.W., Rothstein, S.J., Zhang, S., Bi, Y., Xie, C., 2011. Genome wide identification of microRNAs in response to low nitrogen availability in maize leaves and roots. PLoS One 6, e28009.

Yang, Z., Ebright, Y., Yu, B., Chen, X., 2006. HEN1 recognises 21−24 nt small RNA duplexes and deposits a methyl group onto the 2'OH of the 3'terminal nucleotide. Nucleic Acids Res. 34 (2), 667−675.

Yamaguchi, A., Abe, M., 2012. Regulation of reproductive development by non-coding RNA in *Arabidopsis*: to flower or not to flower. J. Plant Res. 125, 693−704.

Yishai, S., Gilgi, F., Guy, F., Gali, N., Shoshy, A., Ofer, B., Hanah, M., 2007. Regulation of gene expression by small non - coding RNAs:a quantitative view. Mol. Syst. Biol. 3, 138.

Yu, B., Yang, Z.Y., Li, j, 2005. Methylation as a crucial step in plant microRNA biogenesis. Science 307 (5711), 932−935.

Zhang, B., 2015. MicroRNA: a new target for improving plant tolerance to abiotic stress. J. Exp. Bot. 66, 1749−1761.

Zhang, T., Zhao, Y.L., Zhao, J.H., Wang, S., Chen, S.R., Wang, Y.Y., Hua, C.L., Ding, S.W., Guo, H.S., 2016. Cotton plants export microRNAs to inhibit virulence gene expression in a fungal pathogen. Nat. Plants. 2, 16153.

Zhang, W., Xu, Y., Huan, Q., Chong, K., 2009. Deep sequencing of *Brachypodium* small RNAs at the global, genome level identifies microRNAs involved in cold stress response. BMC Genom. 449.

Zhang, X., Zhou, Z., Zhang, J., Zhang, Y., Han, Q., Hu, T., Xiu, X., Liu, H., Li, H., Ye, Z., 2011. Overexpression of sly-miR156a in tomato result in multiple vegetative and reproductive trait alterations and partial phenocopy of sfp mutant. FEBS Lett. 585, 435−439.

Zhou, G.K., Kubo, M., Zhong, R., Demura, J.H., 2007. Overexpression of miR156 affects apical meristem formation, organ polarity establishment and vascular development in *Arabidopsis*. Plant Cell Physiol. 48, 391−404.

Zho, Y., Wang, S., Wu, W., LI, L., Jiang, T., Zheng, B., 2018. Clearance of maternal barriers by paternal miR159 to initiate endosperm nuclear division in arabidopsis. Nat. Commun. 9.

Zhu, Q.H., Helliwell, C.A., 2011. Regulation of flowering time and flowering patterning by miR172. J. Exp. Bot. 62, 487−495.

CHAPTER 12

Characteristic features of genome and pathogenic factors of phytoplasmas

Ai Endo and Kenro Oshima

Department of Clinical Plant Science, Faculty of Bioscience and Applied Chemistry, Hosei University, Koganei, Tokyo, Japan

1. Characteristic features of phytoplasma genomes

Phytoplasmas ('*Candidatus* Phytoplasma') are obligate intracellular plant pathogens in the class Mollicutes (Maejima et al., 2014a). Phytoplasma-infected plants exhibit a wide range of unique symptoms, including flower malformation, yellowing, dwarfing, witches' broom, purple top, and phloem necrosis (Oshima, 2021). They are transmitted by insect vectors such as leafhoppers and psyllids (Hoshi et al., 2007; Oshima et al., 2019) and can infect more than a 1000 plant species worldwide (Hogenhout et al., 2008).

Genome analyses have been very useful for analyses of the feature of plant pathogens. In 2004, the first complete genome sequence of a mildly pathogenic line of the '*Candidatus* Phytoplasma asteris' onion yellows strain was determined (Oshima et al., 2004). In the past decade, whole genome sequences and draft genome sequences have been determined for several phytoplasma strains (Bai et al., 2006; Kube et al., 2008; Tran-Nguyen et al., 2008; Oshima et al., 2013; Wang et al., 2018a), enabling better understanding of the phylogenetic position (Oshima and Nishida, 2007; Cho et al., 2020), virulence (Oshima et al., 2007; Himeno et al., 2014), infection (Miura et al., 2012), transmissibility (Oshima et al., 2001; Ishii et al., 2009a, 2009b), host responses (Ehya et al., 2013; Fan et al., 2015; Shao et al., 2016; Xue et al., 2019), and host interaction (Suzuki et al., 2006; Neriya et al., 2014; Nijo et al., 2017). Genome sequencing has also contributed to the development of phytoplasma detection methods (Sugawara et al., 2012; Ikten et al., 2016).

Although phytoplasma genomes contain genes for basic cellular functions such as DNA replication, transcription, translation, and protein translocation (Jung et al., 2003; Miyata et al., 2003; Miura et al., 2015), they lack genes for amino acid biosynthesis, fatty acid biosynthesis, the tricarboxylic acid cycle, and oxidative phosphorylation. The phytoplasma genome encodes even fewer metabolic functional proteins than mycoplasma genomes, which were previously thought to have the minimum possible gene set (Mushegian and Koonin, 1996). For example, phytoplasma genomes lack the pentose phosphate pathway genes and genes encoding F_1F_0-type ATP synthase (Oshima et al., 2004). ATP synthesis in phytoplasmas may be dependent on glycolysis or malate-related pathway instead of ATP synthase (Oshima et al., 2007; Saigo et al., 2014). Interestingly, phytoplasmas harbor multiple copies of transporter-related genes not found in mycoplasmas. These genomic features suggest that phytoplasmas are highly dependent on metabolic compounds from their hosts (Oshima et al., 2013). Phytoplasmas alter their expression of metabolic and transporter-related genes in response to host switching between plant and insect (Oshima et al., 2011).

Phytoplasma genomes contain clusters of repeated gene sequences called potential mobile units (PMUs) (Bai et al., 2006), which consist of similar genes organized in a conserved order (Arashida et al., 2008). The PMU exists as linear chromosomal and circular extrachromosomal elements in 'Ca. P. asteris' aster yellows witches' broom strain (AY-WB) (Toruño et al., 2010), suggesting that it has the ability to transpose within the genome. Phylogenetic analysis of genes in PMUs indicates that horizontal transfer may have occurred between divergent phytoplasma lineages (Chung et al., 2013; Ku et al., 2013).

Recent advances in sequencing technology enhance genome sequencing of phytoplasmas (Saccardo et al., 2012; Andersen et al., 2013; Chung et al., 2013; Chen et al., 2014; Lee et al., 2015; Quaglino et al., 2015; Cho et al., 2019; Kirdat et al., 2020). Comparisons among phytoplasma genomes revealed the differentiations in gene content and metabolic capacity (Kube et al., 2012; Oshima et al., 2013). However, it is still difficult to obtain pure phytoplasma DNA because the organisms cannot be cultured efficiently in vitro. Total DNA from phytoplasma-infected hosts contains much more host DNA than phytoplasma DNA, so that it is very inefficient to sequence phytoplasma genome using total DNA. The methyl-CpG binding domain (MBD) protein is useful for the enrichment of bacterial DNA by efficient separation of prokaryotic and eukaryotic DNA (Feehery et al., 2013). It has been demonstrated that the MBD-mediated enrichment method facilitates enrichment of phytoplasma DNA from infected plants for next-generation sequencing of phytoplasma genome (Kirdat et al., 2021; Nijo et al., 2021).

The commonly adopted classification system for phytoplasmas has been based on the sequence of 16S rRNA gene or the restriction fragment length polymorphism analysis (Lee et al., 1993; Namba et al., 1993; Wei et al., 2007). However, with the increased availability of genomic sequences, it is believed that comparison of bacteria at the whole genome is highly useful in microbial phylogenetic research (Oshima and Nishida, 2007; Oshima et al., 2012). It has been demonstrated that the comparison of average nucleotide identity can classify phytoplasmas into multiple distinct taxonomic units equivalent to species in other bacteria (Cho et al., 2020). Comparative studies based on the complete sequences of phytoplasma genomes would form the basis for phylogeny and taxonomy.

2. Characteristic features of pathogenic factors

Since phytoplasmas are cell wall—less and reside inside host cells, their secreted proteins function in the cytoplasm of the host plant or insect cell and are predicted to have some important roles in host—parasite interactions and/or virulence. Although the molecular mechanisms behind the symptoms are not fully understood, several secreted proteins of phytoplasmas, called effectors, have been shown to induce the symptoms (Sugio et al., 2011; MacLean et al., 2011; Strohmayer et al., 2019).

A small secreted peptide, TENGU, encoded by onion yellows phytoplasma, was the first phytoplasma pathogenic factor identified that affects plant morphology (Hoshi et al., 2009). TENGU is predicted to be translated as a 70 amino acid preprotein, with a 32 amino acid signal peptide at its N-terminus. The C-terminal 38 amino acids of TENGU are secreted into plant cytoplasm via the Sec system (Kakizawa et al., 2001), a bacterial conserved protein translocation system on the cellular membrane, where the N-terminal signal peptide is cleaved. Transient or transgenic expression of this 38 amino acid peptide, the putative secreted region of TENGU in *Nicotiana benthamiana* and *Arabidopsis thaliana*, results in a short and bushy phenotype similar to the symptoms of phytoplasma-infected plants (Hoshi et al., 2009). Interestingly, the N-terminal 11 amino acids of the secreted region of TENGU are sufficient for its function (Sugawara et al., 2013). It has been also shown that TENGU acts as an inducer of sterility (Minato et al., 2014). Transgenic expression of TENGU induced developmental defects in *A. thaliana* similar to those observed in auxin response factor 6 (ARF6) and ARF8 double mutants. The levels of the ARF6 and ARF8 genes were significantly decreased in TENGU-expressing transgenic plants, and jasmonic acid (JA) and auxin syntheses were decreased in tengu-transgenic buds. These findings suggest that TENGU inhibits the jasmonic acid (JA) and auxin biosynthesis pathways by repression of ARF6 and ARF8, resulting in impaired flower maturation (Minato et al., 2014).

SAP11 is a secretory protein that affects plant morphogenesis, which was identified from AY-WB phytoplasma (Bai et al., 2009). SAP11-expressing plants showed crinkled leaves and produced more stems (Sugio et al., 2011). SAP11 and its homologs interact with and destabilize class II CIN-CINNATA (CIN)-related TEOSINTE BRANCHED1, CYCLOIEDA, PROLIFERATING CELL FACTORs (TCPs) transcription factors (Sugio et al., 2014; Janik et al., 2017; Chang et al., 2018; Wang et al., 2018b), leading in turn to downregulation of JA biosynthesis (Sugio et al., 2011). Transgenic *N. benthamiana* lines expressing the secreted effector SAP11 exhibit an altered aroma phenotype, which is correlated with defects in the development of glandular trichomes and the biosynthesis of 3-isobutyl-2-methoxypyrazine (Tan et al., 2016). The fecundity of insect vectors was increased on the SAP11-expressing plants compared to normal plants, which may be advantageous for the fitness of phytoplasmas and their insect vectors (Sugio et al., 2011). It has been also reported that SAP11 suppresses salicylic acid−mediated defense responses and enhances the growth of a bacterial pathogen (Lu et al., 2014).

Flower malformation, such as phyllody and virescence, is a unique symptom of phytoplasma infection (Himeno et al., 2011). It has been reported that a novel gene family of phytoplasma effectors, designated phyllody-inducing genes or the phyllogen family, induces flower malformation in several plants (MacLean et al., 2011; Maejima et al., 2014b; Yang et al., 2015). Phyllogen was shown to cause phyllody phenotypes in several eudicot species belonging to three different families (Kitazawa et al., 2017). Phylogenetic analyses based on determined phyllogen gene sequences suggest that phyllogen genes are horizontally transferred among phytoplasmas (Iwabuchi et al., 2020). Phyllogen family proteins, such as SAP54 and PHYL, target the products of floral homeotic genes that constitute the floral quartet model (Maejima et al., 2014b; MacLean et al., 2014; Singh and Lakhanpau, 2020), which in turn encode MADS domain transcription factors (MTFs). Phyllogens recognize A- and E-class MTFs of angiosperms and degrade them in a proteasome-dependent manner (Maejima et al., 2014b, 2015; MacLean et al., 2014; Kitazawa et al., 2017). Recently, it has been reported that phyllogen and another secreted effector (SAP05) induces proteasome dependent but ubiquitin-independent degradation of target proteins by functionally mimicking ubiquitin (Huang et al., 2021; Kitazawa et al., 2022). Crystal structure analyses have revealed that the K domain of MTFs, a phyllogen-binding region (MacLean et al., 2014), have two α-helices with conserved hydrophobic residues which are important for the tetramerization of MTFs (Puranik et al., 2014). Two phyllogens (PHYL$_{OY}$ and PHYL$_{PnWB}$, phyllogens of '*Ca.* P. asteris' onion yellows and '*Ca.* P. aurantifolia' peanut witches' broom strains, respectively; also have similar structures based on two α-helices which are important for phyllody-inducing activity (Iwabuchi et al., 2019; Liao et al., 2019). Further structural insights into the interaction between phyllogen family proteins and MTFs will help elucidate the molecular mechanisms how phytoplasmas manipulate plant hosts.

References

Andersen, M.T., Liefting, L.W., Havukkala, I., Beever, R.E., 2013. Comparison of the complete genome sequence of two closely related isolates of 'Candidatus Phytoplasma australiense' reveals genome plasticity. BMC Genom. 14, 1–15.

Arashida, R., Kakizawa, S., Hoshi, A., Ishii, Y., Jung, H-Y., Kagiwada, S., Yamaji, Y., Oshima, K., Namba, S., 2008. Heterogeneic dynamics of the structures of multiple gene clusters in two pathogenetically different lines originating from the same phytoplasma. DNA Cell Biol. 27, 209–217.

Bai, X., Correa, V.R., Toruño, T.Y., Ammar, E.-D., Kamoun, S., Hogenhout, S.A., 2009. AY-WB phytoplasma secretes a protein that targets plant cell nuclei. Mol. Plant Microbe Interact. 22, 18–30.

Bai, X., Zhang, J., Ewing, A., Miller, S.A., Jancso Radek, A., Shevchenko, D.V., Tsukerman, K., Walunas, T., Lapidus, A., Campbell, J.W., 2006. Living with genome instability: the adaptation of phytoplasmas to diverse environments of their insect and plant hosts. J. Bacteriol. 188, 3682–3696.

Chang, S.H., Tan, C.M., Wu, C-T., Lin, T-H., Jiang, S-Y., Liu, R-C., Tsai, M-C., Su, L-W., Yang, J-Y., 2018. Alterations of plant architecture and phase transition by the phytoplasma virulence factor SAP11. J. Exp. Bot. 69, 5389–5401.

Chen, W., Li, Y., Wang, Q., Wang, N., Wu, Y., 2014. Comparative genome analysis of wheat blue dwarf phytoplasma, an obligate pathogen that causes wheat blue dwarf disease in China. PLoS One 9, e96436.

Cho, S-T., Kung, H-J., Huang, W., Hogenhout, S.A., Kuo, C-H., 2020. Species boundaries and molecular markers for the classification of 16SrI phytoplasmas inferred by genome analysis. Front. Microbiol. 11, 1531.

Cho, S-T., Lin, C-P., Kuo, C-H., 2019. Genomic characterization of the periwinkle leaf yellowing (PLY) phytoplasmas in Taiwan. Front. Microbiol. 10, 2194.

Chung, W-C., Chen, L-L., Lo, W-S., Lin, C-P., Kuo, C-H., 2013. Comparative analysis of the peanut witches' broom phytoplasma genome reveals horizontal transfer of potential mobile units and effectors. PLoS One 8, e62770.

Ehya, F., Monavarfeshani, A., Mohseni Fard, E., Karimi Farsad, L., Khayam Nekouei, M., Mardi, M., Salekdeh, G.H., 2013. Phytoplasma-responsive microRNAs modulate hormonal, nutritional, and stress signaling pathways in Mexican lime trees. PLoS One 8, e66372.

Fan, G., Cao, X., Niu, S., Deng, M., Zhao, Z., Dong, Y., 2015. Transcriptome, microRNA, and degradome analyses of the gene expression of paulownia with phytoplasma. BMC Genom. 16, 1–15.

Feehery, G.R., Yigit, E., Oyola, S.O., Langhorst, B.W., Schmidt, V.T., Stewart, F.J., Dimalanta, E.T., Amaral-Zettler, L.A., Davis, T., Quail, M.A., 2013. A method for selectively enriching microbial DNA from contaminating vertebrate host DNA. PLoS One 8, e76096.

Himeno, M., Kitazawa, Y., Yoshida, T., Maejima, K., Yamaji, Y., Oshima, K., Namba, S., 2014. Purple top symptoms are associated with reduction of leaf cell death in phytoplasma-infected plants. Sci. Rep. 4, 4111.

Himeno, M., Neriya, Y., Minato, N., Miura, C., Sugawara, K., Ishii, Y., Yamaji, Y., Kakizawa, S., Oshima, K., Namba, S., 2011. Unique morphological changes in plant pathogenic phytoplasma-infected petunia flowers are related to transcriptional regulation of floral homeotic genes in an organ-specific manner. Plant J. 67, 971–979.

Hogenhout, S.A., Oshima, K., Ammar, E.D., Kakizawa, S., Kingdom, H.N., Namba, S., 2008. Phytoplasmas: bacteria that manipulate plants and insects. Mol. Plant Pathol. 9, 403–423.

Hoshi, A., Ishii, Y., Kakizawa, S., Oshima, K., Namba, S., 2007. Host-parasite interaction of phytoplasmas from a molecular biological perspective. Bull. Insectol. 60, 105–107.

Hoshi, A., Oshima, K., Kakizawa, S., Ishii, Y., Ozeki, J., Hashimoto, M., Komatsu, K., Kagiwada, S., Yamaji, Y., Namba, S., 2009. A unique virulence factor for proliferation and dwarfism in plants identified from a phytopathogenic bacterium. Proc. Natl. Acad. Sci. U. S. A. 106, 6416−6421.

Huang, W., MacLean, A.M., Sugio, A., Maqbool, A., Busscher, M., Cho, S.T., Kamoun, S., Kuo, C.H., Immink, R.G.H., Hogenhout, S.A., 2021. Parasitic modulation of host development by ubiquitin-independent protein degradation. Cell 184, 5201−5214.

Ikten, C., Ustun, R., Catal, M., Yol, E., Uzun, B., 2016. Multiplex real-time qPCR assay for simultaneous and sensitive detection of phytoplasmas in sesame plants and insect vectors. PLoS One 11, e0155891.

Ishii, Y., Kakizawa, S., Hoshi, A., Maejima, K., Kagiwada, S., Yamaji, Y., Oshima, K., Namba, S., 2009a. In the non-insect-transmissible line of onion yellows phytoplasma (OY-NIM), the plasmid-encoded transmembrane protein ORF3 lacks the major promoter region. Microbiology 155, 2058−2067.

Ishii, Y., Oshima, K., Kakizawa, S., Hoshi, A., Maejima, K., Kagiwada, S., Yamaji, Y., Namba, S., 2009b. Process of reductive evolution during 10 years in plasmids of a non-insect-transmissible phytoplasma. Gene 446, 51−57.

Iwabuchi, N., Kitazawa, Y., Maejima, K., Koinuma, H., Miyazaki, A., Matsumoto, O., Suzuki, T., Nijo, T., Oshima, K., Namba, S., Yamaji, Y., 2020. Functional variation in phyllogen, a phyllody-inducing phytoplasma effector family, attributable to a single amino acid polymorphism. Mol. Plant Pathol. 21, 1322−1336.

Iwabuchi, N., Maejima, K., Kitazawa, Y., Miyatake, H., Nishikawa, M., Tokuda, R., Koinuma, H., Miyazaki, A., Nijo, T., Oshima, K., Yamaji, Y., Namba, S., 2019. Crystal structure of phyllogen, a phyllody-inducing effector protein of phytoplasma. Biochem. Biophys. Res. Commun. 513, 952−957.

Janik, K., Mithöfer, A., Raffeiner, M., Stellmach, H., Hause, B., Schlink, K., 2017. An effector of apple proliferation phytoplasma targets TCP transcription factors—a generalized virulence strategy of phytoplasma? Mol. Plant Pathol. 18, 435−442.

Jung, H-Y., Miyata, S.I., Oshima, K., Kakizawa, S., Nishigawa, H., Wei, W., Suzuki, S., Ugaki, M., Hibi, T., Namba, S., 2003. First complete nucleotide sequence and heterologous gene organization of the two rRNA operons in the phytoplasma genome. DNA Cell Biol. 22, 209−215.

Kakizawa, S., Oshima, K., Kuboyama, T., Nishigawa, H., Jung, H-Y., Sawayanagi, T., Tsuchizaki, T., Miyata, S., Ugaki, M., Namba, S., 2001. Cloning and expression analysis of phytoplasma protein translocation genes. Mol. Plant Microbe Interact. 14, 1043−1050.

Kirdat, K., Tiwarekar, B., Thorat, V., Narawade, N., Dhotre, D., Sathe, S., Shouche, Y., Yadav, A., 2020. Draft genome sequences of two phytoplasma strains associated with sugarcane grassy shoot (SCGS) and Bermuda grass white leaf (BGWL) diseases. Mol. Plant Microbe Interact. 33, 715−717.

Kirdat, K., Tiwarekar, B., Thorat, V., Sathe, S., Shouche, Y., Yadav, A., 2021. 'Candidatus Phytoplasma sacchari', a novel taxon-associated with sugarcane grassy shoot (SCGS) disease. Int. J. Syst. Evol. Microbiol. 71, 004591.

Kitazawa, Y., Iwabuchi, N., Himeno, M., Sasano, M., Koinuma, H., Nijo, T., Tomomitsu, T., Yoshida, T., Okano, Y., Yoshikawa, N., Maejima, K., Oshima, K., Namba, S., 2017. Phytoplasma-conserved phyllogen proteins induce phyllody across the Plantae by degrading floral MADS domain proteins. J. Exp. Bot. 68, 2799−2811.

Kitazawa, Y., Iwabuchi, N., Maejima, K., Sasano, M., Matsumoto, O., Koinuma, H., Tokuda, R., Suzuki, M., Oshima, K., Namba, S., Yamaji, Y., 2022. A phytoplasma effector acts as a ubiquitin-like mediator between floral MADS-box proteins and proteasome shuttle proteins. Plant Cell 34, 1709−1723.

Ku, C., Lo, W-S., Kuo, C-H., 2013. Horizontal transfer of potential mobile units in phytoplasmas. Mobile Genet. Elem. 3, e62770.

Kube, M., Mitrovic, J., Duduk, B., Rabus, R., Seemüller, E., 2012. Current view on phytoplasma genomes and encoded metabolism. Sci. World J. 2012, 185942.

Kube, M., Schneider, B., Kuhl, H., Dandekar, T., Heitmann, K., Migdoll, A.M., Reinhardt, R., Seemüller, E., 2008. The linear chromosome of the plant-pathogenic mycoplasma 'Candidatus Phytoplasma mali'. BMC Genom. 9, 1−14.

Lee, I-M., Hammond, R., Davis, R., Gundersen, D., 1993. Universal amplification and analysis of pathogen 16S rDNA for classification and identification of mycoplasmalike organisms. Phytopathology 83, 834–842.

Lee, I-M., Shao, J., Bottner-Parker, K., Gundersen-Rindal, D., Zhao, Y., Davis, R.E., 2015. Draft genome sequence of 'Candidatus Phytoplasma pruni' strain CX, a plant-pathogenic bacterium. Genome Announc. 3 e01117-15.

Liao, Y.T., Lin, S.S., Lin, S.J., Sun, W.T., Shen, B.N., Cheng, H.P., Lin, C.P., Ko, T.P., Chen, Y.F., Wang, H.C., 2019. Structural insights into the interaction between phytoplasmal effector causing phyllody 1 and MADS transcription factors. Plant J. 100, 706–719.

Lu, Y-T., Li, M-Y., Cheng, K-T., Tan, C.M., Su, L-W., Lin, W-Y., Shih, H-T., Chiou, T-J., Yang, J-Y., 2014. Transgenic plants that express the phytoplasma effector SAP11 show altered phosphate starvation and defense responses. Plant Physiol 164, 1456–1469.

MacLean, A.M., Orlovskis, Z., Kowitwanich, K., Zdziarska, A.M., Angenent, G.C., Immink, R.G., Hogenhout, S.A., 2014. Phytoplasma effector SAP54 hijacks plant reproduction by degrading MADS-box proteins and promotes insect colonization in a RAD23-dependent manner. PLoS Biol. 12, e1001835.

MacLean, A.M., Sugio, A., Makarova, O.V., Findlay, K.C., Grieve, V.M., Tóth, R., Nicolaisen, M., Hogenhout, S.A., 2011. Phytoplasma effector SAP54 induces indeterminate leaf-like flower development in *Arabidopsis* plants. Plant Physiol 157, 831–841.

Maejima, K., Iwai, R., Himeno, M., Komatsu, K., Kitazawa, Y., Fujita, N., Ishikawa, K., Fukuoka, M., Minato, N., Yamaji, Y., Oshima, K., Namba, S., 2014b. Recognition of floral homeotic MADS domain transcription factors by a phytoplasmal effector, phyllogen, induces phyllody. Plant J. 78, 541–554.

Maejima, K., Kitazawa, Y., Tomomitsu, T., Yusa, A., Neriya, Y., Himeno, M., Yamaji, Y., Oshima, K., Namba, S., 2015. Degradation of class E MADS-domain transcription factors in *Arabidopsis* by a phytoplasmal effector, phyllogen. Plant Signal. Behav. 10, e1042635.

Maejima, K., Oshima, K., Namba, S., 2014a. Exploring the phytoplasmas, plant pathogenic bacteria. J. Gen. Plant Pathol. 80, 210–221.

Minato, N., Himeno, M., Hoshi, A., Maejima, K., Komatsu, K., Takebayashi, Y., Kasahara, H., Yusa, A., Yamaji, Y., Oshima, K., Kamiya, Y., Namba, S., 2014. The phytoplasmal virulence factor TENGU causes plant sterility by downregulating of the jasmonic acid and auxin pathways. Sci. Rep. 4, 7399.

Miura, C., Sugawara, K., Neriya, Y., Minato, N., Keima, T., Himeno, M., Maejima, K., Komatsu, K., Yamaji, Y., Oshima, K., Namba, S., 2012. Functional characterization and gene expression profiling of superoxide dismutase from plant pathogenic phytoplasma. Gene 510, 107–112.

Miura, C., Komatsu, K., Maejima, K., Nijo, T., Kitazawa, Y., Tomomitsu, T., Yusa, A., Himeno, M., Oshima, K., Namba, S., 2015. Functional characterization of the principal sigma factor RpoD of phytoplasmas via an in vitro transcription assay. Sci. Rep. 5, 11893.

Miyata, S., Oshima, K., Kakizawa, S., Nishigawa, H., Jung, H.Y., Kuboyama, T., Ugaki, M., Namba, S., 2003. Two different thymidylate kinase gene homologues, including one that has catalytic activity, are encoded in the onion yellows phytoplasma genome. Microbiology 149, 2243–2250.

Mushegian, A.R., Koonin, E.V., 1996. A minimal gene set for cellular life derived by comparison of complete bacterial genomes. Proc. Natl. Acad. Sci. U. S. A. 93, 10268–10273.

Namba, S., Oyaizu, H., Kato, S., Iwanami, S., Tsuchizaki, T., 1993. Phylogenetic diversity of phytopathogenic mycoplasmalike organisms. Int. J. Syst. Evol. Microbiol. 43, 461–467.

Neriya, Y., Maejima, K., Nijo, T., Tomomitsu, T., Yusa, A., Himeno, M., Netsu, O., Hamamoto, H., Oshima, K., Namba, S., 2014. Onion yellow phytoplasma P38 protein plays a role in adhesion to the hosts. FEMS Microbiol. Lett. 361, 115–122.

Nijo, T., Neriya, Y., Koinuma, H., Iwabuchi, N., Kitazawa, Y., Tanno, K., Okano, Y., Maejima, K., Yamaji, Y., Oshima, K., Namba, S., 2017. Genome-wide analysis of the transcription start sites and promoter motifs of phytoplasmas. DNA Cell Biol. 36, 1081–1092.

References

Nijo, T., Iwabuchi, N., Tokuda, R., Suzuki, T., Matsumoto, O., Miyazaki, A., Maejima, K., Oshima, K., Namba, S., Yamaji, Y., 2021. Enrichment of phytoplasma genome DNA through a methyl-CpG binding domain-mediated method for efficient genome sequencing. J. Gen. Plant Pathol. 87, 154–163.

Oshima, K., 2021. Molecular biological study on the survival strategy of phytoplasma. J. Gen. Plant Pathol. 87, 403–407.

Oshima, K., Shiomi, T., Kuboyama, T., Sawayanagi, T., Nishigawa, H., Kakizawa, S., Miyata, S., Ugaki, M., Namba, S., 2001. Isolation and characterization of derivative lines of the onion yellows phytoplasma that do not cause stunting or phloem hyperplasia. Phytopathology 91, 1024–1029.

Oshima, K., Kakizawa, S., Nishigawa, H., Jung, H-Y., Wei, W., Suzuki, S., Arashida, R., Nakata, D., Miyata, S., Ugaki, M., Namba, S., 2004. Reductive evolution suggested from the complete genome sequence of a plant-pathogenic phytoplasma. Nat. Genet. 36, 27–29.

Oshima, K., Kakizawa, S., Arashida, R., Ishii, Y., Hoshi, A., Hayashi, Y., Kagiwada, S., Namba, S., 2007. Presence of two glycolytic gene clusters in a severe pathogenic line of 'Candidatus Phytoplasma asteris'. Mol. Plant Pathol. 8, 481–489.

Oshima, K., Ishii, Y., Kakizawa, S., Sugawara, K., Neriya, Y., Himeno, M., Minato, N., Miura, C., Shiraishi, T., Yamaji, Y., Namba, S., 2011. Dramatic transcriptional changes in an intracellular parasite enable host switching between plant and insect. PLoS One 6, e23242.

Oshima, K., Chiba, Y., Igarashi, Y., Arai, H., Ishii, M., 2012. Phylogenetic position of *Aquificales* based on the whole genome sequences of six *Aquificales* species. Int. J. Evol. Biol. 2012, 859264.

Oshima, K., Maejima, K., Namba, S., 2013. Genomic and evolutionary aspects of phytoplasmas. Front. Microbiol. 4, 230.

Oshima, K., Maejima, K., Namba, S., 2019. Plant-insect host switching mechanism. In: Phytoplasmas: Plant Pathogenic Bacteria-III. Springer, Singapore, pp. 57–68.

Oshima, K., Nishida, H., 2007. Phylogenetic relationships among mycoplasmas based on the whole genomic information. J. Mol. Evol. 65, 249–258.

Puranik, S., Acajjaoui, S., Conn, S., Costa, L., Conn, V., Vial, A., Marcellin, R., Melzer, R., Brown, E., Hart, D., 2014. Structural basis for the oligomerization of the MADS domain transcription factor SEPALLATA3 in *Arabidopsis*. Plant Cell 26, 3603–3615.

Quaglino, F., Kube, M., Jawhari, M., Abou-Jawdah, Y., Siewert, C., Choueiri, E., Sobh, H., Casati, P., Tedeschi, R., Lova, M.M., 2015. 'Candidatus Phytoplasma phoenicium' fassociated with almond witches' broom disease: from draft genome to genetic diversity among strain populations. BMC Microbiol. 15, 1–15.

Saccardo, F., Martini, M., Palmano, S., Ermacora, P., Scortichini, M., Loi, N., Firrao, G., 2012. Genome drafts of four phytoplasma strains of the ribosomal group 16SrIII. Microbiology 158, 2805–2814.

Saigo, M., Golic, A., Alvarez, C.E., Andreo, C.S., Hogenhout, S.A., Mussi, M.A., Drincovich, M.F., 2014. Metabolic regulation of phytoplasma malic enzyme and phosphotransacetylase supports the use of malate as an energy source in these plant pathogens. Microbiology 160, 2794–2806.

Shao, F., Zhang, Q., Liu, H., Lu, S., Qiu, D., 2016. Genome-wide identification and analysis of microRNAs involved in witches' broom phytoplasma response in *Ziziphus jujuba*. PLoS One 11, e0166099.

Singh, A., Lakhanpaul, S., 2020. Detection, characterization and evolutionary aspects of S54LP of SP (SAP54 like protein of sesame phyllody): a phytoplasma effector molecule associated with phyllody development in sesame (*Sesamum indicum* L.). Physiol. Mol. Biol. Plants 26, 445–458.

Strohmayer, A., Moser, M., Si-Ammour, A., Krczal, G., Boonrod, K., 2019. 'Candidatus Phytoplasma mali' genome encodes a protein that functions as an E3 ubiquitin ligase and could inhibit plant basal defense. Mol. Plant Microbe Interact. 32, 1487–1495.

Sugawara, K., Himeno, M., Keima, T., Kitazawa, Y., Maejima, K., Oshima, K., Namba, S., 2012. Rapid and reliable detection of phytoplasma by loop-mediated isothermal amplification targeting a housekeeping gene. J. Gen. Plant Pathol. 78, 389–397.

Sugawara, K., Honma, Y., Komatsu, K., Himeno, M., Oshima, K., Namba, S., 2013. The alteration of plant morphology by small peptides released from the proteolytic processing of the bacterial peptide TENGU. Plant Physiol 162, 2005−2014.

Sugio, A., MacLean, A.M., Grieve, V.M., Hogenhout, S.A., 2011. Phytoplasma protein effector SAP11 enhances insect vector reproduction by manipulating plant development and defense hormone biosynthesis. Proc. Natl. Acad. Sci. U. S. A. 108, E1254−E1263.

Sugio, A., MacLean, A.M., Hogenhout, S.A., 2014. The small phytoplasma virulence effector SAP11 contains distinct domains required for nuclear targeting and CIN-TCP binding and destabilization. New Phytol. 202, 838−848.

Suzuki, S., Oshima, K., Kakizawa, S., Arashida, R., Jung, H-Y., Yamaji, Y., Nishigawa, H., Ugaki, M., Namba, S., 2006. Interaction between the membrane protein of a pathogen and insect microfilament complex determines insect-vector specificity. Proc. Natl. Acad. Sci. U. S. A. 103, 4252−4257.

Tan, C.M., Li, C-H., Tsao, N-W., Su, L-W., Lu, Y-T., Chang, S.H., Lin, Y.Y., Liou, J-C., Hsieh, L-C., Yu, J-Z., 2016. Phytoplasma SAP11 alters 3-isobutyl-2-methoxypyrazine biosynthesis in *Nicotiana benthamiana* by suppressing NbOMT1. J. Exp. Bot. 67, 4415−4425.

Toruño, T.Y., Seruga Musić, M., Simi, S., Nicolaisen, M., Hogenhout, S.A., 2010. Phytoplasma PMU1 exists as linear chromosomal and circular extrachromosomal elements and has enhanced expression in insect vectors compared with plant hosts. Mol. Microbiol. 77, 1406−1415.

Tran-Nguyen, L.T., Kube, M., Schneider, B., Reinhardt, R., Gibb, K.S., 2008. Comparative genome analysis of '*Candidatus* Phytoplasma australiense' (subgroup tuf-Australia I; rp-A) and '*Ca.* Phytoplasma asteris' strains OY-M and AY-WB. J. Bacteriol. 190, 3979−3991.

Wang, J., Song, L., Jiao, Q., Yang, S., Gao, R., Lu, X., Zhou, G., 2018a. Comparative genome analysis of jujube witches'-broom phytoplasma, an obligate pathogen that causes jujube witches' broom disease. BMC Genom. 19, 1−12.

Wang, N., Yang, H., Yin, Z., Liu, W., Sun, L., Wu, Y., 2018b. Phytoplasma effector SWP1 induces witches' broom symptom by destabilizing the TCP transcription factor BRANCHED1. Mol. Plant Pathol. 19, 2623−2634.

Wei, W., Davis, R.E., Lee, I-M., Zhao, Y., 2007. Computer-simulated RFLP analysis of 16S rRNA genes: identification of ten new phytoplasma groups. Int. J. Syst. Evol. Microbiol. 57, 1855−1867.

Xue, C., Li, H., Liu, Z., Wang, L., Zhao, Y., Wei, X., Fang, H., Liu, M., Zhao, J., 2019. Genome-wide analysis of the WRKY gene family and their positive responses to phytoplasma invasion in Chinese jujube. BMC Genom. 20, 1−14.

Yang, C-Y., Huang, Y-H., Lin, C-P., Lin, Y-Y., Hsu, H-C., Wang, C-N., Liu, L-Y.D., Shen, B-N., Lin, S-S., 2015. MicroRNA 396-targeted short vegetative phase is required to repress flowering and is related to the development of abnormal flower symptoms by the phyllody symptoms 1 effector. Plant Physiol 168, 1702−1716.

Index

Note: Page numbers followed by 'f' indicate figures those followed by 't' indicate tables and 'b' indicate boxes.

A

Abiotic resistance, 153–154
Acquisition access period (AAP), 46
Alfalfa witches'-broom phytoplasma, 28–29, 55–56
Allium sativum leaf lectin, 128–129
Almond witches'-broom (AlmWB), 23, 58
Antibiotics, 101–102
Apple proliferation, 149–150
Asian citrus psyllid (ACP), 39
Auchenorrhyncha, biological controls, 131
Austroagallia sinuate, 59
Average nucleotide identity (ANI), 76
Axenic culture, 67

B

Bermuda grass white leaves disease, 59
Big bud disease, 27–28
Biocontrol, 112
Biological controls, 131
Biosecurity, 95, 125
Bois Noir (BN) disease, 154
Brinjal little leaf (BLL) phytoplasma
    Hishimonus phycitis vector, 51
    natural resistance, 152

C

Cabbage yellow disease, 57
Carrot witches'-broom, 56
Cetyltrimetyl ammonium bromide (CTAB) method, 8–9
Chemotherapy
    antibiotics, 138
    in vitro, 141–142
Chip graft inoculation assay (CGIA), 27–28
Chloramphenicol, 138
Citrus greening bacterium, graft transmission, 26–27
Clean propagation material, 128
Coding sequences (CDSs), 73
Colladonus montanus, 129
Conservative pesticides, 127
Covering plants, 130–131
Critical-point drying technique, 7
Cross-boundary movement
    National Plant Protection Organizations (NPPOs), 86
    plant quarantine. *See* Plant quarantine
    seed material, 85
Cryofixation, 7–8
Cryoscanning microscopy, 8
Cryotherapy, 101, 142
Cultural control, 31, 103–104

D

Destructive Insects and Pests (DIP) Act, 89–92
Diagnostic methods
    historical background, 2–3
    microscopic techniques, 1, 4
        fluorescence microscopy, 5–6
        light microscopy, 4–5
        transmission electron microscopy, 6–8
    molecular-based disease diagnosis, 1–2
    molecular diagnosis, 8–13
    symptomatic, 3–4
Diaphorina citri transmission
    management, 39
    Witches'-broom disease of lime (WBDL) phytoplasma transmission, 39
Dicer-like 1 (DCL1) protien, 168
Directorate of Plant Protection, Quarantine and Storage (DPPQS), 92–93
Diseases management approach
    antibiotics, 101–102
    biocontrol, 112
    cryotherapy, 101
    cultural control, 103–104
    eradication program, 98–100
    grapevine yellows, 111
    grapevine yellows (GY) detection surveys, 98–100
    healthy planting material, 100–101
    host resistance and tolerance, 104–108
    induced resistance, 108–109
    insect vector control strategies, 102–103
    quarantine, 98–100, 99t–100t
    recovery, 108
    witches'-broom diseases of lime, 109–110
DNA barcodes, 13
DNA-based molecular markers, 53
DNA extraction, 8–9
Dodder transmission, 29–30, 30f
Domestic quarantine, 93–94
Dwarfism, 179–181

E

Effectors, 155
Elicitors, 153–154
Elimination methods, phytoplasma
    Candidatus phytoplasma species, 137
    16S rDNA PCR amplification, 137
    in vitro chemotherapy, 141–142
    in vitro culture
        meristem preparation, 141
        mulberry dwarf phytoplasma elimination, 140
        tissue culture methods, 140
    in vivo methods
        chemotherapy, 138–139
        thermotherapy, 139–140
Endophyte, 154
Endophytic bacteria, 112
Enzyme-linked immunosorbent assay (ELISA), 2–3
Epifluorescence, 5
European stone fruit yellow (ESFY), natural resistance, 151

F

Faba bean phyllody, 56
Floral quartet model, 189–190
Flower and fruit development, microRNAs, 169
Flower malformation, 189–190
Fluorescence microscopy, 5–6

G

Galanthus nivalis agglutinin (GNA), 128–129
Genetic resistance, 153
Genome
    cellular functions, 187–188
    methyl-CpG binding domain (MBD) protein, 188
    potential mobile units (PMUs), 188
Genomic studies
    axenic culture, 67
    effectors and secreted proteins, 74–75
    gene content and metabolism, 73–74
    genome sequencing projects, 69
    historical perspective, 69–72
    mobile genetic elements, 75–76
    next-generation sequencing (NGS) technologies, 69
    phytoplasma strains and genome sequences, 70t–71t
    taxonomy, 76
Grafting methods, 21
Graft inoculation, 21, 40–41
Graft transmission
    Dodder transmission, 29–30, 30f
    herbaceous plants
        alfalfa witches'-broom phytoplasma, 28–29
        chip graft inoculation assay (CGIA), 27–28
        citrus greening bacterium, 26–27
        leaf disc graft transmission, 26–27
        plug grafting method, 28f, 25–26
        wedge grafting procedure, 26f
    micropropagation, 29
    Witches'-broom disease of lime (WBDL) phytoplasma transmission, 40–41
    woody plants
        almond witches'-broom (AlmWB), 23
        apricot tree, 24–25
        Candidatus Phytoplasma pruni, 23
        Candidatus Phytoplasma pyri, 22–23
        infection source, 22–23
        inoculation grafting, 24f
        jujube witches-broom (JWB) disease, 25
        patch grafting, 23
        periwinkle, 25
        Sophora japonica trees, 25
Graft union, 22f
Grapevine yellows (GY)
    detection surveys, 98–100
    management approach, 111

H

Habitat management, 130
Healthy planting material, 100–101
Hemiptera, 45
Hemipteran insect vector
    management, 127–133
        alternative host's and weeds control, 129–130
        biological controls, 131
        clean propagation material, 128
        covering plants, 130–131
        habitat management, 130
        insecticides, 127
        resistant plants, 128–129
        symbiotic control, 132–133
    suborders, 125–126
    taxonomy and ecology, 125–126
HEN 1 protein, 168
High-resolution melt analysis (HRMA), 12–13
Hishimonus phycitis
    biology, 51–52
    distribution, 49
    dorsal and lateral view, 49
    identification, 50–51
    taxonomy, 49
    transmission
        host preference, 38
        host range, 38
Hishimonus sellatus, 52
Host resistance and tolerance

durability, 107−108
elm yellows (EY) phytoplasma infections, 106−107
genotype resistance, 105
molecular DNA methods, 105
molecular mechanisms, 105
quantitative PCR (qPCR) assay, 105
resistance screening, 107
resistant genotype, 106−107
resistant plants, 104
rootstocks, 105−106
woody plants, 104
X-disease resistance, 105
Hot water therapy (HWT), 139−140

I

ICAR-National Bureau of Plant Genetic Resources (ICAR-NBPGR), 92−93
Immunodominant protein (IDP), 97
Immunofluorescence, 5−6
Immunosorbent and immunoelectron microscopy, 7
Induced resistance, 108−109
    effectors and plant immune system, 155
    inducers, 153−154
    recovery phenomenon, 154−155
Insect-exclusion screening (IES), 131
Insecticides, 102, 127
Insect-proof screening, 102
Insect vectors, 47t−49t
    control strategies, 102−103
        FD epidemiological model, 103
        insecticides, 102
        insect-proof screening, 102
        physical prevention, 102
        symbiont-based strategy, 102−103
    Hishimonus spp
        biological controls, 52
        chemical control, 52
        DNA-based molecular markers, 53
        Hishimonus phycitis, 49−52
        Hishimonus sellatus, 52
        microbial symbionts, 53
        symbiotic control, 53−54
        yeast-like symbiont (YLS), 54
    insect acquisition, 46
    malaise trap, 46−47
    management, 59−60
        alternative host's and weeds control, 129−130
        biological controls, 131
        clean propagation material, 128
        covering plants, 130−131
        habitat management, 130
        insecticides, 127
        resistant plants, 128−129
        symbiotic control, 132−133
    Neoaliturus spp
        Neoaliturus fenestratus, 58
        Neoaliturus haematoceps, 56−58
        Neoaliturus tenellus, 58
    Orosius albicinctus vectors
        Alfalfa witches'-broom, 55−56
        carrot witches'-broom, 56
        cucumber and squash phyllody, 54−55
        faba bean phyllody, 56
        Limonium spp, 56
        management, 56
        petunia witches'-broom, 56
        sesame phyllody, 55
        taxonomy and distribution, 54
    transmission experiments, 47
Integrated pest management (IPM), 130
International Standards for Phytosanitary Measures (ISPMs), 87−88

J

Jujube witches-broom (JWB) disease, 25

K

Kaolin, 127

L

Latent period (LP), insect acquisition, 46
Leaf development, microRNAs, 169
Leaf disc graft transmission, 26−27, 27f
Leafhoppers, 131
Lettuce and wild lettuce phyllody, 58
Light microscopy, 4−5
Limonium latifolium witches'-broom, 58
Loop-mediated isothermal amplification (LAMP), 3, 11−12

M

Macropsinae, 126
Macrosteles quadripunctulatus, 46
Membracoidea, 126
Meristem tip culture, 31, 100−101
Methyl-CpG binding domain (MBD) protein, 188
Microarray, 12−13
MicroRNAs
    biogenesis, 168
    biogenesis and target, 173t−174t
    discovery, 167
    disease symptoms, 173t−174t
    phytoplasma infection
        anthers, defective development, 176

MicroRNAs (*Continued*)
    anthocyanin accumulation, leaves, 176–178
    dwarfism, 179–181
    miR156, 170–175
    miR159, 175–176
    miR172, 170–175
    Witches'-broom and phyllody, 178–179
  target gene expression regulation, 168
  upregulation/downregulation, 171t–173t
  vegetative and reproductive development
    abiotic and biotic stress, 170
    flower and fruit development, 169
    leaf development, 169
    root and shoot development, 169
    vascular development, 169
Microscopic diagnosis, 4
Minor phytoplasma vectors, 58–59
Mobile genetic elements (MGE), 75–76
Molecular-based disease diagnosis, 1–2
Molecular diagnosis
  DNA barcoding, 13
  DNA extraction, 8–9
  microarray, 12–13
  PCR techniques, 10–11
  primer selection, 9–10
Molecular memory phenomena, 155
Monophagous vectors, 126
Mulberry dwarf disease, 59
Multilocus sequence analysis (MLSA), 76
Mycoplasma-like organism (MLO), 45
Mycoplasma-like pleomorphic particles, 2

N

National Plant Pests Diagnostic Network, 95
National Plant Protection Organizations (NPPOs), 86
Natural grafting, 32
Natural resistance
  economically important crops
    brinjal little leaf, 152
    jujube witches' broom, 152–153
    sesame phyllody, 152
  pome and stone fruits, 149
  temperate fruit trees
    apple proliferation, 149–150
    European stone fruit yellow (ESFY), 151
    pear decline, 150–151
Neoaliturus fenestratus vector, 58
Neoaliturus haematoceps vector, 56–58
Neoaliturus tenellus vector, 58
Nested PCR, 10–11
Next-generation sequencing (NGS) technologies, 69, 72
Noncoding RNAs, 167

O

Orosius albicinctus vectors
  Alfalfa witches'-broom, 55–56
  carrot witches'-broom, 56
  cucumber and squash phyllody, 54–55
  faba bean phyllody, 56
  Limonium spp, 56
  management, 56
  petunia witches'-broom, 56
  sesame phyllody, 55
  taxonomy and distribution, 54
Oxytetracycline, 101, 138–139

P

Pathogenic factors, 188–190
Pathosystem, 147
Pear decline
  colonization pattern, 150
  natural resistance screening, 150
  rootstocks, 150–151
  seedlings, Pyrus species, 150
Petunia witches'-broom, 56
Phylogenetic analyses, 189–190
Phytoplasma cell membranes, 97
Plant health management, 86
Planthoppers, 131
Plant immune system, 155
Plant quarantine
  International scenario
    imports and exports, 86–88
    sanitary and phytosanitary (SPS) measures, 86–88
  national domestic quarantine, 93–94
  National scenario
    exports, 93
    imports, 89–93
    regulated pest, Asia, 89, 90t
Plants, Fruits, and Seeds (PFS) order, 89–92
Plasma-activated water (PAW), 139
Plug grafting method, 28f, 25–26
Polymerase chain reaction (PCR), 137
  real-time, 11
  reverse transcription, 11
Polyphagous vectors, 126
Potential mobile units (PMUs), 188
Primers, 9–10
Proutista moesta, 59

R

Rapeseed phyllody, 57
Real-time PCR, 11
Regulated phytoplasma, 99t–100t

Index

Resistance
    current status
        India, 156
        Iran, 156−157
        Japan, 157−158
        Turkey, 158
    induced resistance, 153−155
    natural resistance, 149−151
    transgenic, 155−156
Resistant plants, 128−129
Restriction fragment length polymorphism (RFLP) analysis, 76
Reverse transcription-PCR, 11
RNA-induced silencing complex (RISC complex), 168
Root and shoot development, microRNAs, 169
Rootstocks, 105−106

S

Salivary glands, phytoplasma transmission, 46
Sandalwood, 58
Sanitary and phytosanitary (SPS) measures, 86−87
Sanitation techniques, 137
SAP11, 189
Scanning electron microscopy, 7
Scion and stock, 22f
Scleroracus flavopictus, 59
Seed transmission, witches'-broom disease of lime (WBDL) phytoplasma
Sesame phyllody disease, 55, 57−58
    Hishimonus phycitis vector, 51
    natural resistance, 152
Shoot-tip cryotherapy, 142
Sophora japonica trees, 25
Stem-cutting culture, 31, 100−101
Sugarcane grassy shoot (SCGS) disease, 58
Sweet potato little leaf phytoplasma, 31
Symbiotic control, 132−133
Symptomatic diagnosis, 3−4

T

Tetracycline, 101, 138
Thermotherapy, 31, 100−101, 139−140
Thin section procedure, 6−7
Tissue culture techniques, 31, 100−101
Transcription factors (TFs), 168
Transgenic resistance, 155−156
Transmission electron microscopy (TEM), 2−3
    critical-point drying technique, 7
    cryofixation, 7−8
    cryoscanning microscopy, 8
    immunosorbent and immunoelectron microscopy, 7
    scanning electron microscopy, 7
    thin section procedure, 6−7

V

Vascular development, microRNAs, 169
Vectorial capacity, 132

W

Wedge grafting procedure, 26f
Weeds control, 129−130
Witches'-broom disease of lime (WBDL) phytoplasma
    experimental hosts, 40−41
    graft transmission, 40−41
    Hishimonus phycitis vector, 50−51
    management approach, 109−110
    seed transmission, 40
    vector transmission
        Diaphorina citri, 39
        Hishimonus phycitis, 37−38

Y

Yeast-like symbiont (YLS), 54, 133
Yellow dwarf disease, 59

CPI Antony Rowe
Eastbourne, UK
April 21, 2023